Smart Nanomaterials Technology

Nanotechnology is a rapidly growing scientific field and has attracted a great interest over the last few years because of its abundant applications in different fields like biology, physics and chemistry. This science deals with the production of minute particles called nanomaterials having dimensions between 1 and 100 nm which may serve as building blocks for various physical and biological systems. On the other hand, there is the class of smart materials where the material that can stimuli by external factors and results a new kind of functional properties. The combination of these two classes forms a new class of smart nanomaterials, which produces unique functional material properties and a great opportunity to larger span of application. Smart nanomaterials have been employed by researchers to use it effectively in agricultural production, soil improvement, disease management, energy and environment, medical science, pharmaceuticals, engineering, food, animal husbandry and forestry sectors.

This book series in Smart Nanomaterials Technology aims to comprehensively cover topics in the fabrication, synthesis and application of these materials for applications in the following fields:

- Energy Systems—Renewable energy, energy storage (supercapacitors and electrochemical cells), hydrogen storage, photocatalytic water splitting for hydrogen production
- Biomedical—controlled release of drugs, treatment of various diseases, biosensors,
- Agricultural—agricultural production, soil improvement, disease management, animal feed, egg, milk and meat production/processing,
- Forestry—wood preservation, protection, disease management
- Environment—wastewater treatment, separation of hazardous contaminants from wastewater, indoor air filters

Azamal Husen
Editor

Nanomaterials from Agricultural and Horticultural Products

Springer

Editor
Azamal Husen
Wolaita Sodo University
Wolaita, Ethiopia

ISSN 3004-8273 ISSN 3004-8281 (electronic)
Smart Nanomaterials Technology
ISBN 978-981-99-3434-8 ISBN 978-981-99-3435-5 (eBook)
https://doi.org/10.1007/978-981-99-3435-5

This Springer imprint is published by the registered company Springer Nature Singapore Pte Ltd.
The registered company address is: 152 Beach Road, #21-01/04 Gateway East, Singapore 189721, Singapore

Love you Ammi

Preface

Agricultural and horticultural based product or production acts as the primary pillar of the developing economy. The major agricultural products are broadly grouped into foods, fibers, fuels, and raw materials, while horticultural crops include fruits, vegetables, medicinal, aromatic, and ornamental plants. Global population has increased many folds, and thus higher production of agricultural, as well as horticultural products, and, at the same time, waste material production are also accelerated. Thus, sustainable utilization of each kind of waste material is more desirable. In this concern, the past decade has witnessed a phenomenal rise in nanotechnology research due to its wide range of applications in almost every area of science and technology.

The unique features of nanoparticles and/or nanomaterials make them suitable for such kind of wide range of applications, for instance, in food science and technology, agriculture and forestry sectors, cosmetics products, medical science, and materials science. These particles are thought to have been present on earth naturally since their origin in the form of soil, water, volcanic dust, and minerals. Besides their natural origin, they have been also synthesized by using physical, chemical, and biological means. Quite often, it has been noticed that the nanoparticles and/or nanomaterials when manufactured using biological means are eco-friendly and cost-effective. Thus, various kinds of metal and metal-oxide nanomaterials obtained from agricultural or horticultural sources and/or waste products are explored, together with their specific applications. The use of nanoparticles in the food sector involves food processing and preservation and food packaging. In agriculture, nanomaterials are being utilized for the production of nano-fertilizers, pesticides, herbicides, sensors, and so on. In medicine, nanomaterials involve the production of various antibacterial, antifungal, anti-plasmodial, anti-inflammatory, anticancer, antiviral, antidiabetic, and antioxidant agents. They are also useful for the early detection of life-threatening diseases such as cancer. Besides, nanomaterials have also been used for bioremediation due to their capacity to degrade various pollutants such as organic dyes and chemicals. Thus, given the diverse scope of nanoscience, and sustainable use of agricultural and horticultural products, different countries are investing in the discipline of nanotechnology to obtain useful products for various purposes. At the same time, the current developments and safety issues over the use of nanomaterials cannot be ignored and have been discussed in this book.

Taken together, the book in hand covers a wide range of topics as mentioned above. It incorporates chapters that the authors have skillfully crafted with clarity and precision, reviewing up-to-date literature with lucid illustrations. The book would cater to the need of graduate students as a textbook and simultaneously be useful for both novices and experienced scientists and/or researchers working in the discipline of agricultural engineering, horticultural engineering, agricultural science, horticultural science, environmental engineering, waste management, nanotechnology, nanobiotechnology, nanoscience, materials science, biotechnology, molecular plant biology, crop biochemistry, biotechnology, and many other interdisciplinary subjects. The book in hand is also helpful to the researcher and scientist working on minimizing environmental pollution, especially in organic waste management. It should also inspire industrialists and policymakers associated with agricultural and horticultural product management.

I extend my sincere thanks to all the contributors for their timely responses and excellent contributions. Finally, my special thanks go to Shagufta, Zaara, Mehwish, and Huzaifa for providing their time and overall extended support to put everything together. I shall be happy to receive comments and criticism, if any, from subject experts and general readers of this book.

Wolaita, Ethiopia Azamal Husen

Contents

About the Editor

Prof. Azamal Husen is a Foreign Delegate at Wolaita Sodo University, Wolaita, Ethiopia. He has served the University of Gondar, Ethiopia, as a Full Professor of Biology, and worked there as the Coordinator of the M.Sc. Program and as the Head of the Department of Biology. He was a Visiting Faculty member of the Forest Research Institute, and the Doon College of Agriculture and Forest at Dehra Dun, India. Dr. Husen's research and teaching experience of 20 years includes biogenic nanomaterial fabrication and application, plant responses to nanomaterials, plant adaptation to harsh environments at the physiological, biochemical, and molecular levels, herbal medicine, and clonal propagation for improvement of tree species. Dr. Husen contributed to R&D projects of World Bank, ICAR, ICFRE, JBIC, etc. To his credit are >200 publications; and he is Editor-in-Chief of the *American Journal of Plant Physiology*. He is also working as Series Editor of *Exploring Medicinal Plants*, published by Taylor & Francis Group, USA; *Plant Biology, Sustainability, and Climate Change*, published by Elsevier, USA; and *Smart Nanomaterials Technology*, published by Springer Nature Singapore Pte Ltd. Singapore.

Nanomaterials and Nanocomposites in Agricultural and Horticultural Sectors

Y. D. Franco-Aguirre, W. Y. Villastrigo-Lopez, M. D. Davila-Medina, M. E. Castañeda-Flores, R. I. Narro Cespedes, S. C. Esparza Gonzalez, R. Herrera-Rodriguez, and A. Sáenz-Galindo

Abstract This chapter presents an overview of the synthesis and production of nanomaterials and nanocomposites from renewable natural sources from agriculture and horticulture, starting with the definitions of nanomaterials and nanocomposites, followed by the various raw materials and their different processes, to carry out synthesis and the chemical reactions involved or the obtaining presenting the different formulations or the mixing involved to obtain nanomaterials and nanocomposites, as well avocado peel and its derivatives materials avocado and its derivatives stand out, different citrus fruits, wine fruits and their derivatives, watermelon, chili and its derivatives, chitosan, starches from various sources such as corn and potatoes, likewise the different phases presented continuous and discontinuous, highlighting the metallic nanoparticles based zinc, silver, copper, gold, among others, as well nanostructures based on carbon, nanotubes, and graphene. Different applications of their corresponding are also presented, concluding with the general perspectives of this type of nanomaterials and nanocomposites from renewable natural sources.

Keywords Nanomaterials · Nanocomposites · Renewable natural sources · Agriculture · Horticulture

---

Y. D. Franco-Aguirre · W. Y. Villastrigo-Lopez · M. D. Davila-Medina · M. E. Castañeda-Flores · R. I. Narro Cespedes · R. Herrera-Rodriguez · A. Sáenz-Galindo (✉)
Facultad de Ciencias Químicas, Universidad Autónoma de Coahuila, Ing. J. Cárdenas Valdez S/N, Col. República, C. P. 25280 Saltillo Coahuila, México
e-mail: aidesaenz@uadec.edu.mx

S. C. Esparza Gonzalez
Facultad de Odontología, Universidad Autónoma de Coahuila, Dra. Cuquita Cepeda de Dávila, C. P. 25125 Saltillo Coahuila, México

A. Husen (ed.), *Nanomaterials from Agricultural and Horticultural Products*, Smart Nanomaterials Technology,
https://doi.org/10.1007/978-981-99-3435-5_1

1 Introducción

Agriculture is part of the life of human beings, it is present in heritage, cultural identity, and gastronomy, and it is not only an economic activity but also a support of humanity. It is considered a primary activity and is essential for any country; it is currently considered a subsector with possibilities of obtaining economic resources. This activity is responsible for providing food, raw materials, and labor to the agro-industrial and services sectors; it is also demanding a large number of essential industrial products for agricultural production, including fertilizer, herbicides, pesticides, and machinery, among others. In this aspect, social agriculture and horticulture are very important. Is importat the improving the processes involved in these áreas. Nanotechnology is an important area that has an impact on various disciplines and sciences; agriculture (Fig. 1) and horticulture are the exceptions; currently, there are important developments and improvements that benefit agriculture, from herbicides, pesticides, planting methodologies, land treatment, and irrigation materials, among others, carbon-based, carbon nanotubes, nanofibers, graphene, fullerene, nanoparticles metallic, nanoparticle metallic oxide, nanoparticles polymeric, etc. [19–25, 32, 33, 35, 43]. The objective of this chapter is to present an overview of the use and applications of different nanomaterials and nanoparticles in agriculture and horticulture sectors, taking into account that agriculture and horticulture play a vital role in the economic development of countries, even more so in less developed countries because the majority of their population depends on them for their subsistence.

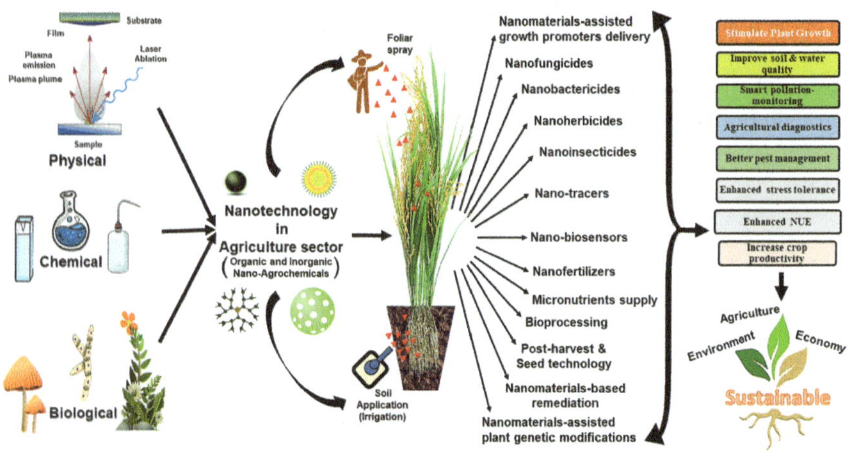

Fig. 1 Various applications of nanotechnology in agricultural sectors (adapted from Sharma et al. [43])

2 Nanomaterials, Nanocomposites, and Nanoparticles

Over the last few years, nanotechnology has developed extensively; currently, several investigations are directly or indirectly related to it. Nanotechnology is the nanometer scale's development, synthesis, characterization, and application of modified devices. The basic and key elements of nanotechnology are nanomaterials, which are defined as those that have dimensions on the nanometer scale (1–100 nm). Its development in agriculture has spread to various fields, such as crop protection, food production, and sewage treatment [29].

In nanotechnology, nanocomposites are used, defined as materials made up of a dispersed phase of nanometric dimensions, which are found inside a matrix, thereby achieving improvements in their properties as a consequence of the large surface area. It is important to obtain a uniform dispersion of the nanomaterial so that it exhibits a significant increase in the properties of nanocomposites [39].

The various potential uses of nanotechnology in agriculture have created great interest, since offering the possibility of improving agricultural production through various strategies for increasing food agrochemical residues and using fewer inputs of energy. The applications of nanotechnology in agriculture are very diverse, mainly highlighting the development of encapsulated nano-pesticides for controlled release, for the production of nano-, macro-, and micronutrients, as well as to make the uses and applications of agrochemicals more efficient [16].

In the case of horticulture, nanomaterials are being used for the treatment of some plant diseases, for the early detection of the pathogens that produce them, for the improvement of the assimilation of essential nutrients, and even to build important nano-biosensors in certain biological processes [48]. Nanotechnology can increase the effectiveness of commercial pesticides and insecticides, reducing the amount of application to the soil of foliage to significantly lower doses than those required conventionally, with the improvement that this implies for ecosystems [49].

Within the raw materials used in the manufacture of nanomaterials, there are residues of plant origin, which are considered as the resulting material that is not intended for consumption, and are obtained from various parts of a plant, such as remains of sugarcane crops, leaves or seeds from legumes, coffee, and shells, among others [12].

These residues are called biomass or lignocellulosic matter and are characterized by having a complex chemical structure that is mainly composed of macromolecules such as cellulose, hemicellulose, and lignin [47].

Lignocellulosic materials are made up of cellulose, which is a natural polysaccharide. It is the most abundant, renewable biopolymer and is part of the cell walls of plants whose proportion ranges 30–50% [15]. It is the most abundant structural component on earth and although it can be produced sustainably from various biomass sources, this polymer, apart from being present in wood, cotton, hemp, and other plants, is also synthesized by other organisms such as Tunicates, Oomycotas, some algae, and bacteria such as *Acetobacter xylinum*; its derivatives are used for a number of applications such as the paper industry, biomaterials, pharmaceuticals, as a thickening agent, cosmetology, and in agriculture [15].

Nanocellulose is a natural fiber, which is made from cellulose materials, and at least one of its dimensions is on the nanometer scale. The greatest efforts to obtain it lie in achieving a nanocellulose with dimensions ranging from 1 to 100 nm; these nanometric dimensions make this biopolymer an interesting material; due to its excellent mechanical properties since at these scales, they present characteristics of low density, high stiffness, high tensile strength, biodegradability, and thermostability [40]. Obtaining cellulose and nanocellulose from lignocellulosic residues has been developed by chemical treatments applying different methods such as acid hydrolysis, alkaline hydrolysis, oxidation agents, ionic liquids, and mechanical treatments such as homogenization, ultrasonication, microfluidization, grinding, and high-pressure heat treatment [40].

2.1 Nanoparticles in Agriculture

Due to the properties that nanoparticles (NPs) possess, which include a high surface area/volume ratio, controlled reléase characteristics, and sorption capacity, they are considered good candidates for the formulation of fertilizer in the agriculture sector [50].

Nanofertilizers are nanometric-sized nutrient particles encapsulated with nanomaterials to facilitate the entry of nutrients into plants. The efficiency of nanofertilizers depends on particle size, soil properties, adsorption, and accumulation by crops [51].

The main use of metallic NPs of Copper (Cu), Iron (Fe), Zin (Zn), Silver (Ag), Titanium (Ti), and others, in agriculture and the food industry, is mainly due to antimicrobial activity. In particular the seeds and plants, it was found a possitive and promoting effect on germination by observing that Nps Zn can penetrate the seed coat and favorably stimulate physiological and biochemical responses. For part NPs, Ag are used as antimicrobial agents, and also applied to panels of refrigerators, as well as in containers or storage containers, packaging lines, and other surfaces intended to come into contact with food [13, 14].

In 2019, Bala et al. used NP ZnO as a nanofertilizer to improve Zn deficiency in rice crops where they carried out the foliar applications of the NP ZnO with four different concentrations with intervals of 15 days, obtaining, as a result, a decrease in soil pH, as an alteration in the length of the nanofertilizer in the face of Zn deficiency [7].

In the same year 2019, Choudhary et al. developed NP chitosan to provide a slow release of Zn ions, promote the growth and yield of corn, and provide protection against leaf spots due to *Curvalaria lunata*. After 35 days of harvest, they made the foliar applications; after 10 days of said applications, they added *Curvalaria lunata* in the leaves, the NPs inhibited mycelial growth of the plant that registered 221.77 cm (day 88 of cultivation) and the maximun weight of 100 grains registered 35.17 g (day 95 of cultivation) [11].

Hussain et al. [27], demonstrated the possible impacts of NP Fe to mitigate the toxic effects of Cadmium (Cd) in the wear. The results observed that the NP Fe at a concentration of 20 mg/kg increased by 54% the length of the spike for both soil and foliar applications. On the other hand, foliar application of 5,10, 15, and 20 mg/L decreased Cd concentrations in roots by 20, 27, 45, and 56%, in grains by 23, 35, 75, and 84%, and in the outbreaks by 11, 26, and 53%; and soil application of 5,10, 15, and 30 mg/kg of NPs reduced roots by 12, 28, 46, and 49%, grains by 23, 30, 72, and 81%, and in outbreaks by 13, 38, and 53%, respectively, concluding that the use of NP Fe increased the length of the spike and decreased the concentrations of Cd; in addition, the foliar application is more efficient than the application in the soil since the absorption of Fe can be affected by the pH [27].

Nandhini et al. [36] evaluated the impact of NP ZnO in the content of saponin for disease resistance of *Mildiú Velloso*. For this treatment, samples of 159, 200, and 250 ppm of NP ZnO, the results show variations in the content Zn with a maximum of 47,055 mg in 200 ppm of NPsZnO. Under greenhouse conditions, improved dry weight (3.13 g) at 250 ppm NP ZnO and plant height (35.06 cm) at a concentration of 250 ppm of NP ZnO; also the concentration of 250 ppm of NP ZnO showed protection of 59% to the plants. In conclusion, the use of NP ZnO can induce resistance effects against *Sclerospora graminicola* [36].

In 2018, Hussain et al. explored and reported the impacts of NP Cd and ZnO on wheat growth and yields. These researchers used a field contaminated with NP Cd and ZnO. In the results, they observed that when applying the different concentrations to the soil, the length of the shoots increased by 10, 23, 33, and 43%, while the foliar application increased by 21, 35, 50, and 55%. Furthermore, Cd concentrations in roots decreased with foliar applications by 25–64%, in grains by 30–77%, and in shoots by 20–77%, but with application on the soli, present a decreases by the roots by 23–60%, in the grains by 16–78%, and in the shoots by 16–78%, and in the shoots between 17 and 68%, respectively. It was concluded that the NP ZnO increased the length of the buds and decreased the amounts of Cd [26].

Ag is very interesting; NP Ag at low concentrations have a positive effect on seed germination and growth promotion in plants. In 2015, Dimkpa et al. reported that NP Ag promote the growth of mustard seedlings (*Brassica juncea*) at concentrations of 25 and 50 mg L^{-1}, reflecting greater root length, dry biomass, and height. However, when using a high concentration of NP Ag (250–500 mg kg^{-1} of soil), the germination and growth of broad bean plants are inhibited. These increases could be related to the endogenous production of phytohormones such as gibberellins and cytokines, which are involved in cell elongation and division. Furthermore, these NPs can improve the efficiency of electron exchange at the cellular level in plants, which could reduce the formation of reactive oxygen species [13, 14].

In 2015, Ibrahim et al. found that the optimal conditions to obtain NP Ag, from banana peel extract, were 1.75 mM silver nitrate, 20.4 mg dry matter, and an incubation time of 72 h, achieving a size of 23.7 nm of NP Ag and a higher % improvement of bacterial inhibition for the pathogen *Pseudomonas aeruginosa,* since inhibition halos of 20 mm were obtained [28].

In 2018, Rodriguez et al. developed a method for the synthesis of TiO_2 based on nanocatalysts using orange peel as a sacrificial template; this material was modified by varying the Ruthenium (Ru) concentrations for the preparation of the catalyst, which was used in the transformation of levulinic acid to N-heterocycles in a continuous flow reactor; it was found that the higher the Ru content, the higher the catalytic yield. Therefore, the orange peel represents an innovative, sustainable, and circular economy option for the synthesis of these materials [41].

3 Nanomaterials and Nanocomposites with Nanostructure of Carbon

One of the most important sectors in the world economy is agriculture, as it produces food and provides raw materials for multiple industries [8]. Population growth together with the increase in natural disasters such as floods, droughts, extreme variability in temperature or rainfall, and the depletion of natural resources often poses a threat to food security that currently demands agricultural development that is viable, efficient, and sustainable [38]. In addition, the population rate will change radically in the coming decades. The world population is expected to exceed 9.7 billion people by 2050, the Food and Agriculture Organization of the United Nations (FAO) estimates that by 2050, agriculture will need to produce more than twice as much food to meet global demand [46].

Consequently, today the search for new technologies that optimize agro-industrial processes has taken on great importance. Several nanomaterials have been introduced with great potential to revolutionize the agriculture industry, seeking to increase food quality and safety, crop growth, and monitor environmental conditions [30].

Nanomaterials are dispersed or solid structures with dimensions from 1 to 100 nm that have at least a one-dimensional structure at the nanometer level. These materials can be synthesized by different methods (chemical, physical, and biological) and using different nanoscale materials such as single- or multiple-walled carbon nanotubes, nanofibers, nanoclays, graphene, AgO, Zn, silicon, and Zn Ti, among others [42].

However, in recent years, carbon-based nanomaterials have shown wide agriculture applicability [1]. Due to their small size and larger surface area, carbon-based nanomaterials have specific characteristics such as high strength, low density, strong hydrophobicity, and biodegradability, unlike common materials, which have facilitated their applications as plant growth regulators and promoters, for improvement of plants by genetic means, as nano-pesticides, nanofertilizers, nanomonitoring sensors, and for detection of pathogens, among others [9].

Carbon nanomaterials have been reported to influence plant physiology in many positive ways, such as enhancement of seed germination, growth shoot and root length, biochemical content enzyme activity, defense system, and many other metabolic activities that in turn improve the productivity of the plants [44]. And in turn, the use of nanotechnologies in the cultivation of plants gives rise to advantages such as improved stability and dispersion of active ingredients, precise delivery of agrochemicals, reduction of residual contamination, and reduction of the cost of labor in different applications, providing ecological and sustainable agents by reducing the number of fertigation compounds and controlled release, in addition to being more sustainable with the environment, therefore, the study of the different applications of this type is of the great importance of nanomaterials [6].

3.1 Nano-biosensors

Nano-biosensors are modified analytical devices that consist of a sensitized element of biological origin linked to another physical–chemical transducer [34]. They have a wide area of applications in agriculture and horticulture, such as control of temperature, quality, fertility, microenvironment and soil moisture, intervention in microbiological growth, irrigation and safety sensors in agronomy, control of residual pesticides, fertilizer, and toxins, and plant pathological monitoring [18].

Nano-biosensors have been developed using graphene-peptide oxide to detect the protease of bacteria such as *Escherichia coli,* which has shown a strategy with the potential for the detection of multiple bacterial species [10].

3.2 Nanofertilizers

Nanofertilizers are particles encapsulated within nanoporous substances, coated with thin nanoscale polymer films [4]. Through the use of nanofertilizers, it has been possible to optimize the use of nutrients, reduce soil toxicity, minimize overdose effects, and reduce application, in addition to improving efficiency in the use of inputs and minimizing costs [2].

Carbon nanotubes provide tools to modify biochemical processes, and the effect of carbon nanotubes on plant cells (especially for crop plants) has shown greater water absorption due to the penetration of carbon nanotubes in the seeds. It has also been used as a vehicle to deliver molecules to seeds to protect plants against disease and promote growth [17].

3.3 Nano-pesticides

Nano-biopesticides are chemical pesticides or derive biologically active compounds that are integrated with nanoparticles and fused into a suitable polymer. However, there are also active pesticides that are reduced by a metal salt to form metal nanoparticles that are homogenized into a suitable polymer. This type of nanotechnology has multiple advantages, such as the release control and biodegradability of the pesticide and the polymer [5].

Multiwall-graft-polycitric acid carbon nanotubes (MWCNT-g-PCA) were used as a hybrid material to encapsulate pesticides; the effect of the encapsulated pesticide against *Alternaria alternata* was studied, showing that the nano-pesticides had an extraordinarily superior effect in inhibiting *Alternaria alternata* compared to commercial pesticides [3].

3.4 Growth Regulators and Promoters

All plants naturally produce hormones that regulate their metabolism, growth, and development. However, there are synthetic biomolecules, which have the ability to carry out similar processes in a controlled manner, such as the formation and growth of roots, shoots, buds, flowers, and fruits. However, there are conditions that can affect its effectiveness such as the type of plant, type of stimulant, amount of stimulant applied, time of application, growth stage, and place of application of the stimulant [45].

Various investigations have shown that carbon nanoparticles can stimulate the germination and growth of some plants, such as barley, chickpeas, corn, broccoli, and onions, since the addition of these nanomaterials stimulates photosynthesis, activating the antioxidant system, and increasing the number of lateral roots that help assimilate nutrients [31].

3.5 Disadvantage of Using Nanomaterials

Nanomaterials in the nanometric size have changed the visiόn of the world since today it is one of the most important and applicable breaches of science in different aspects. However, the size of these materials could propitiate some disadvantages such as an increase in the level of toxicity due to the fact that these particles easily have a cellular penetration that, depending the chemical composition of the material or nanomaterial, would be damage it can cause [37].

4 Conclusion

Agriculture and horticulture are two áreas of great importance for human beings, as well as for the development of any country, developed, underdeveloped, or developing; the feeding of human beings is mostly based on agriculture. Therefore, there is a need to promote development and continuous improvement in these áreas. Taking into account the above, nanotechnology offers great advantages to improve, innovate, implement, design, and apply different methodologies, products, and processes in this area, from the design, formulation, and obtaining of a product with characteristics of fertilizer, nutrient, pesticide, and herbicide, among others, to obtaining materials to help the growth of plants or fruits. The development of numerous nanomaterials, such as metal or metal-oxide nanoparticles, as well as carbon-based nanomaterials, are widely used in these areas, demonstrating that they are a viable and reliable alternative in agriculture and horticulture. However, it is important to take into account different factors such as the concentrations and the controlled use of these nanomaterials in order not to affect the crop and the food.

Acknowledgements The authors thank the CONACyT for supporting the project SEP-CB-2017-2018 A1-S-44977 and Universidad Autónoma de Coahuila, Posgrado en Ciencia y Tecnología de Materiales.

References

1. Aacharya R, Chhipa H (2020) Nanocarbon fertilizers: implications of carbon nanomaterials in sustainable agriculture production. In: Carbon nanomaterials for agri-food and environmental applications. Elsevier, pp 297–321. https://doi.org/10.1016/B978-0-12-819786-8.00015-3
2. Abd El-Azeim MM, Sherif MA, Hussien MS, Tantawy IAA, Bashandy SO (2020) Impacts of nano- and non-nanofertilizers on potato quality and productivity. Acta Ecol Sin. https://doi.org/10.1016/j.chnaes.2019.12.007
3. Abd-Elsalam KA, Prasad R (eds) (2018) Nanobiotechnology applications in plant protection. Nanotechnology in the life sciences. https://doi.org/10.1007/978-3-319-91161-8
4. Abdelsalam NR, Kandil EE, Al-Msari MA, Al-Jaddadi MA, Ali HM, Salem MZ, Elshikh MS (2019) Effect of foliar application of NPK nanoparticle fertilization on yield and genotoxicity in wheat (Triticum aestivum L.). Sci Total Environ 653:1128–1139. https://doi.org/10.1016/j.scitotenv.2018.11.023
5. Agathokleous E, Feng Z, Iavicoli I, Calabrese EJ (2020) Nano-pesticides: a great challenge for biodiversity? The need for a broader perspective. Nano Today 30:100808. https://doi.org/10.1016/j.nantod.2019.100808
6. An C, Sun C, Li N, Huang B, Jiang J, Shen Y, ..., Wang Y (2022) Nanomaterials and nanotechnology for the delivery of agrochemicals: strategies towards sustainable agriculture. J Nanobiotech 20(1):1–19. https://doi.org/10.1186/s12951-021-01214-7
7. Bala R, Kalia A, Dhaliwal SS (2019) Evaluation of efficacy of ZnO nanoparticles as eemedial zinc nanofertilizer for rice. J Soil Sci Plant Nutr 19:379–389. https://doi.org/10.1007/s42729-019-00040-z
8. Barros MV, Salvador R, de Francisco AC, Piekarski CM (2020) Mapping of research lines on circular economy practices in agriculture: from waste to energy. Renew Sustain Energy Rev 131:109958. https://doi.org/10.1016/j.rser.2020.109958
9. Chaudhry N, Dwivedi S, Chaudhry V, Singh A, Saquib Q, Azam A, Musarrat J (2018) Bio-inspired nanomaterials in agriculture and food: current status, foreseen applications and challenges. Microb Pathog 123:196–200. https://doi.org/10.1016/j.micpath.2018.07.013
10. Chen J, Nugen SR (2019) Detection of protease and engineered phage-infected bacteria using peptide-graphene oxide nanosensors. Anal Bioanal Chem 411(12):2487–2492. https://doi.org/10.1007/s00216-019-01766-6
11. Choudhary RC, Kumaraswamy RV, Kumari S, Sharma SS, Pal A, Raliya R, Biswas P, Saharan V (2019) Zinc encapsulated chitosan nanoparticle to promote maize crop yield. Int J Biol Macromol 127:126–135. https://doi.org/10.1016/j.ijbiomac.2018.12.274

12. Corbin KR, Byrt CS, Bauer S, DeBolt S, Chambers D, Holtum JA, Karem G, Henderson M, Lahnstein J, Beahan CT, Bacic A, Fincher GB, Betts N, Burton R (2015) Prospecting for energy-rich renewable raw materials: agave leaf case study. PLoS One 10(8). https://doi.org/10.1371/journal.pone.0135382
13. Dimkpa CO, McLean JE, David W, Britt DW, Anderson AJ (2015) Nano CuO and interaction with nano-ZnO or soil bacterium provide evidence for the interference of nanoparticles in metal nutrition of plants. Ecotoxicology 24:119–129. https://doi.org/10.1007/s10646-014-1364-x
14. Dimkpa C, McLean J, Britt D, Anderson A (2015) Nano-CuO and interaction with nano-ZnO or soil bacterium provide evidence for the interference of nanoparticles in metal nutrition of plants. Ecotoxicology 24:119–129. https://doi.org/10.1007/s10646-014-1364-x
15. Ewulonu CM, Liu X, Wu M, Yong H (2019) Lignin-containing cellulose nanomaterials: a promising new nanomaterial for numerous applications. J Bioresourc Bioprod 4(1):3–10. https://doi.org/10.21967/jbb.v4i1.186
16. Guha T, Gopal G, Kundu R, Mukherjee A (2020) Nanocomposites for delivering agrochemicals: a comprehensive review. J Agricult Food Chem 68:3691–3702. https://doi.org/10.1021/acs.jafc.9b06982
17. Hadj Alouane MH, Ahmed F, Hussein El Semary NA, Aldayel MF, Alhaweti FH, Nasr O (2021) Application of optimised nanocarbon materials and biofertilisers as a potent superfertiliser: towards sustainable agriculture production. Sci Adv Mater 13(5):812–819. https://doi.org/10.1166/sam.2021.3948
18. Hakeem KR, Pirzadah TB (eds) (2020) Nanobiotechnology in agriculture. Nanotechnology in the life sciences. https://doi.org/10.1007/978-3-030-39978-8
19. Husen A (2020a) Interactions of metal and metal-oxide nanomaterials with agricultural crops: an overview. In: Husen A, Jawaid M (eds) Nanomaterials for agriculture and forestry applications. Elsevier Inc., Cambridge, MA, USA, pp 167–197. https://doi.org/10.1016/B978-0-12-817852-2.00007-X
20. Husen A (2020b) Carbon-based nanomaterials and their interactions with agricultural crops. In: Husen A, Jawaid M (eds). Elsevier Inc., All rights reserved. Nanomaterials for agriculture and forestry applications. Cambridge, MA, USA, pp 199–218. https://doi.org/10.1016/B978-0-12-817852-2.00008-1
21. Husen A (2022) Engineered nanomaterials for sustainable agricultural production, soil improvement and stress management. Elsevier Inc., Cambridge, MA, USA
22. Husen A Iqbal M (2019) Nanomaterials and plant potential: an overview. In: Husen A, Iqbal M (eds) Nanomaterials and plant potential. Springer International Publishing AG, Cham, pp 3–29. https://doi.org/10.1007/978-3-030-05569-1_1
23. Husen A, Jawaid M (2020) Nanomaterials for agriculture and forestry applications. Elsevier Inc., Cambridge, MA, USA. ISBN 9780128178522, https://doi.org/10.1016/C2018-0-02349-X
24. Husen A, Siddiqi KS (2014) Phytosynthesis of nanoparticles: concept, controversy and application. Nanoscale Res Lett 9(229):1–24
25. Husen A, Siddiqi KS (2014) Carbon and fullerene nanomaterials in plant system. J Nanobiotech 12(16):1–10
26. Hussain A, Ali S, Rizwan M, Rehman M, Javed M, Imran M, Chatha S, Nazir R (2018) Zinc oxide nanoparticles alter the wheat physiological response and reduce the cadmium uptake by plants. Environ Pollut 242:1518–1526. https://doi.org/10.1016/j.envpol.2018.08.036
27. Hussain A, Ali S, Rizwan M, Rehman M, Qayyum M, Wang H, Rinklebe J (2019) Responses of wheat (*Triticum aestivum*) plants grown in a Cd contaminated soil to the application of iron oxide nanoparticles. Ecotoxicol Environ Saf 173:156–164. https://doi.org/10.1016/j.ecoenv.2019.01.118
28. Ibrahim HM (2015) Green synthesis and characterization of silver nanoparticles using banana peel extract and their antimicrobial activity against representative microorganisms. J Radiat Res Appl Sci 8(3):265–275. https://doi.org/10.1016/j.jrras.2015.01.007
29. Idumah CI, Zurina M, Ogbu J, Ndem U, Igba EC (2019) A review on innovations in polymeric nanocomposite packaging materials and electrical sensors for food and agriculture. Compos Interf. https://doi.org/10.1080/09276440.2019.1600972

30. Ioannou A, Gohari G, Papaphilippou P, Panahirad S, Akbari A, Dadpour MR, ..., Fotopoulos V (2020) Advanced nanomaterials in agriculture under a changing climate: the way to the future? Environ Exp Botany 176:104048. https://doi.org/10.1016/j.envexpbot.2020.104048
31. Khodakovskaya MV, Kim B-S, Kim JN, Alimohammadi M, Dervishi E, Thikra Mustafa T, Cernigla CE (2013) Carbon nanotubes as plant growth regulators: effects on tomato growth, reproductive system, and soil microbial community. Nano Small 1:115–123. https://doi.org/10.1002/smll.201201225
32. Kumar A, Choudhary A, Kaur H, Guha S, Mehta S, Husen A (2022) Potential applications of engineered nanoparticles in plant disease management: a critical update. Chemosphere 295:133798. ISBN: 0045-6535, https://doi.org/10.1016/j.chemosphere.2022.133798
33. Kumar A, Choudhary A, Kaur H, Mehta S, Husen A (2021) Smart nanomaterial and nanocomposite with advanced agrochemical activities. Nanosc Res Lett 16(156):1–26. ISBN: 1556-276X, https://doi.org/10.1186/s11671-021-03612-0
34. Mandal N, Adhikary S, Rakshit R (2020) Nanobiosensors: recent developments in soil health assessment. Soil Anal: Recent Trends Appl 285–304. https://doi.org/10.1007/978-981-15-2039-6_15
35. Mondal R, Dam P, Chakraborty J, Paret ML, Katı A, Altuntas S, Sarkar R, Ghorai S, Mandal AK, Husen A (2022) Potential of nanobiosensor in sustainable agriculture: the state-of-art. Heliyon. https://doi.org/10.2139/ssrn.4134873
36. Nandhini M, Rajinis SB, Udayashankar AC, Niranjana SR, Lund S, Shetty HS, Prakash HS (2019) Biofabricated zinc oxide nanoparticles as an ecofriendly alternative for growth promotion and management of downy mildew of pearl millet. Crop Prot 121:103–112. https://doi.org/10.1016/j.cropro.2019.03.015
37. Patra JK, Fraceto LF, Das G, Campos EVR (eds) (2020) Green nanoparticles. Nanotechnology in the life sciences. https://doi.org/10.1007/978-3-030-39246-8
38. Pawlak K, Kołodziejczak M (2020) The role of agriculture in ensuring food security in developing countries: considerations in the context of the problem of sustainable food production. Sustainability 12(13):5488. https://doi.org/10.3390/su12135488
39. Pelto J, Verho T, Ronkainen H, Kaunisto K, Metsajoki J, Seitsonen J, Karttunen M (2019) Matriz morphology and the particle dispersión in HDPE nanocomposites with enhanced wear resistance. Polym Test 77:105897. https://doi.org/10.1016/j.polymertesting.2019.105897
40. Phanthong P, Reubroycharoen P, Hao X, Xu G, Abudula A, Guan G (2018) Nanocellulose: extraction and application. Carbon Resourc Conv 1:32–43. https://doi.org/10.1016/j.crcon.2018.05.004
41. Rodriguez-Padron D, Puente-Santiago AR, Balu AM, Romero AA, Muñoz-Batista MJ, Luque R (2018) Benign-by-design orange peel-templated nanocatalysts for continuous flow conversion of levulinic acid to N-heterocycles. ACS Sustain Chem Eng 6:16637–16644. https://doi.org/10.1021/acssuschemeng.8b03896
42. Saleem H, Zaidi SJ (2020) Recent developments in the application of nanomaterials in agroecosystems. Nanomaterials 10(12):2411. https://doi.org/10.3390/nano10122411
43. Sharma P, Pandey V, Sharma MMM, Patra A, Singh B, Mehta S, Husen A (2021) A review on biosensors and nanosensors application in agroecosystems. Nano Res Lett 16:136. https://doi.org/10.1186/s11671-021-03593-0
44. Singh M, Sengupta A, Zeller K, Skoptsov G, Vander Wal RL (2019) Effect of hydrogen concentration on graphene synthesis using microwave-driven plasma-mediated methane cracking. Carbon 143:802–813. https://doi.org/10.1016/j.carbon.2018.11.082
45. Small CC, Degenhardt D (2018) Plant growth regulators for enhancing revegetation success in reclamation: a review. Ecol Eng 118:43–51. https://doi.org/10.1016/j.ecoleng.2018.04.010
46. Thurlow J, Dorosh P, Davis B (2019) Demographic change, agriculture, and rural poverty. Sustain Food Agricult 31–53. https://doi.org/10.1016/B978-0-12-812134-4.00003-0
47. Urbaniec K, Bakker RR (2015) Biomass residues as raw material for dark hydrogen fermentation—a review. Int J Hydrogen Energy 40(9):3648–3658. https://doi.org/10.1016/j.ijhydene.2015.01.073

48. Vimala V, Clarke SK, Urvinder KS (2016) Pesticides detection using acetylcholinesterase nanobiosensor. Biosens J 5:1–4. https://doi.org/10.4172/2090-4967.1000133
49. Xue JZ, Luo L, Li P, Cui Y, Wu Q (2014) A residue-free green synergistic antifungal nanotechnology for pesticide thiram by ZnO nanoparticles. Sci Rep 4:1–9. https://doi.org/10.1038/srep05408
50. Zulfiqar F, Navarro M, Ashraf M, Akram N, Munné-Bosch S (2019) Nanofertilizer use for sustainable agriculture: advantages and limitations. Plant Sci 289. https://doi.org/10.1016/j.plantsci.2019.110270
51. Zuverza-Mena N, Martínez-Fernández D, Du W, Hernandez Viezcas J, Bonilla-Bird N, López-Moreno M, Komárek M, Peralta-Videa J, Gardea-Torresdey J (2017) Exposure of engineered nanomaterials to plants: insights into the physiological and biochemical responses—a review. Plant Physiol Biochem 110:236–264. https://doi.org/10.1016/j.plaphy.2016.05.037

Synthesis of Metal Nanoparticles from Vegetables and Their Waste Materials for Diverse Application

Shivam Sharma, Anuj Choudhary, Viveka Katoch, D. R. Chaudhary, Radhika Sharma, Antul Kumar, Payal Sharma, Satyakam Guha, Anand Sonkar, and Sahil Mehta

Abstract Researchers are gaining interest globally in the synthesis of metallic nanoparticles (NPs) due to their unique properties and limitless reach. The risk of conventional chemical and physical methods for synthesis is well aware of its toxic and various side effects on the environment. Therefore, the shift toward the green mode of synthesis is an urgent need in the current wave of sustainability scenarios. The bio-molecules in the plant extracts are showing immense foundation potential in the synthesis of metal NPs. A large amount of vegetable waste is generated on a regular basis and established as an emerging source of bio-molecules for the synthesis of nanomaterials. The metallic NPs synthesized from vegetable wastes are showing high stability, reactivity, and biocompatibility to increase the effectiveness of the end application. The present chapter discusses the green synthesis of NPs from vegetables, their applications, and the role of factors such as pH, temperature, reactant concentration, and reaction time affecting the morphological properties of metallic NPs. The variability in the morphology of NPs with the aid of various extracts of vegetables is going to be discussed.

S. Sharma · D. R. Chaudhary · P. Sharma
Department of Vegetable Science and Floriculture, CSK HPKV, Himachal Pradesh, Palampur 176062, India

A. Choudhary
Department of Biology and Environmental Sciences, CSK HPKV, Palampur 176062, India

V. Katoch
Department of Seed Science and Technology, CSK HPKV, Himachal Pradesh, Palampur 176062, India

R. Sharma
Department of Soil Science, Punjab Agricultural University, Ludhiana, Punjab 141004, India

A. Kumar
Department of Botany, Punjab Agriculture University, Ludhiana, Punjab 141004, India

S. Guha · A. Sonkar · S. Mehta (✉)
Department of Botany, Hans Raj College, University of Delhi, New Delhi 110007, India
e-mail: sahilmehtasm21@hrc.du.ac.in

A. Husen (ed.), *Nanomaterials from Agricultural and Horticultural Products*, Smart Nanomaterials Technology,
https://doi.org/10.1007/978-981-99-3435-5_2

Keywords Green synthesis · Sustainability · Biocompatibility · NPs

Abbreviations

Human cervical carcinoma cells	HeLa
Human epidermoid larynx carcinoma	HEp 2
Methylene blue	MB
Methylene orange	MO
Nanoparticles	NPs

1 Introduction

Nanotechnology is the novel and most attractive globalized concept of advanced science. The focus on green nanotechnology is currently an ideal and in-progress approach for minimizing the effects of manufacturing nanomaterials and associated method problems [82]. The structural attributes (shape and size) are affected by chemical conditions, pH, and temperature. The utilization of nanoparticles (NPs) is advantageous, i.e. they have a very small size with a higher surface area as compared to bulk form [49]. Nano-based materials pose diverse potent alternatives for their use in biological aspects [1, 35, 36, 38, 40, 46, 56, 57, 70, 88]. NPs are utilized more frequently in medical, environmental, and biological fields due to their biological absorption, tumor targeting, bioavailability, bioactivity, biocompatibility, antimicrobial and anti-inflammatory action, and effective drug delivery, [52, 89]. The applications of NPs range from electronics, medicine, sensors, information technology, materials chemistry, biomedical, agriculture, energy, catalysis, optical, and environment [10, 37, 41, 93]. The biological methods are based on the plants, fungi, algae, bacteria, and viruses for the NP biosynthesis [13, 39, 90]. Two kinds of approaches are implemented for nanomaterial synthesis: top-down strategy (large structures are broken by physical, chemical, and biological energy into smaller pieces) and bottom-up strategy (where the material is generated into large nanostructure using biological, physical, and chemical reactions) [20]. Chemical and physical approaches are operational for the NP synthesis. These involve the use of toxic chemicals such as stabilizers, organic solvents, and reducing agents. A clean, reliable, eco-friendly, and biologically appropriate technique is needed for the synthesis of NPs [14, 41, 47, 55].

The diversity in plants has been widely considered for the variability potential of phytoconstituents mainly in leaves involving aldehydes, flavones, ketones, terpenoids, amides, phenols, ascorbic acids, and carboxylic acids. These help to reduce the metallic component of metal NPs. In this chapter, we reviewed the plant-based biosynthesis of NPs, vegetables as nanofactories, factors affecting the biological synthesis, different kinds of NP synthesis from vegetables, and the application of NPs.

2 Plant-Based Synthesis of NPs

Plants have a broad range of phytoconstituents such as terpenoids, alkaloids, steroids, flavonoids, and others such as reducing agents for the production of NPs. *Acalypha indica, Centella asiatica, Ficus benghalensis, Parthenium hysterophorus, Passiflora foetida, Plumbago zeylanica, Sapindus rarak,* and *Zingiber officinale* are the plants that have recently gained attention in the light of plant-based green synthesis of NPs [42, 80, 83, 99]. The extract of plants has greater significance over microbial green synthesis because it is a single-step process, cost-effective, and non-pathogenic [66, 84]. Such NPs have unique properties such as physical, electrical, chemical, thermal, optical, and magnetic properties as compared to bulk materials with several applications [24]. It helps in the exclusion of dangerous by-products and supports fine-tuning the size of NPs. The extract of *Camellia sinensis* was used to generate the spherical-shaped and irregular clustered configurations of iron oxide NPs. The compound phyllanthin extracted from *Phyllanthus amarus* can be utilized to generate silver and gold NPs. The study is based on a single constituent of the plant extract to generate metallic NPs and is considered as unique as compared to whole plant exploitation. The size and shape of generated NPs have been influenced by the phyllanthin concentration. The high concentration of phyllanthin was used to produce hexagonal AuNPs, while the low concentration of phyllanthin was used for hexagonal and triangular NPs production. Polysaccharides, cellulose, chitosan, dextran, hyaluronic acid, alginic acid, and soluble starch have been harnessed for the synthesis of Au and AgNPs successfully [17, 51, 62, 78]. These extracted compounds recommended the use of these less toxic compounds and favor the formation of different nanocomposites with various metals. The extract of lemon grass on incubation with Au tetrachloride solution leads to the generation of liquid-like nanotriangles due to the aggregation of spherical gold NPs [85]. The surface of NPs forms a complex with ketones or aldehydes available in the plant extracts to contribute to the fluidity. ZnONPs were produced from the extracts of leaf of *Calotropis gigantea, Coriandrum sativum, C. sinensis, Acalypha indica,* and *Hibiscus rosa-sinensis.* The TiO NPs were produced from the extracts of *Eclipta prostrata* and *Jatropha curcas.* Similarly, the oxides of Cu were produced from the leaf extracts of *Aloe sylvestris* and *A. barbadensis* [45]. The metal of the NPs was observed to be stabilized and reduced by the phytoconstituents, and macromolecules were reported [80]. Macro-molecules are the redox intermediates for the reduction of metals and capped mediators for non-agglomeration, and the modification of post surface for NPs. Moreover, the generated NPs are pollutant-free and appropriate for physiologically mediated uses [75]. Several experimental studies on the plant components to generate ZnO from the herbal extracts of *Vitex trifolia, Ocimum basilicum,*

Swertia chirayita, Catharanthus alata, C. roseus, Scadoxus multiflorus, Eichhornia crassipes, and *Agathosma betulina* [32]. Similarly, the SeNPs were synthesized from various plant species for impacted application [42].

3 Role of Various Factors in Green Synthesis of Metal NPs

Morphological features of NPs can be modified by several factors such as temperature, pH, reactant concentration, and time of reaction as shown in Fig. 1. Such factors are decisive for understanding the impact of environmental factors for the NP synthesis as they have a role in the optimization of metallic NPs in biological procedures [33, 60].

3.1 Temperature

The reaction temperature is critical for the determination of the size, shape, and yield of synthesized NPs with the aid of plants [43]. For instance, the peel extract of sweet orange was allowed to generate the particle with 35 nm of average size at 25 °C. When the temperature was increased to 60 °C, the average NP size was decreased to 10 nm [50]. The stable AgNPs were produced from the leaf extract of persimmon at 25–95 °C of reaction temperature [92]. Alteration in the reaction temperature acted

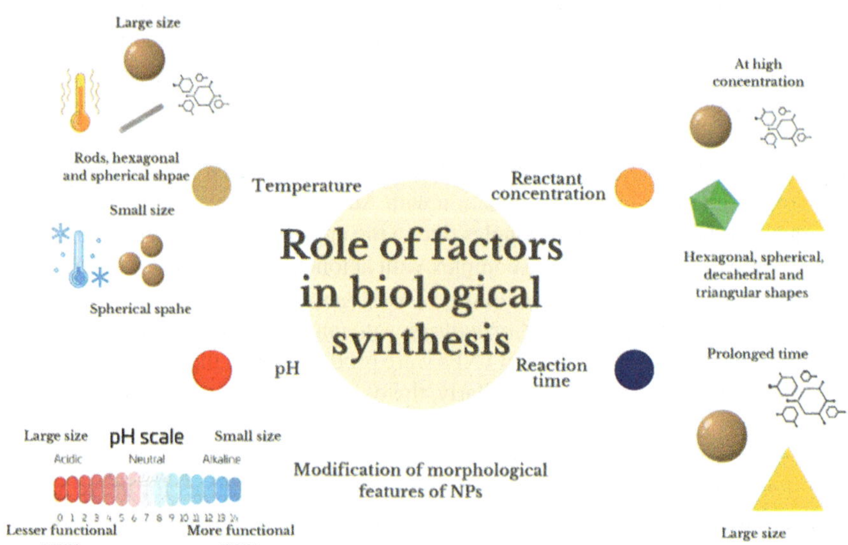

Fig. 1 Representation of morphological refining of metallic nanoparticles under certain factors

as determining factor in the size and shape of AuNPs from the oat biomass. At high temperatures, the enhancement in the rate of AuNPs was observed. At lower temperatures, the spherical shape of AuNPs was produced while at higher temperatures, the plate-like and rod-like NPs were produced [29]. AgNPs were generated as spherical shapes at 4 °C, whereas the different shapes such as rods, hexagonal, and spherical were obtained at 60 °C. And after 5–7 days at 60 °C, the size of AgNPs prepared exceeded from nano- to micro-scale [7].

3.2 pH

The pH of the reaction medium is one of the crucial factors in the NP formations. The change in pH alters the shape and size of NPs as the large-sized NPs are generated in the acidic medium [81]. For instance, the rod shape AuNPs are synthesized with the biomass of *Avena sativa* at 2 pH of 25–85 nm of a size range, while it is reduced to 5–20 nm at 3–4 pH [9]. Moreover, functional group accessibility for the nucleation of a particle in the extract was comparatively better at 3 or 4 values of pH than at 2 values of pH where lesser functional groups were available for particle aggregation. The enhancement in the number of spherical AgNPs was obtained by the bark extract of *Cinnamon zeylanicum* at pH value (pH > 5). The increase in the particle size was reported at acidic pH; PdNPs showed 15–20 nm of particle size at pH < 5, while it was 20–25 nm in an acidic medium [59]. AgNP synthesis was produced by green synthesis procedure using the leaf extract of *Lawsonia inermis* under different temperatures and pH. There were mono-dispersive AgNPs obtained at a 9 pH value [54].

3.3 Reactant Concentration

Metallic NP formation is influenced by the reactant concentration of bio-molecules of the extract [4]. The shape of biologically synthesized Ag and Au NPs was affected by the biomass concentration used by the leaf extract of camphor [34]. The addition of chloroauric acid precursor to the extract was led to the growing concentration for the spherical-shaped NP synthesis despite triangular-shaped NPs. Alteration regarding the ratio of spherical to triangular NPs in the chloroaurate ion-rich medium was due to carbonyl compounds while using the leaf extract of *Aloe vera* [18]. Hexagonal, spherical, triangular, and decahedral shapes of silver NPs were generated by the different concentrations of leaf extract of *Plectranthus amboinicus* [74].

3.4 Reaction Time

The reaction time is the key factor in the synthesis of NPs. The color change was rapidly reported within 2 min with the extract of Pineapple for the synthesis of AgNPs and the faster decrease in the aqueous solution of $AgNO_3$, thus forming the NPs within 2 min. The synthesized NPs have 12 nm of average spherical size [3]. The leaf extract of the *Chenopodium album* was used for the generation of Au and AgNPs. After a span of 15 min, the NPs were formed and it continued for 120 min, and lesser number of NPs having larger sizes were produced [26]. Similarly, the alteration in the size of the particle was observed from the 10–35 nm range with the prolongation in time from 30 min to 4 h with the help of leaf extract of *Azadirachta indica* and $AgNO_3$ solution [95]. The leaf extract of *Ennab* was studied at different times for the stable formation of NPs [6].

4 Vegetables as Nanofactories

Even in the scenario of the twenty-first century, increase in vegetable yield has become a challenge for breeders and researchers to feed our ever-growing population. Vegetables are consumed raw or cooked and are the richest source of vitamins and minerals with anti-oxidant properties [65]. Vegetables are preferred on a daily basis by people not only in India but all over the world. The 7.2 billion current global population is estimated to increase by 1 billion within the upcoming 3 years, reaching 8.1 billion in 2025 and 9.6 billion in 2050. As the world population continues to grow, the rising demand for vegetable production is significant. This continuous increase in vegetable production has also led to a 60% increase in the wastage of vegetables in terms of stale/peels (Surendra et al. 2016b). The management of vegetable waste is very much difficult due to its wastage. As a biodegradable material, it is dumped freely on soil which in return will deplete its fertility and quality. Moreover, these peels of vegetable crops due to improper management accumulate on the surface of the soil which is then further responsible for air, water, and land/soil pollution [61]. These pollutions lead to global warming which is the major reason behind the sudden change in global climatic conditions as rightly said by IPCC [77]. That is, the global temperature is raised by 2 °C in the past 10 years.

Even today, the biodegradation or management of vegetable waste remains under concern. Therefore, there is a huge need to develop eco-friendly techniques or products which can be used in a certain way with huge benefits which will ultimately resolve the problem of dumping/management of vegetable waste [22]. Significant efforts have been made by many researchers to have newer products from vegetable waste that have both social and economic importance [28]. These products have many active constituents including polyphenols (gallic acid and caffeic acid), amino acids, minerals, and vitamins. Nowadays, the synthesis of nano-based materials/nanoparticles with the aid of vegetable-derived waste has shown the potential to solve the many associated problems [79]. Vegetable waste is efficiently exploited in the production of nano-based materials with various functional groups therefore vegetable waste is also referred to as nanofactories. Moreover, vegetable waste reduces the NP toxicity, easily replaces the toxic reducing agents, and tunes the morphology and surface-associated functional groups. Thereby, the exploitation in end applications is especially in biological-based applications [87]. This technique will save our environment and many lives of the biotic community that may be plants, humans, animals, birds, etc. Worldwide, the techniques of developing nanoparticles or nanomaterials have been adopted by people to have proper management of vegetable wastage.

5 Biosynthesis of Different Nanoparticles

The biosynthesis of nanoparticles or nanomaterials might be simple, eco-friendly, and economically viable. A huge number of metals and metal-oxides have been biosynthesized by the bottom-up approach using vegetable waste in which the atoms come together and assemble making the building block of precursor (Table 1). Thereafter, these nanoparticles have been used in various applications like the removal of pollutants from water and antimicrobial activity [79]. The green synthesis of nanomaterials/NPs depends on many experimental steps like features of vegetable waste, the concentration of vegetable waste in a reaction, metal ion concentration present in a reaction, and temperature, pH, and contact time of reaction mixture. In the early stage, the metal ions are mixed with the vegetable wastes extracted to produce the nanoparticles [19]. Generally, vegetable waste acts as a reducing agent in a reaction process. When the formation of metal nanoparticles takes place, the nucleation starts, and then the growth of those nanoparticles occurs. When the growth ends, the stabilization or encapsulation of NPs happens, which leads to biocompatibility, reactivity, and stability. Thereafter, these nanoparticles can be used in endless applications. Wastage of many vegetable crops has been used to develop nanoparticles having anti-bacterial, antifungal, and many more properties.

The waste extract of Cauliflower was used to biosynthesize AgNPs and then further test the potential application of AgNPs in photo-catalytic methylene blue (MB) dye degradation and Hg^{2+} bio-sensing. It was observed that AgNPs can act as an amazing tool for degrading MB dye in detecting Hg-based pollutants and industrial effluents [48]. The CuO NPs have been synthesized by using cauliflower, potatoes, and peas waste. Cauliflower has shown high degradation of Methylene Blue as compared to Potatoes and Peas peels [97]. Zinc oxide NPs were synthesized by using *Allium cepa* peel waste, and the study showed that zinc oxide NPs have been recommended as a nano-based application of nutrient source for agriculture [67]. Magnetic iron oxide nanoparticle green synthesis from the iron (III) chloride has been done by use of aqueous extract of peel of *Brassica oleracea* var. *capitata* sub. var. *rubra*. The solution prepared of iron oxide nanoparticles was characterized

Table 1 Biosynthesis of metal nanoparticles from various vegetable extracts

Name	NPs	Part used	Size (nm)	Morphology	Phyto-constituents	References
Watermelon	Ag	Fresh leaf	31	Spherical	Polyphenols	[86]
Ivy gourd	Ag	Leaf extract	20–30	Cubic	Polyphenols	[11]
Yam	Ag	Tuber	8–20	Nanorod, triangle, hexagonal and spherical	Flavonoids	[31]
Pepper	Ag	Leaf extract	5–60	Spherical and cubic (biphase)	Proteins	[64]
Drumstick	Ag	Fresh stem bark	40	Spherical and pentagonal	Phenols	[98]
Beetroot	Ag	Root	15	Spherical	Phenols	[15]
Bitter gourd	Ag	Juice extract	600	Spherical and rod	Polyphenols	[23]
Bottle gourd	Au	Peel	40–50	Cubic	Polyphenols	[58]
Tomato, brinjal, cucumber, and bottle and sponge gourd	Fe	Peel	20	Spherical	Proteins	[87]
Drumstick	Zn	Peel	40–45	Hexagon	Phenol and amine	[93]
Cauliflower, peas, and potato	Cu	Peel	22.20, 24.70, 31.60	Flat, cubic	–	[97]
Green vegetable waste	Fe	Peel	10–90; 10–70	Spherical	Polysaccharides	[72, 73]
Broccoli	Zn	Broccoli extract	14–17	Hexagonal	Flavonoid (quercetin)	[76]
Carrot	Ag	Tops	2–25	Spherical and pseudo-spherical	Ascorbic, gallic, and chlorogenic acids	[30]
Brinjal	Ag	Green calyx	8–10, 8–12	Spherical	Phenolic hydroxyl moieties	[12]
Cauliflower	Ag	Extract	35.08	Cubic	Proteins and flavanones	[48]
Tomato	Ag	Fruit extract	9.58–72.69	Cubic	Phenol/alcohol	[68]

(continued)

Table 1 (continued)

Name	NPs	Part used	Size (nm)	Morphology	Phyto-constituents	References
Pumpkin	Zn	Seed extract	28.07	Hexagonal	Polyphenols, terpenoids, and proteins	[5]
Cabbage	Mn	Leaves	10.70	Spherical and ellipsoidal	Flavonoids, alkaloids, and proteins	[8]
Okra	Ag	Flowers	5.52–31.96	Spherical	Phenols	[21]
Red cabbage	Fe	Leaves	300–500	Spherical	–	[27]
Spinach	Fe	Leaves	10–70	Amorphism	Phenol	[96]
Radish	Ag	Seeds	5–20	Spherical	Free amino and carboxyl group	[53]
Onion	Zn	Peel	20–80	Spherical	Amide and hydroxyl group	[67]

with ultraviolet-visible spectroscopy, Fourier transform infrared spectroscopy, scanning electron microscopy, energy-dispersive X-ray spectroscopy, and zeta potential (Fig. 2). The results showed that it makes a potential chemotherapeutic agent for breast cancer treatment [27].

Many NPs have been used for dye degradation from the water like ZnO, AuO, and AgO. These metals and metal-oxide nanoparticles showed toxicity which leaches out into the environment and kills many animals and micro-organisms. Therefore, many vegetable wastes are exploited to generate metal and metal-oxide NPs to manage and reduce the bad effects of the wastes. Many experiments have been conducted to fulfill the need of today's scenario to have proper management of vegetable waste such as making diverse products including paints, sunscreens, and textiles, therefore, ZnONPs are important materials. In the study, the photo-catalytic effect of produced zinc oxide NPs from *Moringa oleifera* was studied on the degradation of harmful crystal violet dye in the environment. ZnONPs gave excellent results in the case of anti-bacterial, and hemolytic activities and against *Alternaria solani* and *Sclerotium rolfsii* strains (antifungal activity) [94]. AuNPs are synthesized by using vegetable extract of *Sterculia acuminata* and examined against 4-nitrophenol, MB, direct blue 24, and methylene orange (MO) dyes. The study showed that the AuNPs degraded all the used dyes. Methylene blue and methylene orange degrade within a period of 12 min but direct blue 24 dyes and 4-nitrophenol took 18 and 36 min, respectively for degradation [16].

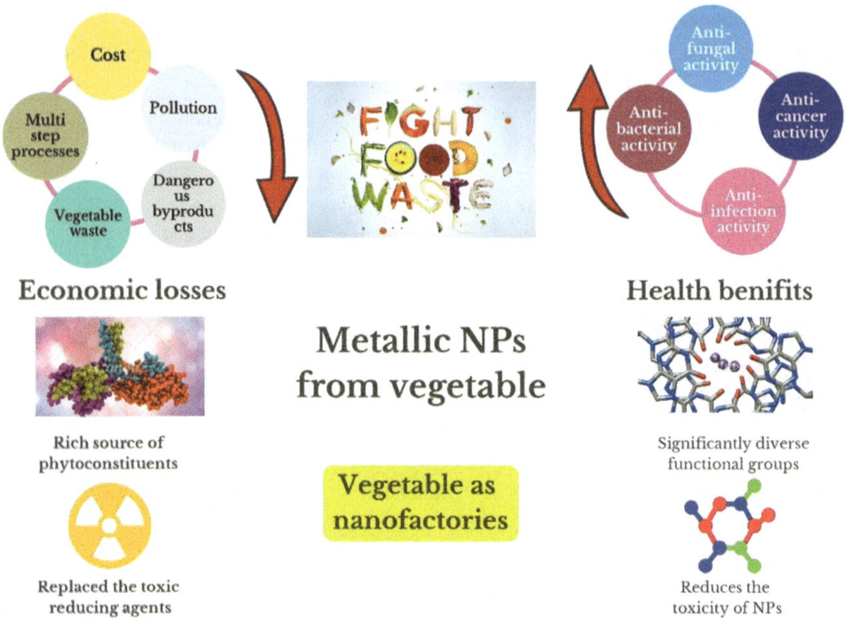

Fig. 2 Overview of metallic nanoparticles from vegetable waste

6 Applications

There are many applications observed for nanoparticles/nanomaterials, but the major ones are they act as an anti-bacterial and antifungal agent and they also help to clean polluted water which is an essential need in the current scenario. There are many vegetable crops whose waste is being used to synthesize different nanoparticles which have very less toxicity levels [19]. Table 2 shows various vegetable crops involved, different nanoparticles produced, and their applications. The metals/metal oxides like Cu, Zn, Ag, AuO, ZnO, CuO, and AgO have anti-bacterial or antifungal activity, but the metals which are biosynthesized have both properties as well as are less toxic as compared to the former [2].

The antimicrobial activity can be explained by many researchers like in a study where vegetable waste was prepared by using five different plants such as *Cucumis sativus, Lagenaria siceraria, Solanum lycopersicum, S. melongena,* and *Luffa cylindrica.* By using this source, silver nanoparticles (AgNPs) were synthesized and found to be anti-bacterial against *K. pneumoniae* and *E. coli* [87]. *Musa acuminata* and *Spinacia oleracea* were selected for the experimental study, and an aqueous extract of both was prepared for the FeNP synthesis, and their anti-bacterial activity against MTTC 1133 of *Bacillus subtilis* and MTTC 62 of *Escherichia coli* can be exploited as anti-bacterial activity against the pathogenic bacteria and shows non-toxicity in nature [96]. AgNPs were produced from the vegetable waste of *Beta*

Table 2 Potential application of nanoparticles derived from vegetables

NPs	Source	Potential role	Descriptions	References
Ag	*Citrullus colocynthis*	Anti-cancer activity	Reduced the viability of Human epidermoid larynx carcinoma (HEp 2) to half	[86]
Ag	*Coccinia grandis*	Anti-bacterial activity	Against the *E. coli*	[11]
Ag	*Dioscorea bulbifera*	Antimicrobial activity	Against the *E. coli*	[31]
Ag	*Moringa oleifera*	Anti-cancer activity	Triggered apoptosis in Human cervical carcinoma cells (HeLa)	[98]
Ag	*Beta vulgaris*	Antimicrobial and catalytic activity	–	[15]
Au	*Sterculia acuminata* vegetable extract	Dye degradation	Degraded the methylene blue, orange, 4-nitrophenol dye, and direct blue 24 dyes	[16]
Ag	*Moringa oleifera* and *Murraya koenigii*	Anti-bacterial activity	Prevent growth of *S. aureus* strain	[44]
Ag	*Lagenaria siceraria*	Antioxidant activity	–	[58]
Ag	*Bauhinia tomentosa* vegetable extract	Anti-cancer activity	Inhibit the cancer cell lines	[71]
Ag	*Lagenaria siceraria, Luffa cylindrica, Solanum lycopersicum, Solanum melongena,* and *Cucumis sativus*	Anti-bacterial activity	Against the *E. coli* and *K. pneumonia*	[87]
ZnO	*Moringa oleifera*	Dye degradation and antimicrobial activity	Degrade the crystal violet dye inhibit several microbial growth	[93]
Cu	Cauliflower, Potatoes, and Peas	Dye degradation	Induce degradation of Methylene Blue	[97]
Ag	Vegetable waste extract	Antimicrobial activity	Against *Klebsiella* species and *Staphylococcus* species	[72]
Au	Vegetable waste extract	Antimicrobial activity	Against *Klebsiella* species and *Staphylococcus* species	[73]
Ag	Vegetable plant waste extract	Anti-bacterial activity	Inhibit the *E. coli* and *S. aureus*	[69]
Zn	*Pithecellobium dulce* peel waste	Antifungal activity	Inhibit the *Aspergillus flavus* and *Aspergillus niger*	[63]

(continued)

Table 2 (continued)

NPs	Source	Potential role	Descriptions	References
Ag	*Trigonella foenum-graecum*	Dye degradation	Degrade the highly reactive blue 9 and reactive yellow 186	[91]
Ag	*Citrus limetta* and vegetable extract	Anti-bacterial activity	Inhibit *Escherichia coli*, *Staphylococcus aureus Micrococcus luteus* and *Candida* sp.	[25]
Ag	*Daucus carota*	Anti-bacterial activity	Against *E. coli* and *S. epidermidis*	[30]
Ag	*Solanum melongena*	Anti-bacterial activity	Antimicrobial against *E. coli*	[12]
Ag	*Brassica oleracea* var. *botrytis*	–	Detecting mercury-based pollutants	[48]
Mn	*Brassica oleracea* var *capitata*	Anti-bacterial activity	Against *S. aureus*, *E. coli*, and *S. typhi*	[8]
Ag	*Abelmoschus esculentus*	Anti-bacterial and anti-cancer activity	Anti-bacterial against *Bacillus subtilis* and anti-cancerous against the TERT-4 and A-549 cell lines	[21]
Fe	*Brassica oleracea* var. *capitata* sub. var. *rubra*	Anti-cancer activity	Breast cancer treatment	[27]
Fe	*Spinacia oleracea* and *Musa acuminata*	Anti-bacterial activity	Against pathogenic bacteria *Bacillus subtilis* (MTTC 1133) and *Escherichia coli* (MTTC 62)	[96]
Ag	*Ipomoea batatas L* var. *Rancing*	Anti-bacterial activity	Against *Staphylococcus aureus*	[100]
Ag	*Raphanus sativus*	Anti-cancer activity	Against the colon carcinoma cell line HCT-15	[53]

vulgaris, *Lepidium aucheri*, and *Petroselinum crispum* and observed that they have strong anti-bacterial properties against *S. aureus* and *E. coli* [69].

7 Conclusion

The green synthesis at the nanoscale has proved to be of immense potential. It suffered from some shortcomings involving material selection, product quality, synthesis condition, and application. These shortcomings pose challenges for adopting them on an industrial scale and large-scale implementation of green synthesized metallic NPs. Several plant materials have been implemented for the green synthesis of NPs, and research has been made to exploit the local abundance of plant materials. Various kinds of natural-based extracts have been utilized as effective resources for material synthesis and fabrication. Plant extracts have higher efficiency as stabilizing and reducing agents for the biosynthesis of controlled size, shape, structures, and other related features. The improvement of eco-friendly and reliable processes for metal NPs is the major walkthrough in the field of nanotechnology. The progress in these approaches is still under challenge and at the developmental stage. Purification and separation of NPs are other important parameters that need immense exploration. Metallic NPs produced by plant extracts are highly stable as compared to extracts of various organisms. Future direction in the research and development of green synthesis NPs should channelize the lab work toward the integration into industrial scale by traditional health and environmental issues. Several vegetable extracts are shown to regulate the shape and size of NPs resulting in high efficiency in the light of end applications. Conclusively, the NPs generated from the vegetable is an attractive applicant with various applications.

References

1. Abd Elkodous M, El-Sayyad GS, Abdelrahman IY, El-Bastawisy HS, Mosallam FM, Nasser HA, El-Bata AI (2019) Therapeutic and diagnostic potential of nanomaterials for enhanced biomedical applications. Colloids Surf B 180:411–428
2. Afreen S, Talreja N, Chauhan D, Ashfaq M (2020) Polymer/metal/carbon-based hybrid materials for the detection of heavy metal ions. In: Abd-Elsalam KA (ed) Multifunctional hybrid nanomaterials for sustainable agri-food and ecosystems. Elsevier
3. Ahmad N (2012) Green synthesis of silver nanoparticles using extracts of *Ananas comosus*. Green Sustain Chem 2:141–147
4. Ahmed SF, Mofijur M, Rafa N, Chowdhury AT, Chowdhury S, Nahrin M, Islam AS, Ong HC (2022) Green approaches in synthesizing nanomaterials for environmental nano bioremediation: technological advancements, applications, benefits and challenges. Environ Res 204:111967
5. AIP Conference Proceedings 2349, 020010 (2021); https://doi.org/10.1063/5.0051826
6. Aljabali AA, Akkam Y, Al Zoubi MS, Al-Batayneh KM, Al-Trad B, Abo Alrob O, Alkilany AM, Benamara M, Evans DJ (2018) Synthesis of gold nanoparticles using leaf extract of *Ziziphus zizyphus* and their antimicrobial activity. Nanomat (Basel) 8(3):174
7. Al-Namil DS, Khoury EE, Patra D (2019) Solid-state green synthesis of AgNPs: higher temperature harvests larger AgNPs but smaller size has better catalytic reduction reaction. Sci Rep 9(1):1–9
8. Amatya SP, Shrestha S (2021) Biosynthesis of manganese nanoparticles (MnNPs) from *Brassica oleraceae* (cabbage leaves) and its antibacterial activity. Asian J Chem Sci 9(1):1–11
9. Armendariz V, Herrera I, Jose-Yacaman M, Troiani H, Santiago P, Gardea-Torresdey JL (2004) Size controlled gold nanoparticle formation by *Avena sativa* biomass: use of plants in nanobiotechnology. J Nanoparticle Res 6:377–382
10. Arokiyaraj S, Kumar VD, Elakya V, Kamala T, Park SK, Ragam M, Vincent S (2015) Biosynthesized silver nanoparticles using floral extract of *Chrysanthemum indicum*-potential for malaria vector control. Environ Sci Pollut Res 22(13):9759–9765
11. Arunachalam R, Dhanasingh S, Kalimuthu B, Uthirappan M, Rose C, Mandal AB (2012) Phytosynthesis of silver nanoparticles using *Coccinia grandis* leaf extract and its application in the photocatalytic degradation. Colloids Surf B94:226–230
12. AttiaGH AHS, Orabi MAA, Gaara AH, El Raey MA (2020) Antimicrobial activity of silver and zinc nanoparticles mediated by eggplant green calyx. Int J Pharmacol 16:236–243

13. Bachheti RK, Abate L, Deepti, Bachheti A, Madhusudhan A, Husen H (2021) Algae-, fungi-, and yeast-mediated biological synthesis of nanoparticles and their various biomedical applications. In: Kharisov B, Kharissova O (eds) Handbook of greener synthesis of nanomaterials and compounds. Fundamental principles and methods, vol 1. Elsevier Inc., 50 Hampshire St., 5th Floor, Cambridge, MA 02139, USA, pp 701–734
14. Bachheti RK, Gupta V, Husen A, Joshi A, Konwarh R (2019) Green synthesis of iron oxide nanoparticles: cutting edge technology and multifaceted applications. In: Husen A, Iqbal M (eds) N*anomaterials* and plant potential, vol 11, 6330th edn. Springer International Publishing AG, Gewerbestrasse, Cham, pp 239–259. ISBN: 978-3-030-05569-1. https://doi.org/10.1007/978-3-030-05569-1_9
15. Bindhu M, Umadevi M (2015) Antibacterial and catalytic activities of green synthesized silver nanoparticles. Spectrochim Acta A 135:373–378
16. Bogireddy NKR, Anand MKK (2015) Gold nanoparticlessynthesis by *Sterculia acuminata* extract and its catalytic efficiency in alleviating different organic dyes. J Mol Liq 211:868–875
17. Cai J, Kimura S, Wada M, Kuga S (2009) Nanoporous cellulose as metal nanoparticles support. Biomacromole 10:87–94
18. Chandran S, Chaudhary M, Pasricha R, Ahmad A, Sastry M (2006) Synthesis of gold nanotriangles and silver nanoparticles using aloevera plant extract. Biotechnol Prog 22:577–583
19. Chauhan D, Afreen S, Talreja N, Ashfaq M (2020) multifunctional copper polymer-based nanocomposite for environmental and agricultural applications. In: Abd-Elsalam KA (ed) Multifunctional hybrid nanomaterials for sustainable agri-food and ecosystems. Elsevier
20. Das RK, Pachapur VL, Lonappan L, Naghdi M, Pulicharla R, Maiti S, Cledon M, Dalila LM, Sarma SJ, Brar SK (2017) Biological synthesis of metallic nanoparticles: plants, animals and microbial aspects. Nanotechnol Environ Eng 2:18
21. Devanesan S, AlSalhi MS (2021) Green synthesis of silver nanoparticles using the flower extract of *Abelmoschus esculentus* for cytotoxicity and antimicrobial studies. Int J Nanomed 16:3343–3356
22. Dhanorkar MN, Sabnis A, Kale SP (2022) Perfect home composting of biodegradable waste resources for reducing carbon foot print and climate change associated with global warming-a microbiological approach. Cardiometry 22:185–190
23. Dhivya G, Rajasimman M (2015) Synthesis of silver nanoparticles using *Momordica charantia* and its applications. J Chem Pharm Res 7(9):107–113
24. Duran N, Seabra A (2012) Metallic oxide nanoparticles: state of the art in biogenic syntheses and their mechanisms. Appl Microbiol Biotechnol 95:275–288
25. Dutta T, Ghosh NN, Das M, Adhikary R, Mandal V, Chattopadhyay AP (2020) Green synthesis of antibacterial and antifungal silver nanoparticles using *Citrus limetta* peel extract: experimental and theoretical studies. J Environ Chem Eng 8(4):104019
26. Dwivedi A, Gopal K (2010) Biosynthesis of silver and gold nanoparticles using *Chenopodium album* leaf extract. Colloids Surf A Physicochem Eng Asp 369:27–33
27. Erdogan O, Pasa S, Demirbolat GM, Çevik Ö (2021) Green biosynthesis, characterization, and cytotoxic effect of magnetic iron nanoparticles using *Brassica Oleracea* var *capitata* sub var *rubra* (red cabbage) aqueous peel extract. Turk J Chem 45:1086–1096
28. Ganesh KS, Sridhar A, Vishali S (2022) Utilization of fruit and vegetable waste to produce value-added products: conventional utilization and emerging opportunities-a review. Chemosphere 287(3):132221
29. Gericke M, Pinches A (2006) Biological synthesis of metal nanoparticles. Hydrometallurgy 83:132–140
30. Ghosh PR, Fawcett D, Platten M, Sharma SB, Fosu-Nyarko J, Poinern GEJ (2020) Sustainable green chemical synthesis of discrete, well-dispersed silver nanoparticles with bacteriostatic properties from carrot extracts aided by polyvinylpyrrolidone. AIMS Mater Sci 7(3):269–287
31. Ghosh S, Patil S, Ahire M, Kitture R, Kale S, Pardesi K, Cameotra SS, Bellare J, Dhavale DD, Jabgunde A, Chopade BA (2012) Synthesis of silver nanoparticles using *Dioscorea bulbifera* tuber extract and evaluation of its synergistic potential in combination with antimicrobial agents. Int J Nanomed 7:483

32. Gupta M, Tomar RS, Kaushik S, Mishra RK, Sharma D (2018) Effective antimicrobial activity of green ZnO nano particles of *Catharanthus roseus*. Front Microbiol 9:2030
33. Gur T, Meydan I, Seckin H, Bekmezci M, Sen F (2022) Green synthesis, characterization and bioactivity of biogenic zinc oxide nanoparticles. Environ Res 204:111897
34. Huang J, Li Q, Sun D, Lu Y, Su Y, Yang X, Wang H, Wang Y, Shao W, He N, Hong J (2007) Biosynthesis of silver and gold nanoparticles by novel sundried *Cinnamomum camphora* leaf. Nanotechnology 18:105104
35. Husen A (2019a) Natural product-based fabrication of zinc oxide nanoparticles and their application. In: Husen A, Iqbal M (eds) N*anomaterials* and plant potential, vol 11, 6330th edn. Springer International Publishing AG, Gewerbestrasse, Cham, pp 193–291. https://doi.org/10.1007/978-3-030-05569-1_7
36. Husen A (2019b) Medicinal plant-product based fabrication nanoparticles (Au and Ag) and their anticancer effect. In: Kintzios SE, Barberaki M, Flampouri (eds) Plants that fight cancer, 2nd edn. Taylor & Francis/CRC Press, pp 133–147
37. Husen A (2022) Engineered nanomaterials for sustainable agricultural production, soil improvement and stress management. Elsevier Inc., 50 Hampshire St., 5th Floor, Cambridge, MA 02139, USA
38. Husen A Iqbal M (2019a) N*anomaterials*and plant potential: an overview. In: Husen A, Iqbal M (eds) N*anomaterials*and plant potential, vol 11, 6330th edn. Springer International Publishing AG, Gewerbestrasse, Cham, pp 3–29. https://doi.org/10.1007/978-3-030-05569-1_1
39. Husen A, Iqbal M (2019b). N*anomaterials*and plant potential: an overview. In: Husen A, Iqbal M (eds) N*anomaterials*and plant potential, vol 11, 6330th edn. Springer International Publishing AG, Gewerbestrasse, Cham, pp 3–29. https://doi.org/10.1007/978-3-030-05569-1_1
40. Husen A, Jawaid M (2020) Nanomaterialsfor agriculture and forestry applications. Elsevier Inc., 50 Hampshire St., 5th Floor, Cambridge, MA 02139, USA. ISBN: 9780128178522. https://doi.org/10.1016/C2018-0-02349-X
41. Husen A, Siddiqi KS (2014) Phytosynthesis of nanoparticles: concept, controversy and application. Nanoscale Res Lett 9(229):1–24
42. Ikram M, Javed B, Raja NI, Mashwani ZUR (2021) Biomedical potential of plant-based selenium nanoparticles: a comprehensive review on therapeutic and mechanistic aspects. Int J Nanomed 16:249
43. Jamkhande PG, Ghule NW, Bamer AH, Kalaskar MG (2019) Metal nanoparticles synthesis: an overview on methods of preparation, advantages and disadvantages, and applications. J Drug Deliv Sci Technol 53:101174
44. Jaybhaye SV (2015) Antimicrobial activity of silver nanoparticles synthesized from waste vegetable fibers. Mater Today Proceed 2(9, Part A):4323–4327
45. Jeevanandam J, Chan YS, Danquah MK (2016) Biosynthesis of metal and metal oxide nanoparticles. Chem Bio Eng Rev 3(2):55–67
46. Jin-Chul K, Madhusudhan A, Husen A (2021) Smart nanomaterials in biomedical applications, vol 11, 6330th edn. Springer Nature Switzerland AG, Gewerbestrasse, Cham, Switzerland. https://doi.org/10.1007/978-3-030-84262-8
47. Joshi A, Sharma A, Bachheti RK, Husen A, Mishra VK (2019) Plant-mediated synthesis of copper oxide nanoparticles and their biological applications. In: Husen A, Iqbal M (eds) N*anomaterials* and plant potential, vol 11, 6330th edn. Springer International Publishing AG, Gewerbestrasse, Cham, pp 221–237. ISBN 978-3-030-05569-1. https://doi.org/10.1007/978-3-030-05569-1_8
48. Kadam J, Dhawal P, Barve S, Kakodkar S (2020) Green synthesis of silver nanoparticles using cauliflower waste and their multifaceted applications in photocatalytic degradation of methylene blue dye and Hg^{2+} biosensing. SN Appl Sci 2:738
49. Kaur S, Roy A (2021) Bioremediation of heavy metals from wastewater using nanomaterials. Environ Dev Sustain 23(7):9617–9640

50. Kaviya S, Santhanalakshmi J, Viswanathan B, Muthumary J, Srinivasan K (2011) Biosynthesis of silver nanoparticles using *Citrus sinensis* peel extract and its antibacterial activity. Spectrochim Acta A Mol Biomol Spectrosc 79:54–598
51. Kemp MM, Kumar A, Mousa S, Park TJ, Ajayan P, Kubotera N, Mousa SA, Linhardt RJ (2009) Synthesis of gold and silver nanoparticles stabilized with glycosaminoglycans having distinctive biological activities. Biomacromol 10:589–595
52. Khalith SM, Anirud RR, Ramalingam R, Karuppannan SK, Dowlath MJH, Pandion K, Arunachalam KD (2021) Synthesis and characterization of magnetite carbon nanocomposite from agro waste as chromium adsorbent for effluent treatment. Environ Res 202:111669
53. Khare S, Singh RK, Prakash O (2022) Green synthesis, characterization and biocompatibility evaluation of silvernanoparticles using radish seeds. Results Chem 4:100447
54. Kredy HM (2018) The effect of pH, temperature on the green synthesis and biochemical activities of silver nanoparticles from *Lawsoniainermis* extract. J Pharm Sci Res 10(8):2022–2026
55. Kulkarni N, Muddapur U (2014) Biosynthesis of metal nanoparticles: a review. J Nanotechnol 2014:510246
56. Kumar A, Choudhary A, Kaur H, Guha S, Mehta S, Husen A (2022) Potential applications of engineered nanoparticles in plant disease management: a critical update. Chemosphere 295:133798. https://doi.org/10.1016/j.chemosphere.2022.133798
57. Kumar A, Choudhary A, Kaur H, Mehta S, Husen A (2021) Smart nanomaterial and nanocomposite with advanced agrochemical activities. Nanoscale Res Lett 16(156):1–26. https://doi.org/10.1186/s11671-021-03612-0
58. Kumar V, Hussainb PR, Chatterjeea S, Variyara PS (2015) Evaluation of in vitro antioxidant activity and characterization of phenolic compounds of bottle gourd towards the green synthesis of gold nanoparticles and its bio-efficacy. Int J Food Nutr Saf 6(3):125–149
59. Kumar V, Yadav S (2009) Plant-mediated synthesis of silver and gold nanoparticles and their applications. J Chem Technol Biotechnol 84:151–157
60. Lee KX, Shameli K, Yew YP, Teow SY, Jahangirian H, Rafiee-Moghaddam R, Webster TJ (2020) Recent developments in the facile bio-synthesis of gold nanoparticles (AuNPs) and their biomedical applications. Int J Nanomed 15:275–300
61. Liu X, Gu S, Yang S, Deng J, Xu J (2021) Heavy metals in soil-vegetable system around e-waste site and the health risk assessment. Sci Total Environ 20:146438
62. Ma Y, Li N, Yang C, Yang X (2005) One-step synthesis of amino-dextranprotected gold and silver nanoparticles and its application in biosensors. Anal Bioanal Chem 382:1044–1048
63. Madhumitha G, Fowsiya J, Gupta N, Kumar A, Singh M (2019) Green synthesis, characterization and antifungal and photocatalytic activity of *Pithecellobium dulce* peelmediated ZnO nanoparticles. J Phys Chem Solids 127:43–51
64. Mallikarjuna K, Sushma NJ, Narasimha G, Manoj L, Raju BD (2014) Phytochemical fabrication and characterization of silver nanoparticles by using pepper leaf broth. Arab J Chem 7:1099–1103
65. Mekonnen TW, Gerrano AS, Mbuma NW, Labuschagne MT (2022) Breeding of vegetable cowpea for nutrition and climate resilience in Sub-Saharan Africa: progress, opportunities, and challenges. Plants 11:1583
66. Mittal S, Roy A (2021) Fungus and plant-mediated synthesis of metallic nanoparticles and their application in degradation of dyes. Elsevier, Photocatalytic Degradation of Dyes, pp 287–308
67. Modi S, Yadav VK, Choudhary N, Alswieleh AM, Sharma AK, Bhardwaj AK, Khan SH, Yadav KK, Cheon J-K, Jeon B-H (2022) Onion peel waste mediated-green synthesis of zinc oxide nanoparticles and their phytotoxicity on mung bean and wheat plant growth. Materials 15:2393
68. Mohamed M, Faraj K, Al-Jobori K (2020) Green synthesis of silver nanoparticles using tomato (*Lycopersicon esculentum*) extract and evaluation of their antifungal activities. Plant Archives 20:5773–5786

69. Mohammad AE, Taher DM (2019) Antimicrobial activity of silver nanoparticles fabricated from some vegetable plants. J Phys Conf Ser 1294:062048
70. Mondal R, Dam P, Chakraborty J, Paret ML, Katı A, Altuntas S, Sarkar R, Ghorai S, Mandal AK, Husen A (2022) Potential of nanobiosensor in sustainable agriculture: the state-of-art. Heliyon. https://doi.org/10.2139/ssrn.4134873
71. Mukundan D, Mohankumar R, Vasanthakumari R (2015) Green synthesis of silver nanoparticles using leaves extract of *Bauhinia tomentosa* and its invitro anticancer potential. Mater Today Proc 2(9, part a):4309–4316
72. Mythili R, Selvankumar T, Kamala-Kannan S, Sudhakar C, Ameen F, Al-Sabri A, Selvam K, Govarthanan M, Kim H (2018) Utilization of market vegetable waste for silver nanoparticle synthesis and its antibacterial activity. Mater Lett 225:101–104
73. Mythili R, Selvankumar T, Srinivasan P, Sengottaiyan A, Sabastinraj J, Ameen F, Al-Sabri A, Kamala-Kannan S, Govarthanan M, Kim H (2018) Biogenic synthesis, characterization and antibacterial activity of gold nanoparticles synthesised from vegetable waste. J Mol Liq 262:318–321
74. Narayanan KB, Sakthivel N (2010) Phytosynthesis of gold nanoparticles using leaf extract of coleus amboinicuslour. Mater Charact 61:1232–1238
75. Naseer M, Aslam U, Khalid B, Chen B (2020) Green route to synthesize zinc oxide nanoparticles using leaf extracts of *Cassia fistula* and *Melia azadarach* and their antibacterial potential. Sci Rep 10(1):1–10
76. Osuntokun J, Onwudiwe DC, Ebenso EE (2019) Green synthesis of ZnO nanoparticles using aqueous *Brassica oleracea* L. var. *italica* and the photocatalytic activity. Green Chem Lett Rev 12(4):444–457
77. Parajuli R, Thoma G, Matlock MD (2019) Environmental sustainability of fruit and vegetable production supply chains in the face of climate change: a review. Sci Total Environ 650(2):2863–2879
78. Park Y, Hong YN, Weyers A, Kim YS, Linhardt R (2011) ChemInform abstract: polysaccharides and phytochemicals: a natural reservoir for the green synthesis of gold and silver nanoparticles. IET Nanobiotechnol 5:69–78
79. Patel R, Mehta M (2019) Green synthesis of silver nanoparticles by using fruit and vegetable waste: a review. J Eng Res Appl 9(3):78–85
80. Peralta-Videa JR, Huang Y, ParsonsJG ZL, Lopez-Moreno L, Hernandez-Viezcas JA, Gardea-Torresdey JL (2016) Plant-based green synthesis of metallic nanoparticles: scientific curiosity or a realistic alternative to chemical synthesis? Nanotechnol Environ Eng 1(1):1–29
81. Rafique M, Tahir R, Gillani SSA, Tahir MB, Shakil M, Iqbal T, Abdellahi MO (2022) Plant-mediated green synthesis of zinc oxide nanoparticles from *SyzygiumCumini* for seed germination and wastewater purification. Int J Eng Res Appl 102(1):23–38
82. Ramalingam M, Kokulnathan T, Tsai PC, Arasu MV, Al-Dhabi NA, Prakasham K, PonnusamyVK (2021) Ultrasonication-assisted synthesis of gold nanoparticles decorated ultrathin graphitic carbon nitride nanosheets as a highly efficient electrocatalyst for sensitive analysis of caffeic acid in food samples. Appl Nanosci 1–12
83. Roy CA, ElzakiA TirthV, Kajoak S, Osman H, Algahtani A, Bilal M (2021) Biological synthesis of nanocatalysts and their applications. Catalysts 11(12):1494
84. RoyA BN (2017) Qualitative analysis of phytocompounds and synthesis of silver nanoparticles from *Centella asiatica*. Innovat Tech Agric 1(2):88–95
85. Sangaru SS, Rai A, Ahmad A, Sastry M (2004) Rapid synthesis of Au, Ag, and bimetallic Au core-Ag shell nanoparticles using neem (*Azadirachta indica*) leaf broth. J Colloid Interface Sci 275:496–502
86. Satyavani K, Gurudeeban S, Ramanathan T, Balasubramanian T (2011) Biomedical potential of silver nanoparticles synthesized from calli cells of *Citrullus colocynthis* (L.) Schrad. J Nanobiotechnology 9:43
87. Sharma K, Kaushik S, Jyoti (2016) Green synthesis of silver nanoparticles by using waste vegetable peel and its antibacterial activities. J Pharm Sci Res 8(5):313–316

88. Sharma P, Pandey V, Sharma MMM, Patra A, Singh B, Mehta S, Husen A (2021) A Review on biosensors and nanosensors application in agroecosystems. Nano Res Lett 16:136. https://doi.org/10.1186/s11671-021-03593-0
89. Siddiqi KS, Husen A (2017) Recent advances in plant-mediated engineered gold nanoparticles and their application in biological system. J Trace Elem Med Biol 40:10–23
90. Siddiqi KS, Husen A, Rao RAK (2018) A review on biosynthesis of silver nanoparticles and their biocidal properties. J Nanobiotechnology 16(14):1–28
91. Singh J, Kumar V, Singh J, Kim S, Rawat K-H, Kukkar M, Tsang DYF (2019) Biogenic synthesis of silver nanoparticles and its photocatalytic applications for removal of organic pollutants in water. J Ind Eng Chem 80:247–257
92. Song J, Kwon EY, Kim BS (2009) Biological synthesis of platinum nanoparticles using Diopyros kaki leaf extract. Bioprocess Biosyst Eng 33:159–164
93. Surendra TV, Roopan SM, Al-Dhabi NA, Arasu MV, Suthindhiran GSK (2016) Vegetable peel waste for the production of ZnO Nanoparticles and its toxicological efficiency, antifungal, hemolytic, and antibacterial activities. Nanoscale Res Lett 11:546
94. Surendra TV, Roopan SM, Arasu MV, Al-Dhabi NA, Rayalu GM (2016) RSM optimized *Moringa oleifera* peel extract for green synthesis of *M. oleifera* capped palladium nanoparticles with antibacterial and hemolytic property. J Photochem Photobiol B 162:550–557
95. Tc P, Raichur A, Chandrasekaran N, Mukherjee A (2011) Kinetic evolution studies of silver nanoparticles in a bio-based green synthesis process. Colloids Surf A Physico Chem Eng Asp 377:212–216
96. Tyagi PK, Gupta S, Tyagi S, Kumar M, Pandiselvam R, Daştan SD, Sharifi-Rad J, Gola D, Arya A (2021) Green synthesis of iron nanoparticles from spinach leaf and banana peel aqueous extracts and evaluation of antibacterial potential. J Nanomater 11
97. UllahH UZ, Fazal A, Irfan M (2017) Use of vegetable waste extracts for controlling microstructure of CuO nanoparticles: green synthesis, characterization, and photocatalytic applications. J Chem 2721798:5
98. Vasanth K, Ilango K, MohanKumar R, Agrawal A, Dubey GP (2014) Anticancer activity of *Moringa oleifera* mediated silver nanoparticles on human cervical carcinoma cells by apoptosis induction. Colloids Surf B 117:354–359
99. VermaY, SinghSK, JatavHS, RajputVD, Minkina T (2021) Interaction of zinc oxide nanoparticles with soil: insights into the chemical and biological properties. Environ Geochem Health 1–14
100. Wibowo A, Tajalla GUN, Marsudi MA, Cooper G, Asri LATW, Liu F, Ardy H, Bartolo PJDS (2021) Green synthesis of silver nanoparticles using extract of cilembusweet potatoes (*Ipomoea batatas* L. var. *Rancing*) as potential filler for 3D printed electroactive and anti-infection scaffolds. Molecules 26(7):2042

Synthesis of Metal-Oxide Nanoparticles from Vegetables and Their Waste Materials for Diverse Applications

A. P. C. Ribeiro, Isabelle Zheng, and M. M. Alves

Abstract Nanotechnology is emerging as an important field of research in the last decades, and nanoparticles (NPs) are the basic element of nanotechnology as building blocks of several nanostructured devices or materials. Metal-oxide NPs are among the several nanomaterials found to have different applications in the fields of science and technology. Conventionally, metal-oxide NPs use different chemical agents as reducing and stabilizing agents with various biological and environmental risks due to the toxicity of used chemicals. Green synthesis is emerging as another safer method for the synthesis of metal/metal-oxide NPs using plants and agricultural wastes as a source of precursor material. The main advantages of using vegetable waste are highly biocompatible, insignificant toxicity, high reactivity, and high stability compared with that of the traditional process. Characterization techniques for metal-oxide NPs are already available and include electronic microscopy (transmission and scanning), diffraction techniques (single crystal X-ray and powder X-ray) as well as UV–vis, Fourier transform infrared spectroscopy, energy-dispersive X-ray spectroscopy (XPS), dynamic light scattering (DLS), and zeta potential measurements. This chapter addresses these issues and presents an updated vision of recent developments in this field. The use of waste has the advantage of recycling materials, promoting CO_2 valorization, and water and land usage.

Keywords Agri-food waste · Biosynthesis · Metal-oxide NPs · Plant potential

1 Introduction

Food waste from households, retail establishments, and the food service industry totals 931 million tonnes/year [45]. The global average of 74 kg per capita of food wasted each year is remarkably similar from lower middle income to high-income

A. P. C. Ribeiro · I. Zheng · M. M. Alves (✉)
Centro de Química Estrutural, Institute of Molecular Sciences, Instituto Superior Técnico, Universidade de Lisboa, Av. Rovisco Pais, 1049 001 Lisboa, Portugal
e-mail: martamalves@tecnico.ulisboa.pt

A. Husen (ed.), *Nanomaterials from Agricultural and Horticultural Products*, Smart Nanomaterials Technology,
https://doi.org/10.1007/978-981-99-3435-5_3

countries [45]. The fact that such considerable amounts of food are produced but not consumed has large negative impacts, namely, at the environmental, social, and economic levels. Estimates suggest that 8–10% of global greenhouse gas emissions are associated with food that is not consumed. Fresh fruit and vegetables contribute to almost 50% of the food waste generated by households [13]. Vegetable wastes are mostly commonly edible spoilage and leftovers, as well as inedible parts such as peels, with inedible waste being prominent. At the retail level, these are estimated to represent a food loss of around 9.7% per capita [24]. Fresh vegetables typically spoiled in supermarkets are presented in Fig. 1a. In an endeavour to make Europe carbon–neutral, and encourage a circular economy, expanding food waste management was identified by the European Union (EU) as a key factor [5] to create sustainable food systems and meet United Nations Sustainable Development Goals and tackle food waste across the food supply chain. Food waste management offers multi-faceted wins for people and the planet, saving money and reducing pressures on land, water, biodiversity, and waste management systems. This potential has until now been woefully underexploited. There are several solutions to tackle food waste (Fig. 1b), one being the reuse of these wastes as raw materials to produce valuable chain materials.

A remarkable example is the reuse of vegetable waste for the synthesis of NPs [34]. The gained prominence in technological advancements of NPs represents an active area of research and a techno-economic sector with full expansion in many application domains [25]. Among the existing NPs, metal-oxide ones are among the widest used manufactured nanomaterials due to their unique physicochemical properties [12]. Among the diverse methods used to synthesize NPs, green synthesis is highlighted as a robust, environmentally friendly, and cost-effective technique where the library of biomolecules available in vegetable waste can be used to improve NPs' properties [20, 21, 23, 40] (Fig. 2). During the green synthesis process of the NPs, biomolecules can be used to control size, shape, and surface properties, essential to tune the physicochemical properties of the synthesized NPs and will act as well as reducing and capping agents to stabilize the formed NPs. Sustainable routes for the synthesis of NPs can be achieved throughout circular economy, where the use of agriculture waste will target the design of safer and low-cost NPs. Herein, we provide an overview of the current status, challenges, and future directions for the utilization of vegetables wastes as suitable inputs for the green synthesis of metal-oxide NPs as added-value products. We conclude with some remarks from the sustainability viewpoint, that emphasize the critical aspects necessary for scale-up manufacturing and deployment of waste-derived nanomaterials.

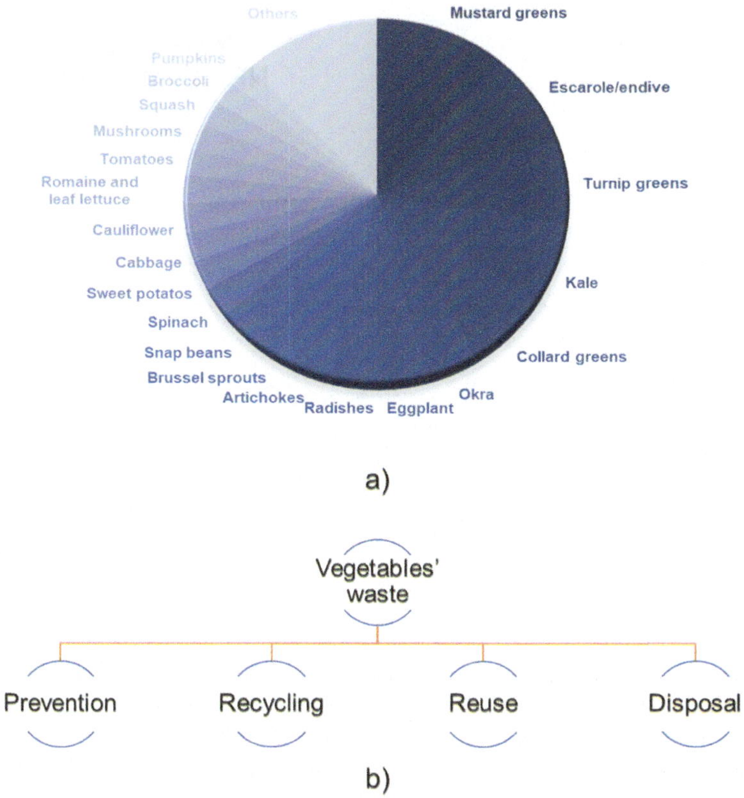

Fig. 1 Vegetable wastes; **a** supermarket loss of fresh vegetables [24] and **b** typical strategies used to mitigate their waste

1.1 Vegetables and Their Waste Materials for Nanoparticles Synthesis

Metal ions reduction occurs in an aqueous-based solution having waste extract and precursor cations. Metal-oxide NPs are usually made by the hydrolysis of metal salts at ambient temperature or temperatures below 100 °C [7]. Biomolecules present in the aqueous extract, namely such as alkaloids, amino acids, enzymes, phenolics, proteins, polysaccharides, tannins, saponins, vitamins and terpenoids, among others, act as reducing agents, whilst others play a role in shaping agents that drive particle size and shape, for instance, rods, spheres, ellipsoids, cylinders, triangular, hexagonal and more [7, 26]. In the final stages of NPs growth, capping agents stabilize the NPs preventing their aggregation [34]. Green routes, when using vegetable extracts tend to be faster, have lower running costs, and can be easily scaled up to industrial amounts of metal-oxide NPs production [34]. Importantly, using vegetable-based sources to manufacture metal-oxide NPs also provides (i) an aqueous-based solvent,

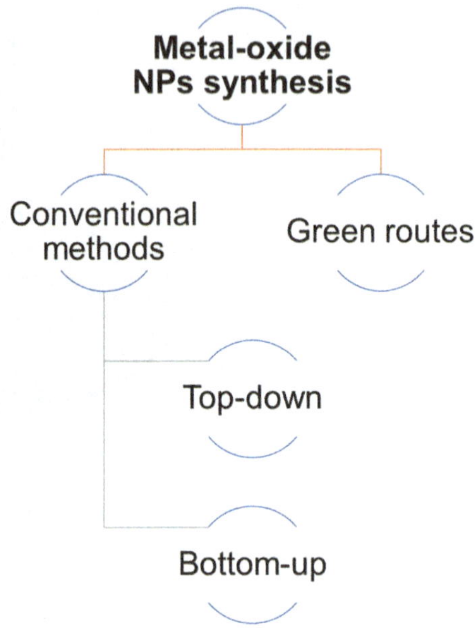

Fig. 2 Scheme of the synthetic routes used for the fabrication of metal-oxide nanoparticles (NPs) fabrication

(ii) nontoxic reducing agents, and (iii) safe stabilizers [34]. While many reports refer to the use of fruits and their wastes [34] (which are out of the scope of this chapter), few have considered the reuse of vegetable waste (Fig. 3).

Bell peppers (*Capsicum annuum* L. var. grossum) besides having an exotic flavour, are an important source of vitamins and various bioactive compounds that are beneficial for health. Bell peppers extract was used to synthesize ZnO nanorods with sizes ranging between 70 and 80 nm [47].

Beetroot (*Beta vulgaris)* is rich in various bioactive phytochemicals, which are beneficial for human health. Beet extracts resulted in ZnO NPs with average sizes of 20 nm [33].

Broccoli (*Brassica oleracea*) is one of the most common cruciferous vegetables in the world. Broccoli florets and sprouts are usually used for consumption, while its other parts, such as stalks and leaves, are wasted during harvesting. Broccoli extracts were utilized in ZnO NPs preparation, which was obtained after calcination [32]. Hexagonal phases were identified for NPs, with an average crystallite size of 14 nm [32]; other work reported average sizes of 47 nm [49]. When using leaf extracts, ZnO NPs presented an average size ranging from 10 to 120 nm, depending on the precursor concentration [49] (Fig. 4a). The synthesis of CaO NPs using aqueous extract of broccoli, resulted in sizes between 38 and 29 nm that were modulated by the volume of extract used as the volume increased the particles' size decreased [31].

Cardamom (*Elettaria cardamomum*), known as the "queen of spice" is an aromatic spice with a rich pool of polyphenolic compounds. Cardamom seed extract-assisted ZnO NPs resulted in homogenous spherical forms with average sizes of 18.7 nm [48].

Fig. 3 Schematic representation of vegetable extracts used to synthesize metal-oxide nanoparticles (NPs)

Carrots (*Daucus carota*) are among the most important horticultural crops ranked in the worldwide top 10 vegetable crops, which exceeds an annual worldwide production of 40 million tonnes. Carrot extract used in the green synthesis of ZnO was reported to yield NPs with average sizes from 16 to 21 nm [27].

Cassava (*Manihot esculenta)* tuber is a major staple food and the world's fourth source of calories. Cassava leaf extract supported the production of monodisperse and hexagonal-shaped MgO NPs with average sizes of 36.7 nm [15].

Coriander (*Coriandrum sativum)* is an edible plant mainly cultivated for fresh leaves and dried seeds. Coriander leaf extract was used to synthesize ZnO NPs resulting in average particle sizes of 24 and 52 nm [18, 41] (Fig. 4b).

Cumin (*Cuminum Cyminum*) has wide usage in the beverage, food, liquor, medicine, perfume, and toiletry industries. Cumin seed extracts were used for the synthesis of Fe_3O_2 NPs used to assist drought resistance in wheat plants [30].

Fig. 4 Scanning electron microscopy (SEM) imaging of the ZnO NPs synthesized with **a** broccoli leaf extract, **b** coriander leaves extract, **c** lupin peel seed extract (authors' unpublished data), **d** okra mucilage extract, **e** onion extract, **f** starch-rich potato extract, **g** pumpkin seed extract, and white radish extract. [adapted from: 11, 35, 36, 41, 43, 49]

Garlic (*Allium sativum)* is one of the oldest plants cultivated for its dietary and medicinal values. Garlic extracts were used for the successful synthesis of 14 nm sized ZnO NPs [42], and 34–70 nm in size with a cubic structure of FeO NPs when using garlic peel extracts [2].

Ginger (*Zingiber officinale*) is a medicinal herb that has been commonly used in food and pharmaceutical products. Ginger extracts were applied for the synthesis of Fe_3O_2 NPs used to assist drought resistance in wheat plants [30].

Long pepper (*Piper longum)* aqueous extracts resulted in ZnO NPs that exhibited hexagonal wurtzite crystalline structure [46].

Lupins are gaining global interest due to their nutritional value, greater sustainability, and low production costs. Furthermore, lupin consumption is related to numerous health benefits including improved bowel function and reduced cholesterol, blood glucose, and glycaemic index [39]. Lupin seed peel extract was used to synthesize ZnO NPs, that resulted in spherically shaped NPs with average sizes of 40 nm (unpublished data, Fig. 4c).

Marjoram (*Origanum majorana*) leaf extracts were used in the green synthesis of CeO_2 NPs rendering spherically shaped particles with an average size of 10–70 nm [29].

Okra *(Abelmoschus esculentus)*, which belongs to the mallow family, is an important vegetable that can be found in almost every country in the world and is generally used in traditional medicine [6]. ZnO NPs were reported via a green synthetic strategy using okra mucilage [35]. The morphology of the NPs was uneven where they comprised heterogeneous particles with uniform spheres that had an average size of 29 nm, and elongated and rod-like structures measuring 70 nm [35] (Fig. 4d). ZnO NPs synthesized from okra leaves had uniformly distributed spherical shapes from 20 to 45 nm [28]. The one-pot green fabrication of $BiFeO_3$ NPs via okra leaf extracts resulted in spherical-shaped particles with a size of around 10 nm [22]. The okra seed extract was also used for the synthesis of TiO_2 resulting in NPs with sizes ranging between 60 and 120 nm [6].

Onion (*Allium cepa*) is a common vegetable, widely consumed all over the world. Onion contains diverse phytochemicals, including organosulfur compounds, phenolic compounds, polysaccharides, and saponins. Onion extracts resulted in spherical NPs of ZnO [36], and hexagonal wurtzite structures, with particle sizes of 21 nm [42] (Fig. 4e). The use of onion peel extracts resulted in the synthesis of Fe_2O_3 NPs with 37–66 nm in size [2].

Parsley (*Petroselinum crispum)* is a globally used common culinary herb. Parsley extracts supported the synthesis of ZnO NPs with 70 nm in size [42].

Potato production is global estimated at 370.4 million tons [16]. Potatoes are one of the most water-efficient crops and produce the greatest number of calories per unit of water input, being most probably one of the crops with good adaptability to climate changes. Potato-derived extracts were used for the synthesis of magnetic NPs with an average size of about 40 nm [10], and ZnO NPs with hexagonal (wurtzite) shape and size of 20 nm [11] (Fig. 4f). Pumpkin is the second-largest global production [16].

Pumpkin (*Cucurbita moschata*) seed extract-based ZnO NPs have a spherical shape and have a relatively uniform particle size distribution with an average particle size of 28 nm [37] and spherical NPs with an average size of 50–60 nm [43] (Fig. 4g).

Rosemary *(Rosmarinus officinalis)* is an aromatic plant that has been utilized as a food additive or ingredient for flavouring purposes. Rosemary extracts were used in the production of magnesium oxide (MgO) nanoflowers with average particle sizes of 20 nm [1].

Spinach (*Spinacia oleracea*) is a highly nutritious leafy vegetable rich in vitamins and mineral elements, with a global production of 30.1 million tonnes in 2019 [16] and the harvested acreage increasing over the years. Spinach leaf extract was used to synthesize CuO NPs [4]. These particles formed rod-like agglomerates, with their width ranging between 10 and 30 nm, and their length extending from 60 to 130 nm [4].

Tomato (*Lycopersicon esculentum*) is cultivated for fresh fruit and processed products. Tomatoes contain many health-promoting compounds including vitamins, carotenoids, and phenolic compounds [17]. Tomato peel extracts were used for the synthesis of SnO_2 NPs [17]. The crystallinity of the NPs grew in a purely tetragonal crystal structure. Different sizes and shape homogeneity were depending on the amount of extract used [17].

White radish (*Raphanus sativus*) is a biennial plant cultivated as a major vegetable crop throughout the world. Extracts of white radish yielded spherical NPs of ZnO around 20–25 nm [36] (Fig. 4h). In this work, the authors compared conventional and green-synthesis methods, and proved that green routes represented a better alternative to the conventional chemical reduction of metal precursors for the formation of ZnO NPs, where an improved distribution of particles' size and shape was achieved [36].

An overview of the reagents used in these vegetable wastes-based synthesis shows that acetate [18, 28, 35–37, 43, 47, 48] and nitrate [4, 11, 15, 22, 27, 29, 33, 42] precursors predominate over chloride [2, 17, 31, 32, 41], and in a less extent sulphate [10, 38, 49] (Fig. 5). This overview clearly shows that there is room to explore other metal precursors in these synthetic routes, namely acetylacetonate and oxalates, among others. When considering the reports for the synthesized NPs, there is clear evidence for an increased interest in ZnO fabrication [11, 18, 27, 28, 32, 33, 35–37, 41–43, 46–48, 49] over the remaining particles, namely iron oxides [2, 10, 30, 38], MgO [1, 15], and then CaO [31], CeO_2 [29], CuO [4], SnO_2 [17], TiO_2 [6] and bimetallic $BiFeO_3$ [22] (Fig. 5).

There is unexplored NPs synthesis that can profit from these green routes, both by yet unexplored vegetables and metal-oxide NPs (Fig. 6). Some examples of NPs are Co_3O_4, NiO, MnO, Mn_3O_4, SiO_2, ZrO_2 and the list continues (Fig. 6). The bimetallic NPs are certainly some with an upcoming interest due to their unique properties, that by differing from those of pure elemental particles result in powerful synergistic properties [8]; some examples of unexplored vegetable-based synthesis

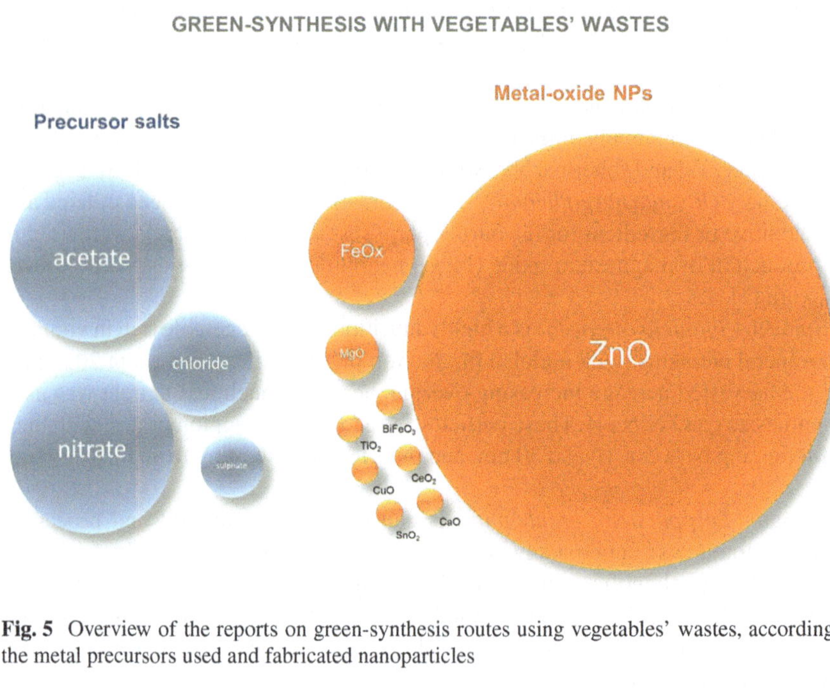

Fig. 5 Overview of the reports on green-synthesis routes using vegetables' wastes, according to the metal precursors used and fabricated nanoparticles

of bimetallic particles are $MnFe_2O_4$, $CoFe_2O_4$, $NiFe_2O_4$, $ZnMn_2O_4$ and many others (Fig. 6).

This area of knowledge, of reusing vegetables' wastes in the green synthesis of metal-oxide NPs, is still in its infancy. The current overview of the state-of-the-art (Figs. 5 and 6) showed us where gaps in the design of the NPs still exist; however, the routes to achieve better and safer particles require many iterations in the overall process (Fig. 7). There are at least four different variables that must be explored: (i) vegetables, (ii) extract, (iii) precursor salts and (iv) additives. The variable vegetables, as extensively described, depends on the nature of the waste (Fig. 3). Within this, the maturation and deterioration states will heavily impact the nature of the phytochemicals present in the extracts (e.g. [9]). Moreover, the presence of Mn and Fe in ZnO powders synthesized by using plant extracts, highlighted by ICP-MS [42], supports that besides the biomolecules, the metals ions present in the

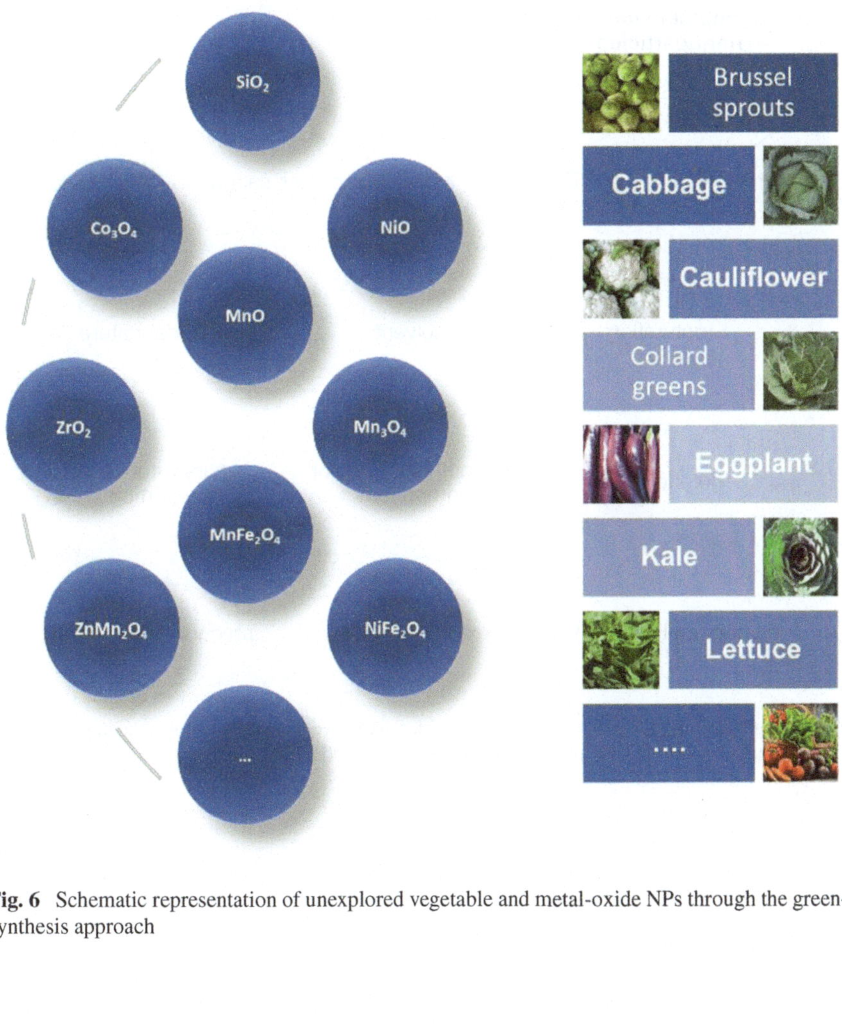

Fig. 6 Schematic representation of unexplored vegetable and metal-oxide NPs through the green-synthesis approach

vegetable extracts can also influence the composition of the as-synthesized NPs. Likewise, the vegetable parts, being these tubers, leaves, peels or a mixture of these will as well modulate the phytochemical pools. In the extraction procedure, many variables can be considered, not only the solvent itself that can go from pure aqueous to an organic mixture will lead to the preferential extraction of phytochemical with a higher affinity for aqueous or organic solvents, or even mixtures. In the same line of thought the chemical nature of the metal precursors, as mentioned above, and their mixtures can themselves impact the synthesis of the metal-oxide NPs. Either from the extract or the precursor sides, the proportion between these two will have a great impact on the final particles' properties. Although little explored, the combination of additives with the vegetables' waste extracts [36] can bring new avenues to these synthetic routes (Fig. 7).

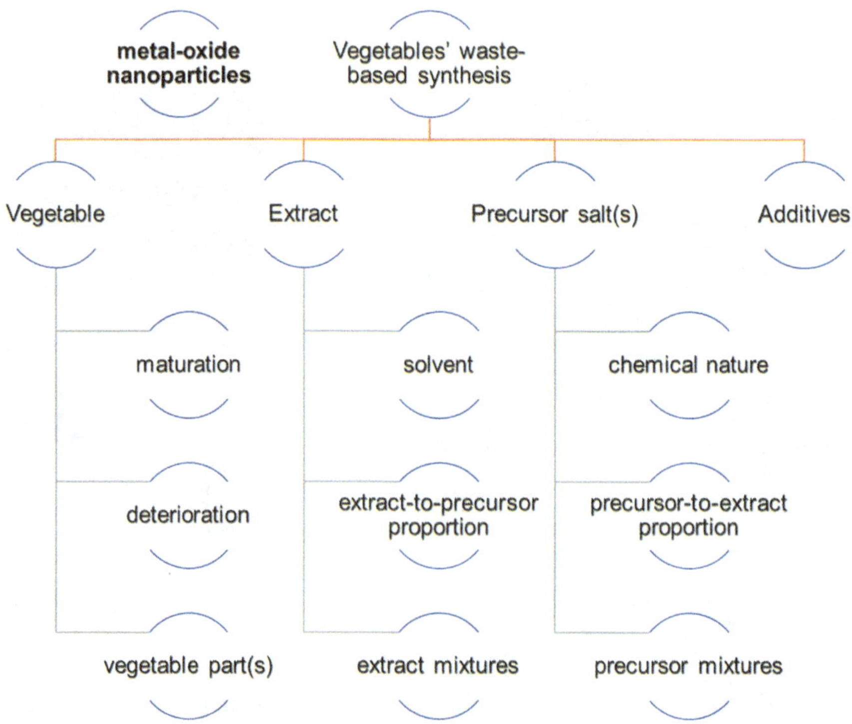

Fig. 7 Overview of the green-synthesis variables using vegetables' wastes that have room to improve the fabrication of the metal-oxide nanoparticles (NPs)

1.2 Green-Synthesized Metal-Oxide Applications

Agricultural nanotechnology emerged in the late 1990s and was quickly adopted in many fields. The biogenic synthesis of NPs using vegetable extracts is also reliable, ecofriendly and cost-effective. The aqueous extract from plants acts as both a reducing and stabilizing agent for NP synthesis. Metal-oxide NPs exhibit a wide range of enhanced physicochemical properties that make them ideal candidates for developing new chemical sensors, and electronic and optoelectronic devices [12, 34]. Vegetable-based food waste, by avoiding the use of hazardous contaminating chemicals, offers an attractive renewable source of molecules and compounds that by being used in metal-oxide NPs synthesis can be applied to pharmaceuticals and biomedical applications. The applications per NPs (Fig. 3) are detailed below.

1.2.1 Calcium Oxide

Calcium oxide (CaO) is an alkaline earth metal oxide and the heaviest among the group of light alkaline earth oxides. CaO NPs are the non-toxic, low-cost, readily available raw material for heterogeneous catalysis and many other applications such as an additive in refractory, adsorbent especially in catalysis, bactericides, in the biomedical field, and as well as a precursor in bioceramics [31]. Nanoparticles of CaO prepared using an aqueous extract of broccoli were used as catalysts for the degradation of bromocresol green (BG) via photo irradiation with ultraviolet light, and the result revealed a degradation efficiency of 60.1% [31]. The exploitation of environmentally friendly materials to degrade such complex aromatic molecular structures are of utmost importance to preserve aquatic life and the food web.

1.2.2 Cerium Oxide

Cerium oxide (CeO) NPs have been pointed to as a neurodegenerative therapeutic agent and indicated for other multiple applications such as catalysts, fuel cells and antioxidants in biological systems [29]. CeO_2 NPs synthesized using marjoram leaf extract were reported to have higher cytotoxic effects against breast cancer compared to normal cells indicating the potential use of these NP as anticancer agents [29], the most common malignancy in women [44].

1.2.3 Copper Oxide

Copper oxide NPs have remarkable properties and are used in the following areas: as solar energy conversion tools, high-temperature superconductors, batteries, catalysis, also gas sensors. Anticancer activities of cupric oxide NPs synthesized using spinach leaves extract were reported against breast cancer cells as well [4].

1.2.4 Iron Oxide

Green synthesis of iron/iron oxide NPs is cost-effective, nontoxic and ecofriendly. Iron oxide NPs possess multi-valent oxidation states and their characteristic structure at the nanoscale plays a major role in catalysis, imaging, biosensors, gene delivery and targeted drug delivery. The magnetic and ferromagnetic properties of these NPs attract various applications in supercapacitors, lithium-ion batteries and electrocatalysis [26]. Magnetic NPs synthesized using potato extract were effectively utilized in the degradation of rhodamine B (RhB) [10, 38], whereas those synthesized by garlic and onion peel extracts resulted in methylene blue (MB) rapid degradation, reaching 90% at 35 min and 97% at 30 min, respectively [2]. Ecofriendly strategies are crucial to assist the degradation of such hazardous and carcinogenic pollutants.

From a completely different perspective, Fe_3O_2 NPs biofabricated by using aqueous extract of ginger and cumin, induced drought tolerance to wheat plants by upregulating biochemical resistance mechanisms [30]. Nanotributes-priming could be considered an efficient method for sustainable food production especially if used in marginal soils [30].

1.2.5 Magnesium Oxide

Magnesium oxide (MgO) is an ecofriendly, economically feasible and industrially important NP due to its unique physicochemical behaviours such as outstanding refractive index, excellent corrosion resistance, high thermal conductivity, low electrical conductivity, physical strength, stability, flame-resistance, dielectric resistance, mechanical strength and excellent optical transparency [3]. Because of the above properties, MgO is used as a semi-conducting material, catalyst in organic transformations, sorbent for organic and inorganic contaminants from wastewater, electrochemical biosensors, photocatalysts and refractory materials [3]. It also possesses good antibacterial, anticancer and antioxidant properties [3]. Biomedical engineering is also an emerging field of study, as MgO is being used for tissue regeneration, implant coatings, bioimaging, wound healing and the development of cancer therapies [3].

Bacteriological tests indicated that MgO nanoflowers produced with rosemary extracts significantly inhibited bacterial growth, biofilm formation, and motility of *Xanthomonas oryzae* pv. *oryzae*, which is the causal agent of bacterial blight disease in rice [1]. The ecofriendly nature of these particles makes these important tools that can be widely used in agricultural fields to suppress bacterial infection and support food production.

1.2.6 Tin Oxide

Tin oxide (SnO_2), like many other metal oxides, has unique optical and electrical properties, such as low resistivity, optical transparency and high specific theoretical capacity. All these properties have generated various applications for SnO_2, such as chemical and gas sensors, transparent conductive electrodes, supercapacitors, solar cells and photocatalysts [17]. SnO_2 NPs produced with 4% of tomato peel extracts presented a UV light photocatalytic degradation rate of MB of around 100% at 120 min. These results are better than those achieved by commercially available SnO_2 NPs [17] and represent an expedited strategy to cope with the current and excessive use of such synthetic dyes in different industries.

1.2.7 Titanium Oxide

Titanium oxide (TiO_2) NPs have been widely used as an environmentally friendly and clean photocatalyst in recent years. TiO_2 material is an important semiconducting transition metal oxide with unique properties such as ease of control, low cost and non-toxic [6]. It also possesses good resistance to chemical erosion, making it suitable for solar cells, chemical sensors and environmental applications [6]. The electrical, magnetic and optical properties of these NPs are better compared to those of their bulk counterparts [6]. TiO_2 exists in both amorphous and crystalline forms, with the most common crystalline polymorphous of anatase, rutile and brookite [6]. Okra seed extract synthesized TiO_2 NPs achieved more than 80% photocatalytic activity, which was quite satisfactory when compared to commercial TiO_2 NPs. The time required for maximum dye degradation was approximately 200–240 min of exposure to irradiation [6]. The authors attributed that the presence of reducing organics such as polyphenolic tannins in Okra seed extract could be the reason for the improved MB degradation efficiency as compared to commercial TiO_2 NPs [6].

1.2.8 Zinc Oxide

Zinc oxide (ZnO) is a distinctive inorganic material exhibiting various properties like piezoelectric, pyroelectric, semiconducting, optoelectronics and catalysis. ZnO is also documented as bio-safe, and biocompatible with various unique applications in biomedical and drug delivery systems [11, 26]. Its compatibility with skin rendered its use as a UV blocker in sunscreen products, whereas other researchers point to it as a strong antimicrobial agent [26]. Currently, ZnO is listed by the US FDA as a GRAS (generally recognized as safe) metal oxide [26]. As such, many studies have been carried out with this important metal oxide.

The photocatalytic efficiency of ZnO NPs against MB obtained via broccoli, carrot, as well as okra mucilage and leaves were reported to reach 74–100% degradation efficiency [27, 28, 32, 35]. Other photodegradation studies conducted in the presence of UV light irradiation indicated that ZnO NPs prepared using garlic extract exhibit the highest efficiency in the photodegradation of MB dye when compared with those synthesized with onion or parsley extracts [42]. The photocatalytic efficiency against phenol red (PR) under UV light irradiation of ZnO obtained via the broccoli extract was reported to be 71% degradation efficiency [32]. ZnO NPs synthesized using coriander leaf extract presented an efficiency of 93% degradation of yellow 186 dye and 96% against anthracene ([18, 41].

Significant anticancer potential of ZnO NPs produced with pumpkin seed extract against a breast cancer cell line was also reported for this metal oxide [37].

The antilarval activity was reported for cardamom seed extract-assisted ZnO NPs acting as a larvicidal agent to the constraint of mosquito vectors [48]. The antibacterial properties of ZnO NPs synthesized with broccoli and beetroot were confirmed against some Gram-negative and Gram-positive bacteria, those produced with cardamom seed extract reduce the bacterial biofilm formation, particularly of Gram-negative bacteria [33, 48, 49]. Moreover, an increase in the efficacy of these NPs with a decrease in their size was also evident [49]. ZnO NPs fabricated with the aqueous extract of bell peeper showed efficacy against multi-drug-resistant strains (MDR) of non-typhoidal *Salmonella (NTS)* spp. [46]. Despite the ZnO NPs prepared using beetroot were found to be inactive towards *Staphylococcus aureus*, their antifungal activity was demonstrated against *Candida albicans* and *Aspergillus niger* fungal stains [33]. These results are of utmost importance as biosafe alternatives urgently need to overcome the increasing antibiotic resistance of pathogenic bacteria and fungi [14, 19].

1.2.9 Bimetal Oxides

Bimetallic NPs have acquired particular importance because of their unique optical, electronic, magnetic, and catalytic properties, which, in most cases, are significantly distinguishable from their monometallic counterparts attributed to the synergistic properties between the two different metal parts.

In particular, multiferroic nanomaterials have recently attracted great interest due to the coexistence of different order parameters in a crystalline phase, and broad applicability in multifunctional, low-power consumption and environmentally friendly devices [50]. The photocatalytic activity of $BiFeO_3$ NPs synthesized from leaves of okra resulted in 94% of the degradation of MB after 2 h [22].

An overview of the applications of the vegetables' synthesized metal-oxide NPs (Fig. 8) supports the unexplored nature of this field, as limited applications for catalysis, biomedical applications and plant breeding have been reported. Other emerging areas of knowledge can profit from this approach, namely for energy and semiconductors applications, among others.

2 Conclusions and Future Perspectives

Increasingly larger amounts of vegetable waste are generating environmental, social and economic problems. Therefore, waste valorization is an attractive strategy for utilizing renewable vegetable-based wastes into valuable marketable-derived products. The use of vegetable extracts to synthesize metal-oxide NPs brings an add-value

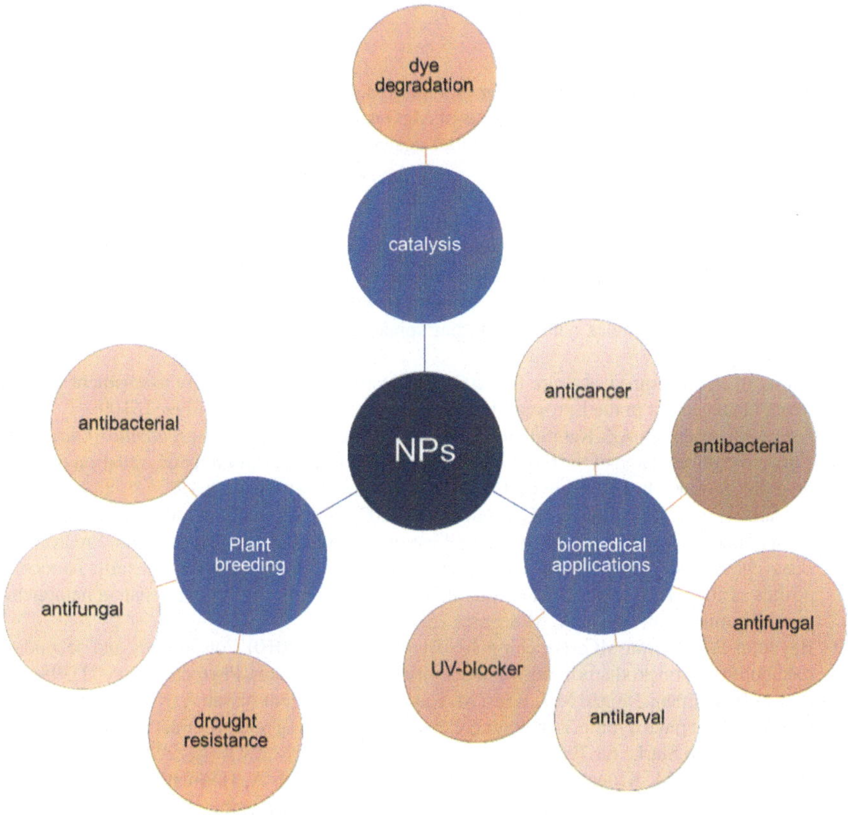

Fig. 8 Overview of the applications of the vegetable green-synthesized metal-oxide nanoparticles (NPs)

to the food chain waste while providing low cost and safety to the newly synthesized nanomaterials. Furthermore, the aqueous-based method can provide reliable, sustainable and green NPs that are less toxic to the environment. The exploitation of this green route method relieves the pressure on human and animal food consumption. Nevertheless, the waste, by its nature, brings a large variability to the biomolecules, which can at some point be detrimental to the reproducibility of the NPs for high precision and sensibility applications. Therefore, overcoming the variations within the green synthesis process to ensure the reproducibility of NPs properties is essential to support this new waste valorization strategy to be adopted. Also, limited scale-up studies have been reported. As such, there is considerable scope and opportunity for further study in this relatively new field of research.

Acknowledgements Authors acknowledge funding to Fundação para a Ciência e a Tecnologia (FCT) for financial support, namely through FCT projects LA/P/0056/2020, UIDB/00100/2020.

References

1. Abdallah Y, Ogunyemi SO, Abdelazez A, Zhang M, Hong X, Ibrahim E, Hossain A, Fouad H, Li B, Chen J (2019) The green synthesis of MgO nano-flowers using rosmarinus officinalis L. (Rosemary) and the Antibacterial Activities against Xanthomonas oryzae pv. oryzae. Biomed Res Int 5620989
2. Abid MA, Abid DA, Aziz WJ, Rashid TM (2021) Iron oxide nanoparticles synthesized using garlic and onion peel extracts rapidly degrade methylene blue dye. Physica B 622:413277
3. Abinaya S, Kavitha HP, Prakash M, Muthukrishnaraj A (2021) Green synthesis of magnesium oxide nanoparticles and its applications: a review. Sustain Chem Pharm 19:100368
4. Al-Jawhari H, Bin-Thiyab H, Elbialy N (2022) In vitro antioxidant and anticancer activities of cupric oxide nanoparticles synthesized using spinach leaves extract. Nano-Struct Nano-Objects 29:100815
5. Albizzati PF, Tonini D, Astrup TF (2021) A quantitative sustainability assessment of food waste management in the European Union. Environ Sci Technol 55:16099–16109
6. Aslam M, Abdullah AZ, Rafatullah M, Fawad A (2022) *Abelmoschus esculentus* (Okra) seed extract for stabilization of the biosynthesized TiO_2 photocatalyst used for degradation of stable organic substance in water. Environ Sci Pollut Res Int 29:41053–41064
7. Aswathi VP, Meera S, Maria CGA, Nidhin M (2022) Green synthesis of nanoparticles from biodegradable waste extracts and their applications: a critical review. Nanotechnol Environ Eng
8. Basavegowda N, Patra JK, Baek K-H (2020) Essential oils and mono/bi/tri-metallic nanocomposites as alternative sources of antimicrobial agents to combat multidrug-resistant pathogenic microorganisms: an overview. Molecules 25:1058
9. Bayoumi SA, Rowan MG, Beeching JR, Blagbrough IS (2010) Constituents and secondary metabolite natural products in fresh and deteriorated cassava roots. Phytochemistry 71:598–604
10. Buazar F, Baghlani-Nejazd MH, Badri M, Kashisaz M, Khaledi-Nasab A, Kroushawi F (2016) Facile one-pot phytosynthesis of magnetic nanoparticles using potato extract and their catalytic activity. Starch Stärke 68:796–804
11. Buazar F, Bavi M, Kroushawi F, Halvani M, Khaledi-Nasab A, Hossieni SA (2015) Potato extract as reducing agent and stabiliser in a facile green one-step synthesis of ZnO nanoparticles. J Exp Nanosci 11:175–184
12. Chavali MS, Nikolova MP (2019) Metal oxide nanoparticles and their applications in nanotechnology. SN Appl Sci 1:607
13. De Laurentiis V, Corrado S, Sala S (2018) Quantifying household waste of fresh fruit and vegetables in the EU. Waste Manag 77:238–251
14. Eleraky NE, Allam A, Hassan SB, Omar MM (2020) Nanomedicine fight against antibacterial resistance: an overview of the recent pharmaceutical innovations. Pharmaceutics 12:142
15. Essien ER, Atasie VN, Okeafor AO, Nwude DO (2020) Biogenic synthesis of magnesium oxide nanoparticles using Manihot esculenta (Crantz) leaf extract. Int Nano Lett 10:43–48
16. Faostat, S.D. (2019). Food and Agriculture Organization of the United Nations. Retrieved from http://www.fao.org/faostat/en/
17. Garrafa-Galvez HE, Nava O, Soto-Robles CA, Vilchis-Nestor AR, Castro-Beltrán A, Luque PA (2019) Green synthesis of SnO2 nanoparticle using Lycopersicon esculentum peel extract. J Mol Struct 1197:354–360
18. Hassan SSMEA, W. I. M., Ali, H. R., Mansour, M. Doi S. M. (2015) Green synthesis and characterization of ZnO nanoparticles for photocatalytic degradation of anthracene. Adv Nat Sci: Nanosci Nanotechnol 6:045012
19. Hendrickson JA, Hu C, Aitken SL, Beyda N (2019) Antifungal resistance: a concerning trend for the present and future. Curr Inf Dis Rep 21:47
20. Husen A (2019a) Natural product-based fabrication of zinc oxide nanoparticles and their application. In: Husen A, Iqbal M (eds.), N*anomaterials* and Plant Potential. Springer International Publishing AG, Gewerbestrasse 11, 6330 Cham pp. 193–291. https://doi.org/10.1007/978-3-030-05569-1_7

21. Husen A (2019b) Medicinal plant-product based fabrication nanoparticles (Au and Ag) and their anticancer effect, In: Kintzios SE, Barberaki M, Flampouri (eds.), Plants that Fight Cancer – Second Edition. Taylor & Francis/CRC Press pp. 133–147
22. Indriyani A, Yulizar Y, Tri Yunarti R, Oky Bagus Apriandanu, D, Marcony Surya R (2021) One-pot green fabrication of $BiFeO_3$ nanoparticles via Abelmoschus esculentus L. leaves extracts for photocatalytic dye degradation. Appl. Surf. Sci. 563:150113
23. Jadoun S, Arif R, Jangid NK, Meena RK (2021) Green synthesis of nanoparticles using plant extracts: a review. Environ Chem Lett 19:355–374
24. Jean C Buzby, HFW, Axtman B, Mickey J (2009) Supermarket loss estimates for fresh fruit, vegetables, meat, poultry, and seafood and their use in the ERS loss-adjusted food availability data. Econ Inform Bull 44
25. Jeevanandam J, Barhoum A, Chan YS, Dufresne A, Danquah MK (2018) Review on nanoparticles and nanostructured materials: history, sources, toxicity and regulations. Beilstein J Nanotechnol 9:1050–1074
26. Kurhade P, Kodape S, Choudhury R (2021) Overview on green synthesis of metallic nanoparticles. Chem Pap 75:5187–5222
27. Luque PA, Nava O, Soto-Robles CA, Vilchis-Nestor AR, Garrafa-Galvez HE, Castro-Beltran A (2018) Effects of *Daucus carota* extract used in green synthesis of zinc oxide nanoparticles. J Mater Sci: Mater Electron 29:17638–17643
28. Mirgane NA, Shivankar VS, Kotwal SB, Wadhawa GC, Sonawale MC (2021) Degradation of dyes using biologically synthesized zinc oxide nanoparticles. Mater Today: Proc 37:849–853
29. Nezhad SA, Ali E-H, Tabrizi MH (2019) Green synthesis of cerium oxide nanoparticle using *Origanum majorana* L. leaf extract, its characterization and biological activities. Appl Organometall Chem 34:e5314
30. Noor R, Yasmin H, Ilyas N, Nosheen A, Hassan MN, Mumtaz S, Khan N, Ahmad A, Ahmad P (2022) Comparative analysis of iron oxide nanoparticles synthesized from ginger (*Zingiber officinale*) and cumin seeds (*Cuminum cyminum*) to induce resistance in wheat against drought stress. Chemosphere 292:133201
31. Osuntokun J, Onwudiwe DC, Ebenso EE (2018) Aqueous extract of broccoli mediated synthesis of CaO nanoparticles and its application in the photocatalytic degradation of bromocrescol green. IET Nanobiotechnol 12:888–894
32. Osuntokun J, Onwudiwe DC, Ebenso EE (2019) Green synthesis of ZnO nanoparticles using aqueous Brassica oleracea L. var. italica and the photocatalytic activity. Green Chem Lett Rev 12:444–457
33. Pillai AM, Sivasankarapillai VS, Rahdar A, Joseph J, Sadeghfar F, Anuf A, R., Rajesh, K., and Kyzas, G.Z. (2020) Green synthesis and characterization of zinc oxide nanoparticles with antibacterial and antifungal activity. J Mol Struct 1211:128107
34. Poinern GEJ, Fawcett D (2019) Chapter 1 –Sustainable utilization of renewable plant-based food wastes for the green synthesis of metal nanoparticles. In: D.J. Henry (ed.), Harnessing Nanoscale Surface Interactions. Elsevier, pp. 1–39
35. Prasad AR, Garvasis J, Oruvil SK, Joseph A (2019) Bio-inspired green synthesis of zinc oxide nanoparticles using *Abelmoschus esculentus* mucilage and selective degradation of cationic dye pollutants. J Phys Chem Solids 127:265–274
36. Saravanan P, Senthilkannan K, Mustafa A, Vimalan M, Bououdina M, Balasubramanian S, Meena M, Tamilselvan S (2021) Dielectric and magnetic properties of *Allium cepa* and *Raphanus sativus* extracts biogenic ZnO nanoparticles. J Mater Sci: Mater Electron 32:590–603
37. Saridewi N, Syaputro HT, Aziz I, Dasumiati D, Kumila BN (2021) Synthesis and characterization of ZnO nanoparticles using pumpkin seed extract (Cucurbita moschata) by the sol-gel method. AIP Conf Proc 2349:020010
38. Sharma RK, Yadav S, Gupta R, Arora G (2019) Synthesis of magnetic nanoparticles using potato extract for dye degradation: a green chemistry experiment. J Chem Educ 96:3038–3044
39. Shrestha S, Hag LVT, Haritos VS, Dhital S (2021) Lupin proteins: structure, isolation and application. Trends Food Sci Technol 116:928–939

40. Siddiqi KS, Husen A (2017) Recent advances in plant-mediated engineered gold nanoparticles and their application in biological system. J Trace Elem Med Biol 40:10–23
41. Singh J, Kaur S, Kaur G, Basu S, Rawat M (2019) Biogenic ZnO nanoparticles: a study of blueshift of optical band gap and photocatalytic degradation of reactive yellow 186 dye under direct sunlight. Green Process Synth 8:272–280
42. Stan M, Popa A, Toloman D, Dehelean A, Lung I, Katona G (2015) Enhanced photocatalytic degradation properties of zinc oxide nanoparticles synthesized by using plant extracts. Mater Sci Semicond Process 39:23–29
43. Tabrez S, Khan AU, Hoque M, Suhail M, Khan MI, Zughaibi TA (2022) Biosynthesis of ZnO NPs from pumpkin seeds' extract and elucidation of its anticancer potential against breast cancer. Nanotechnol Rev 11:2714–2725
44. Tong CWS, Wu M, Cho WCS, To KKW (2018) Recent advances in the treatment of breast cancer. Front Oncol 8
45. Un EP (2021) "FOOD WASTE INDEX REPORT 2021"
46. Unni V, Abishad P, Prasastha Ram V, Niveditha P, Yasur J, John L, Prejit N, Juliet S, Latha C, Vergis J, Kurkure NV, Barbuddhe SB, Rawool DB (2022) Green synthesis, and characterization of zinc oxide nanoparticles using Piper longum catkin extract and its in vitro antimicrobial activity against multi-drug-resistant non-typhoidal Salmonella spp. Inorg. Nano-Metal Chem. 1–9
47. Vijayakumar S, González-Sánchez ZI, Malaikozhundan B, Saravanakumar K, Divya M, Vaseeharan B, Durán-Lara EF, Wang M-H (2021) Biogenic synthesis of rod shaped ZnO nanoparticles using red paprika (*Capsicum annuum* L. var. grossum (L.) Sendt) and their in vitro evaluation. J Cluster Sci 32:1129–1139
48. Vinotha V, Yazhiniprabha M, Raj DS, Mahboob S, Al-Ghanim KA, Al-Misned F, Govindarajan M, Vaseeharan B (2020) Biogenic synthesis of aromatic cardamom-wrapped zinc oxide nanoparticles and their potential antibacterial and mosquito larvicidal activity: an effective eco-friendly approach. J Environ Chem Eng 8:104466
49. Wali SZ, M; Okla MK, Wahidah HA, Tauseef I, Haleem KS, Farid A, Maryam A, Abdelgawad H, Adetunji CO, Akhtar N., Akbar S, Rehman W, Yasir H, Shakira G (2024) *Brassica oleracea* L. (Acephala Group) based zinc oxide nanoparticles and their efficacy as antibacterial agent. Braz. J. Biol. 84:e259351
50. Wang N, Luo X, Han L, Zhang Z, Zhang R, Olin H, Yang Y (2020) Structure, performance, and application of $BiFeO_3$ nanomaterials. Nano-Micro Lett 12:81

Synthesis of Metal Nanoparticles from Fruits and Their Waste Materials for Diverse Applications

Radhika Sharma, Manik Devgan, Arshdeep Kaur, Antul Kumar, Taruna Suthar, Anuj Choudhary, Satyakam Guha, Anand Sonkar, and Sahil Mehta

Abstract In recent times, green fabricated nanomaterials have been used in numerous fields of scientific study substantially. Replacement for the current chemical strategies that contribute significantly to environmental contamination with the metallurgical nanoparticles, biosynthetic pathway has been created that is simple, harmless, economical, and environmentally benign to produce. Fruit peels, which are typically disposed of as waste even by the fruit processing sector, contain a variety of biologically active compounds, including ellagitannins, phenols, anthocyanins, carotenoids, flavonoids, glycosides, tannins, triterpenoids, essential oils, steroids, and vitamin C. All of them have significant health advantages and are anti-carcinogenic. The biosynthesis of metallic NPs such as silver, gold, nickel, iron, copper, aluminium, and platinum has been studied extensively utilizing fruit peel extract. Herein, the book chapter demonstrates the utility of multipurpose, cost-efficient NPs for enhanced medicinal, nutritional, and industrial applications. Additionally, fruit peel NPs uses and biosynthesis techniques have been addressed.

Keywords Green fabrication · Fruit peels · Bio synthesis · Metallic NPs

---

R. Sharma
Department of Soil Science, Punjab Agricultural University, Ludhiana, Punjab, India

M. Devgan · A. Kaur
Department of Plant Breeding and Genetics, Punjab Agricultural University, Ludhiana, Punjab, India

A. Kumar · T. Suthar
Department of Botany, Punjab Agriculture University, Ludhiana, Punjab, India

A. Choudhary
Department of Biology and Environmental Sciences, CSKHPKV, Palampur, India

S. Guha · A. Sonkar · S. Mehta (✉)
Department of Botany, Hansraj College, University of Delhi, New Delhi 110007, India
e-mail: sahilmehtasm21@hrc.du.ac.in

A. Husen (ed.), *Nanomaterials from Agricultural and Horticultural Products*, Smart Nanomaterials Technology,
https://doi.org/10.1007/978-981-99-3435-5_4

Abbreviations

FTIR	Fourier transform infrared
GCE	Glossy carbon electrode
DLS	Dynamic light scattering
MG	Methylglyoxal
AFM	Atomic force microscopy
MO	Methyl orange
SEM	Scanning electron microscopy
SPR	Surface plasmon resonance
TEM	Transmission electron microscopy
XRD	X-Ray diffraction
NZVI	Nanoscale Zero-Valent Iron
NPs	Nanoparticles
Cu	Copper
Zn	Zinc
Fe	Iron
Au	Gold
Ag	Silver
Al	Aluminium
Sn	Stannic
Pt	Platinum

1 Introduction

Nanotechnology is the most widely used technology and has attracted numerous researchers in the development of metal-based nanoparticles in an ecofriendly manner. Nanotechnology involves the 1–100 nm of particle size, manipulation, and synthesis strategy. This field allows the blending of the knowledge of natural science with physics, chemistry, engineering, computational science, biological sciences, and material science in the development of nanostructures. Nanostructures have various applications based on their morphology, distribution, and size. It has applications in several fields involving drug delivery, cosmetics, energy science, health care, optics, space industries, biomedical, chemical industries, catalysis, electronics, non-linear optical devices, environment, photo-electrochemical applications, and single-electron transistors [44–52, 58]. The research interest is shifted toward the synthesis of nanoscale metals with the help of physical, chemical, and green methods of synthesis these days [45].

Chemical and physical methods are under the status of replacement by green synthesis due to the release of harmful or toxic chemicals, large amount of energy consumption, and complex equipment for synthesis [4, 9, 53]. Green synthesis has become the more preferable over the physical and chemical methods in the vision of sustainability [29]. Although, issues are there with the extraction of raw materials, quality of final products, and the reaction time. For instance, the wide availability of raw material is less, particle size is highly homogenous, and a long time is required for synthesis [115, 120].

The potential of nanoparticles has revolutionized the environmental-friendly approaches for generating nanoparticles. The employment of bacteria, fungi, actinomycetes, and plants and where the fruits waste has appeared to be the most economic and latest viable reserve for the synthesis of NPs. Numerous research studies have reported the development of Ag NPs from the waste material of dragon fruit, orange, tangerine, kinnow, banana, papaya, avocado, and pomegranate [1, 53, 89], Au NPs from the mangosteen, pomegranate, mango, banana, and orange [34, 87], and zinc oxide NPs, iron NPs and other NPs [61, 81, 119]. Recently, the identification, extraction, measurement, and evaluation of bioactive compounds from fruit waste is the active thrust area of research [27]. The antioxidant potential of these phytochemicals is very crucial for health benefits [21]. The bioactive compounds are flavonoids, phenols, triterpenoids, tannins, glycosides, steroids, anthocyanins, carotenoids, vitamin C, essential oils, and ellagitannins are observed in the waste material of fruits. In this chapter, we are going to discuss the green synthesis of nanoparticles to put forward the challenges and major issues related to metallic NPs and the perspective in the direction of future target research.

2 Methods of Biosynthesis of NPs

The selection of an ecofriendly green solvent, a good reducing agent, and a safe stabilizing substance are the key requirements for the fabrication of nanoparticles. Nowadays, numerous strategies are being used but primarily these techniques can be divided into two categories: bottom-up and top-down techniques [57]. The paramount distinction between the two approaches is the raw material used to prepare the NPs. The top-down methods start with bulk material whereas atoms or molecules are the starting material in bottom-up approaches. While synthesis of nanoparticles from fruits is presented in Table 1.

2.1 Top-Down Method

With this technique, big portions of materials are broken down into tiny NPs. The size reduction of the initial material through various physical and chemical processes forms the basis for the creation of nanostructures [57, 64]. It covers techniques including mechanical milling, thermal, and laser. Although top-down approaches are simple to use, they are not appropriate for creating irregularly shaped and very small particles. This method's primary drawback is that nanomaterials' chemical and physical characteristics alter [108]

Table 1 Synthesis of NPs from fruits

Fruit common name	Scientific name	Biological extract	Types of NPs synthesized	Reaction temperature/time	Morphology	Size	Color appear	References
Pomegranate	*Punica granatum*	Peel	Zinc oxide	80 °C/ND	Sphere and hexagonal	32–81 nm	Greenish and pinkish-brown color	[111]
Indian gooseberry	*Emblica officinalis*	Pulp	Silver	RT/30 min	Sphere	15 nm	Yellowish brown	[111]
Christ's thorn jujube	*Ziziphus spina-christi*	Pulp	Copper Oxide	80 °C/ND	Sphere	5–20 nm	–	[111]
Papaya	*Caricaya papaya*	Leaves	Fe	–	Irregular	Irregular	–	[106]
Orange	*Citrus × sinensis*	Pulp	Fe	–	Spherical	3–300 nm	–	[61]
Pomegranate	*Punica granatum*	Peel	Gold	2 min	Spherical	25 nm	Purplish red	Naganathan and Thirunavukkarasu [28]
Tomato	*Lycopersicon esculentum*	Peel	Zinc	10 °C/min	–	50 nm	Yellow	[2]
Orange	*Citrus sinensis*	Peel	Zinc	5 h	Spherical	9–25 nm	Yellow to dark brown	[2]
Grapefruit	*Citrus paradise*	Peel	Zinc	30 min	Irregular	Irregular		[2]
Lemon	*Citrus aurantifolia*	Peel	Zinc	30 min	Irregular	Irregular	Yellow to dark brown	[2]
Sweet lime and sweet lemon	*Citrus sinensis and Citrus limetta*	Peel	Gold	Less than 2 h	Spherical	9–46 nm	Yellow to dark brown	Ahmed et al. [5]

(continued)

Table 1 (continued)

Fruit common name	Scientific name	Biological extract	Types of NPs synthesized	Reaction temperature/ time	Morphology	Size	Color appear	References
Wild orange	*Citrus macroptera*	Juice	Gold	40–50° C/ 90 min	Sphere	7–25 nm	Greenish blue	[118]
Pomegranate	*Punica granatum*	Leaves	Fe	–	Spherical	10–30 nm	–	[32]
Pomegranate	*Punica granatum*	Peel	Silver	RT/24 h	Sphere	20–40 nm	Brown	[]
Avocado	*Persea americana*	Avocado oil	Gold	RT/120 min	Spherical, Quasi-Spherical, Decahedral, and triangular	48.8 ± 24.8	Light yellow	[66]
Chebulic myrobalan	*Terminalia chebula*	Whole fruit	Silver	RT/ND	Cubic	25 nm	Brownish red	[3]
Grapes and orange	*Vitis vinifera* and *Citrus sinensis*	Wastes	Gold	20 min	Spherical	10–50 nm	Ruby red	[114]
Pineapple	*Ananas comosus*	Peel	Silver	24 h	Spherical	–	Brown	[3]
Pineapple	*Ananas comosus*	Peel	Silver	RT/24 h	Sphere	–	Yellow	[56]
Avocado	*Persea americana*	Seeds	Silver	RT/24 h	Spherical	35.6 nm	Yellow–brown	Kumar et al. [67]
Peach	*Prunus persica*	Peel	Silver	RT/24 h	Crystalline size	39.9 nm	Reddish brown	[8]
Longan	*Euphoria longana*	Peel	Silver	30 °C/24 h	Cubic	9–32 nm	Brownish to dark black	[64]
Purple mangosteen	*Garcinia mangostana*	Pulp	Zinc Oxide	70–80 °C/ ND	Sphere	21 nm	Bluish-green	Kumar et al. [67]

(continued)

Table 1 (continued)

Fruit common name	Scientific name	Biological extract	Types of NPs synthesized	Reaction temperature/time	Morphology	Size	Color appear	References
Pomelo	*Citrus maxima*	Juice	Zinc Oxide	400 °C/5–10 min	Agglomerated	10–20 nm	Light yellow	[116]
Orange	*Citrus sinensis*	Peel	Silver	90 °C/15 min	Sphere	7.36 nm	Yellow to dark brown	Kumar et al. [67]
Banana	*Musa paradisiaca*	Peel	Silver	30 °C/ND	Sphere	23.7 nm	Brown	[56]
Apple	*Malus pumila*	Pulp	Silver	80 °C/ND	Sphere	30.25 nm	Dark Brown	[116]
Date palm	*Phoenix dactylifera*	Pulp	Silver	55 °C/10 min	Sphere	25–60 nm	Deep red to brown	[116]
Black hawthorn	*Crataegus pentagyna*	Pulp	Silver	RT/2 h	Sphere	25–45 nm	–	Kumar et al. [67]
Fig	*Ficus carica*	Whole fruit	Silver	RT/24 h	Sphere	54–89 nm	Dark brown	[56]
Papaya	*Carica papaya*	Juice	Silver	NS	Sphere	75.68 nm	Yellowish brown	[116]
Caperberry	*Capparis spinosa*	Whole fruit	Copper Oxide	60 °C/24 h		17–41 nm	Amber yellow	Kumar et al. [67]
Pomegranate	*Punica granatum*	Juice	Gold	RT/20 min	Irregular	100 nm	Purple	Kumar et al. [67]
Grapes	*Vitis vinifera*	Peel	Silver	90 °C for 20 min	Crystalline and spherical	10–30 nm	Dark brown	[91]
Orange	*Citrus × sinensis*	Peel	Silver	90 °C for 20 min	Crystalline and spherical	10–30 nm	Yellow to dark brown	[91]
Melon	*Cucumis melo*	Peel	Silver	24 h	Spherical	78.11 mm	Yellowish	[92]

(continued)

Table 1 (continued)

Fruit common name	Scientific name	Biological extract	Types of NPs synthesized	Reaction temperature/time	Morphology	Size	Color appear	References
Watermelon	*Citrullus lanatus*	Rind	Silver	30 min	Spherical	20–140 nm	Yellow to dark brown	[91]
Banana	*Musa* × *paradisiaca*	Waste	Silver	38 min	Spheres	less than 20 nm	Reddish brown	[92]
Naseberry	*Achras sapota*	Fruit latex	Gold	30 min	Spherical	20–40 nm	Purple	[15]
Custard apple	*Annona squamosa*	Peel	Gold	4 h	Spherical	35 nm	Ruby red	[15]
Dragon fruit	*Selenicereus undatus*	Peel	Silver	24 h	Spherical	26 nm	Brown	[3]
Watermelon	*Citrullus lanatus*	Rind	Pd	24 h	Spherical	96 nm	Dark brown	[33]
Banana	*Musa paradisiac*	Peel	Pd	–	Irregular shape	50 nm	Dark brown	[33]
Tangerine	*Citrus reticulata*	Peel	Fe	–	Spherical	50–200 nm	–	[105]
Lime	*Citrus* × *aurantiifolia*	Peel	Fe	–	Cylindrical	3–300 nm	–	[105]
Orange	*Citrus sinensis*	Peel	Mg	3 h	Spherical	29 nm	–	Nejati et al. [90]
Banana	*Musa paradisiac*	Peel	Mn	2 h	Spherical	20- 50 nm	–	Nejati et al. [90]

2.1.1 Mechanical Milling

A particle size reduction method has the potential to manufacture high-purity NPs with improved physical attributes, and improved solubility of the therapeutic components [57, 108]. The alteration of surface characteristics of NPs is the major outcome of this process. Ball milling and mechano-chemical method are widely preferred for the synthesis of nanomaterials. Process variables and the characteristics of the milling powder have an impact on the effectiveness of mechanical milling [18].

2.1.2 Laser Ablation

The laser irradiation employed in the laser ablation method causes the particles to be reduced to the nanoscale dimensions. After being covered by a light coating, the target material is then exposed to pulsed laser irradiation [18]. When a material is exposed to laser energy, it fragments into NPs, which remain in the liquid surrounding the target and create a colloidal solution. The proportionate quantity of ablated molecules and electrons generated depends on the time of exposure and energy emitted by the laser [57].

2.1.3 Ion Sputtering

It involves vaporizing a solid by the sputter method with a beam of ions from inert gas. Recently, employing magnetron sputtering of metal targets, this technique was used to create NPs from a variety of metals. By using this technique, vast amounts of nanocrystal coatings are produced on Si substrates in the form of collinear nanoparticulate streams. Minimal pressure is used during the entire procedure [101].

2.2 Bottom-Up Methods

Bottom-up nanoparticle synthesis relies on combining atoms, molecules, or tiny particles to create nanoscale dimensions from larger molecules. This technique involves first creating nanosized building pieces, which are then put together to create the finished nanoparticle [108].

2.2.1 Solid State Methods

It is classified into physical and chemical vapor deposition methods. In physical vapor deposition, a substance is placed on a surface either as a thin layer or as a nanoparticle [101]. Vacuum technology with strict control, such as thermal material makes the substrate concentrate on the surface physically. Another technique of coating involves the chemical reaction of gaseous molecules carrying atoms necessary for the coating process to produce a thin coating of a material on a surface.

2.2.2 Liquid State Synthesis Methods

Jamkhande et al. [57] classified into further types.

- **Sol–gel method**—In this, gelatin plus colloidal suspension (sol) are used in this technique to create a network in a continuous liquid phase (gel). Colloids are created using the ions of metal alkoxides and alkoxysilanes both a predecessor.
- **Chemical reduction method**—Various chemical methods are used to reduce ionic salt in an acceptable environment while surfactant is present. Metallic NPs are created by reducing a substance in an aqueous phase, like sodium borohydride.
- **Hydrothermal process**—It relies on the depositing of tiny particles triggered by the interaction of aqueous medium vapors with solid material under high temperature and pressure.
- **Solvothermal method—**In the presence of water or other organic solvents including methanol, ethanol, and polyol, the solvothermal process is employed to prepare nanophase. The pressure vessel used to create the reaction permits the solvents (water and alcohol) to be heated above their boiling points.

2.2.3 Gas Phase Methods

It is further divided into spray, laser, and flame pyrolysis. In spray pyrolysis, the heated reactor receives vaporized nanoparticle antecedents when using the spray pyrolysis method. A nebulizer is used to deliver the precursors straight into the heated reactors in the shape of tiny, imperceptible droplets [101]. In laser pyrolysis, the creation of nanomaterials by laser pyrolysis requires the use of laser power. In this technique, homogenous nucleation processes are induced by allowing the precursors to receive laser power. The flame pyrolysis method's basic operating principle is indeed the straight spray of a liquid precursor into a flame to produce nanostructures [57, 108]

2.2.4 Biological Methods

The biological processes for making NPs use a variety of microorganisms, their enzymes, and plant products like isolates and extracts. These approaches have a variety of advantages over traditional physical and chemical processes since they are economical, environmentally friendly, and simple to scale up for mass production [108]. Additionally, green synthesis avoids the use of harmful chemicals, high temperatures, pressures, and energies. Bioreduction and biosorption are the primary methods used NPs synthesis [101].

2.2.5 Other Techniques of Nanoparticle Synthesis

Electrochemical deposition, microwave nanoparticle preparation, supercritical fluid technology, and ultra sound technique for the synthesis of NPs, are more precise and are needed to get over their limitations and ensure proper application at the commercial level [108].

3 Synthesis of Metallic NPs

3.1 Zinc Oxide NPs (ZnO NPs)

Recent research on dragon fruit by using its peel extracts for the environmentally friendly production of zinc oxide NPs was reported by Vishnupriya et al. [124]. To create NPs, dragon fruit peel extract and zinc chloride was combined in a 1:1 ratio. As a result, the color of the reaction medium altered from half-white to pale yellow. In the UV–vis spectral investigation, a distinctive absorbance peak was discovered at 360 nm. The synthesized NPs had an outer thickness of 70 nm, a size variation of 10–100 nm, and a spherical form with agglomeration, according to SEM examination [116, 124]. By mixing 2 g of zinc nitrate with 42.5 mL of each extract, ZnO NPs were created. The mechanism is explained in Fig. 1A. These combinations were again mixed continuously for 60 min, immersed in a water bath at about 60 °C until they each had the consistency of glassy caramel, and then heat treated at 400 °C for 1 h [88]. The final step was to use a mortar to grind the obtained samples into fine white powders, store them separately, and classified them as follows: M1 utilized tomato peel extract, M2 used orange peel extract, M3 used grapefruit peel extract, and M4 used lemon peel extract [88, 116].

In another study, Liu et al. [73] ZnO NPs were made from the extract of *Amomum longiligulare* which is non-toxic, cost-effective, and quick to biodegrade. UV–vis spectra were used to establish the creation of ZnO NPs from *A. longiligulare* fruit extract. Finally, the green-produced Al-ZnO NPs confirm the degradation of dyes as a function of time utilizing Al-ZnO NPs. The Al-ZnO NPs demonstrated photocatalysis rate and MB and MG degradation in a short span. It is also obvious that the degradation ratio and decomposing time are related. The pure aqueous solution from the fruit of the *G. mangostana* was used by [10] to create ZnO NPs, most of which were spherical and had a size distribution of 21 nm. Malachite green color breakdown during solar irradiation was employed to measure the photo-catalytic activity of biosynthesized ZnO NPs. By monitoring absorbance at its typical maximum value of 615 nm, the degree of MG dye degradation was tracked spectrophotometrically. Using the liquid chromatography-mass spectrophotometry approach, degradation products were found. The tiny dimension and high purity of the NPs synthesized ZnO NPs resulted in exceptional photo catalytic activity. Within an area of pharmaceutical

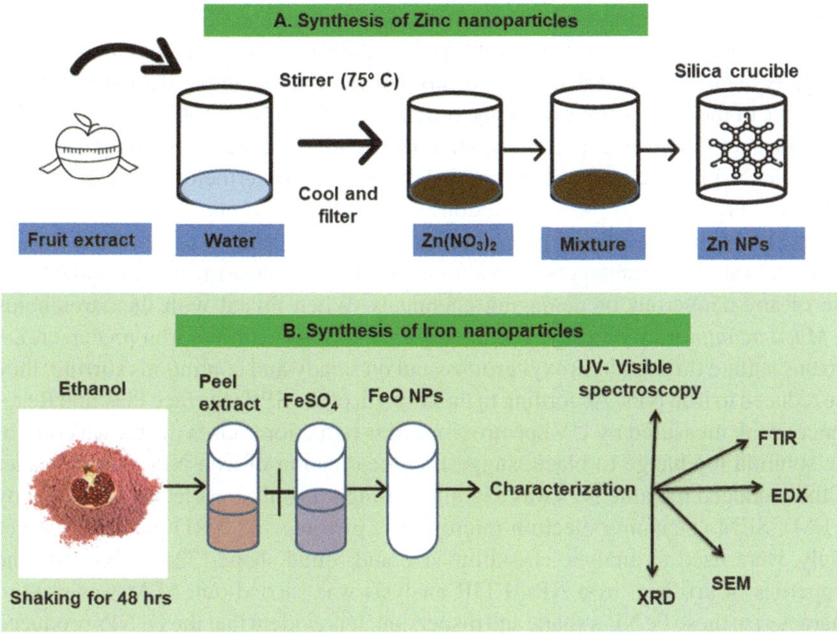

Fig. 1 Green synthesis of Zn NPs (**a**) from fruit extract and Fe NPs (**b**) from fruit pomegranate

application, this environmentally friendly production of ZnO-NPs utilizing hydro-methanolic extracts of *Flueggea leucopyrus* fruit paves the way for the creation of innovative drugs [123]. The secondary metabolites found in the hydro-methanolic extracts of *F. leucopyrus* fruit that are involved in the synthesis of ZnO-NPs were identified utilizing UV–visible, FT-IR spectrum analyses.

3.2 *Iron Oxide NPs (FeO NPs)*

Iron oxide NPs could be produced using Maghemite ($FeCl_3.6H_2O$), a hexahydrate of ferric chloride, using an aqueous extract of dried *Ficus carica* fruits. When these NPs are described, it was discovered that they were spherical in shape and had an average size of 9–4 nm [95]. The peel extract from watermelon rind is a good source of rutin, carotenoids, phytosterols, flavonoids, and other bioactive chemicals with functional groups like carboxylic and hydroxyl groups needed to create zero-valent iron NPs [60]. The mechanism of synthesis of NPs from pomegranate is explained in Fig. 1B. According to the strength of the SPR (surface Plasmon resonance) peak measured by UV spectroscopy, this reduction causes the brown concentration of the water to transform to black, suggesting the creation of iron NPs [32]. The overall size of the produced iron NPs was examined using transmission electron microscopy

(TEM). SEM (Scanning electron microscopy) pictures and XRD (X-Ray Diffraction) patterns, respectively, were employed to analyze crystallinity and round shape. Analysis using the Fourier transform infrared (FTIR) technique demonstrated the existence of the moiety involved in the production of iron NPs [106]. With the help of iron salts that served as a substrate again for the synthesis reaction, the iron nanoconstruct can be produced by reducing and stabilizing the watermelon rind extract. The inclusion turns the light brown solution into a dark brown one, removing any iron oxide NPs from the colloidal particles [105].

The filtrate from banana peel was used to create iron nanostructures without the use of any dangerous or damaging chemicals. When mixed with the carotenoids in *Musa acuminata* with polyphenol, they even engage with the iron precursor, *i.e.* ferrous sulfate through hydroxyl groups, and on steady and continuous stirring, they are reduced to iron NPs. According to the strength of the SPR (Surface Plasmon Resonance) peak measured by UV spectroscopy, this reduction causes the brown color of the solution to change to black, suggesting the creation of iron NPs. The diameter of the produced metallic NPs was examined using transmission electron microscopy (TEM). SEM (Scanning electron microscopy) pictures and XRD patterns, respectively, were used to analyze crystallite size and round shape [32]. To identify the properties of artificial iron NPs, FTIR analysis was carried out. SEM was used to characterize these Fe NP's shape and dispersion. It is evident that the Fe NPs produced utilizing pomegranate peel extracts are polydispersed and have a size ranging from 15 to 34 nm. XRD analysis was used to determine the crystallinity and phase purity. Results from UV–Visible spectroscopy showed that nanostructured materials had fully formed [60]. Iron NPs that are safe, dependable, and affordable are produced when a single iron precursor (ferric chloride) is combined with grape seed proanthocyanidin (a reducing agent and stabilizer). This was demonstrated by the transformation of the yellow solution into a greenish-brown color. Through XRD, DLS, and TEM investigation, structural and physicochemical characterization revealed that the material was crystalline with inverted spinal structure and size distribution. The presence of polyphenol in the production of iron NPs was established by FTIR analysis [106].

FTIR, XRD, UV–visible spectroscopy, SEM, and TEM are examples of high throughput techniques that were used to generate magnetic NPs using *Caricaya papaya* leaves as a reducing and stabilizing agent and ferric chloride salts as a precursor. This was done because of the existence of such specific composition. A significant sensitivity in the visible range of a produced Fe_3O_4 NPs UV–visible spectra demonstrated that perhaps the NPs are durable and evenly disseminated in the solution. As shown by the FTIR results, carboxylate ions strongly coordinate with iron metals and convert iron salts into Fe NPs. These NPs' erratic shape and precise size were visible in SEM and TEM pictures [105]. As both a reduction and stabilizing agent, the extract of the berry fruits was used to create ferromagnetic NPs from ferric salts. The ferric chloride solution and fruit extract were combined to produce black-colored precipitates. The creation of iron NPs, which was confirmed by UV–visible spectroscopy, is indicated by a shift in wavelength to the visible range. According to FTIR data, polyphenols carboxylic and hydroxyl groups serve as a good example of reducing ferric ions and stabilizing the iron NPs that are generated [79]. By mixing a dilute $FeCl_3$ mixture with an extract of *Psidium guajava* leaf, nanoparticles can be created. A UV–visible spectroscopy investigation confirmed that the mixture's yellow color turns black when metallic NPs are generated [38].

3.3 Copper Oxide NPs (Cu NPs)

Ficus carica fruit extracts were used by [95] in an aqueous solution and combined with 1 mM copper sulfate ($CuSO_4 \cdot 5H_2O$) for 12. After that, a solution containing CuO NPs was cleaned by repeatedly centrifuging it for 15 min at 12,000 rpm. They used an X-ray diffractometer to study the crystalline nature, shape, mass of the CuO NPs as well as UV absorbance. In a sample mixture, 150 mL of produced extract from *C. sinensis* leaf was added to a 1 molar solution of $CuCl_2$, which was then heated at 75–85 °C for an hour while being constantly stirred. The reduction of metal ions to metal NPs caused the suspension's hue to change from dark green to bright green [6].

The mixture was spun at 10,000 rpm for 10 min after being allowed to cool at ambient temperature for over two hours. The resulting metal NPs were then cleaned with water and ethanol three to four times to get rid of any contaminants before being baked in a drying oven for 24 h [6]. While the synthesis of copper NPs from papaya is represented in Fig. 2A. The study by Khani et al. [62] looked at the potential of *Ziziphus spina-christi* (L.) fruit extracts as novel reducing agents for the environmentally friendly manufacture of copper NPs. X-ray diffraction, Field emission scanning electron microscopy, Fourier transform infrared spectroscopy, and transmission electron microscopy were used to analyze the biosynthesized Cu-NPs.

Cu-NPs were discovered to be an effective adsorptive nanomaterial for removing crystal violet (CV) from aqueous solutions. The production of CuO NPs from discarded papaya peel and copper (II) nitrate trihydrate salt was shown by Phang et al. [98]. 40 mL of freshly made PPE were initially heated at 70–80 °C. The greenish-blue solution was promptly generated after around 1 g (0.004 mol) of copper (II) nitrate trihydrate was slowly incorporated into the hot papaya peel. The color of a liquid progressively altered from greenish-blue to green after being heated at 70–80 °C with steady mixing. The dark green slurry was again transferred to a ceramic crucible and cooled down to room temperature. The material was then heated to 450 °C in a thermal muffle furnace for two hours.

3.4 Gold NPs (Au NPs)

There is a lot of interest in the green synthesis of Au NPs using fruit extracts that contain phytochemical agents. This ecologically friendly method may enable larger synthesizes and is more cost-effective and biocompatible. It has been extensively

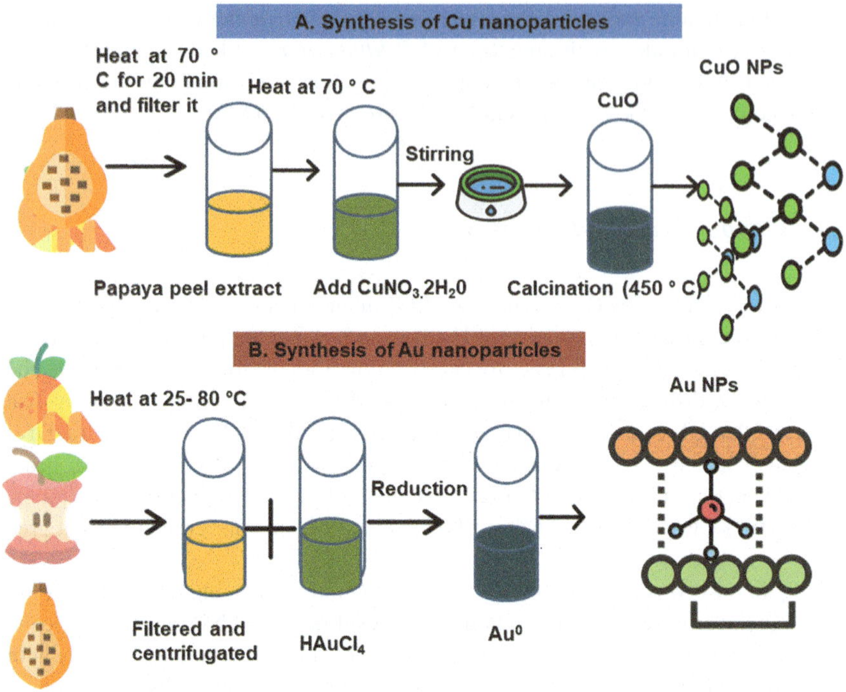

Fig. 2 Green synthesis of Cu NPs (**a**) from papaya and Au NPs (**b**) from different fruit wastes

studied by Baran et al. [20] utilizing environmentally acceptable techniques from the fruit peel of the *Ananas comosus* species. A UV–visible spectrophotometer was used to characterize the collected particles and the NPs had a spherical appearance, 11.61 nm in crystal nanosize, and maximum absorbance at 463 nm. The gram-positive and Gram-negative pathogen bacteria were inhibited by the NPs, which showed promising effectiveness Baran et al. [20]. The synthesis of gold NPs is represented in Fig. 2B.

For the environmentally friendly production of Au NPs, 10 mL of silver nitrate aqueous and 20 mg of the appropriate lyophilized extract were combined (0.1 M). The reduction of Ag^+ ions inside the mixture was observed by UV–Vis spectrophotometry for 20 min at 90 °C. At 5, 10, 15, and 20 min into the reaction, samples of 250 L were obtained, and UV–Vis spectra in the 200–800 nm regions were observed. [114]. In the form of colloidal particles, Au NPs are either red or purple in color (spherical NPs or large-sized NPs). These possess biological, photochemical, electrical, visual, and photodynamic qualities. The distinguishing qualities rely exclusively on its size and appearance [92]. Using debris from either melon or peach, green Au NPs were created by using 100 mL $AuCl_3$ solution with 10 mL of the extract solution, and the reaction mixture was continuously stirred for 24 h at room temperature [91]. *Sapindus mukorossi* fruit pericarp, has been used to create highly crystalline (face-centered cubic structures) Au NPs [47]. In the typical reaction, a 10 mL aqueous extract of soapnut shells was combined with 1 mL of 10 mM $HAuCl_4$ aqueous solution. The color shift proved the bioreduction of the Au ions and took 8 h to complete. Without employing any additional physical, chemical, or biological agents, an aqueous extract of leftover fruits at ambient temperature [28].

By merely reacting 9.7 mL (1 mM) $AuCl_4 \cdot 3H_2O$ with 0.3 mL of extract at room temperature, they were able to create Au NPs. Visual observation of the process was accompanied by UV–visible spectroscopy, and then circular NPs of about 20 nm in size were produced. Additionally, by employing *Citrus sinensis* peel extract as a reducing and capping agent, ultra-small Au NPs have been created. This procedure involved reacting 0.20 mL of fresh peel extract with 0.75 mL (2 mM) of $HAuCl_4$ solution in a water bath at 100 °C for 5 min [3].

3.5 Palladium NPs (Pd NPs)

Pd NPs were created by utilizing an aqueous suspension comprising an extraction from *Musa paradisiaca* (banana) peel. The resultant nanomaterials used to have a median size of approximately 50 nm, became crystalline and had an amorphous form [33]. Parallel to this, [69] created Pd NPs with a mean particle size of 96 nm but used extraction from watermelon rind. Using the leaf of *Prunus yedoensis*, [77] demonstrated the synthesis of spherical Pd NPs (50–150 nm). The environmentally friendly procedure involved swirling Pd (II) and leaf extract continuously for 30 min at 80 °C and pH 7. Patil [95] taking into account these advantages of successful green synthesis, created Pd NPs by treating an aqueous extract of fresh *Ficus carica* fruits with 1 mL of 1 mM H_2PdCl_4 for 3 h.

Additionally, graphene oxide reduction sheets coated with Pd NPs were shown to function as a catalyst in Suzuki cross-coupling reactions where phenyl boronic acid is coupled with an organo-halide. They optimized the reaction to get a 98% yield using base K_2CO_3 heated to 200 °C and 0.96 Pd NP/rGO (%).Hemmati et al. [43] used strawberries to make Pd NPs. First, 500 mg of Fe_3O_4 NPs were dissolved in a volume of 100 ml of water and sonicated for 20 min at 60 °C. The strawberry extract was then added to the mixture. An aqueous liquid containing Na_2PdCl_4 was then slowly added into the reaction solution (0.04 g, 50 ml, the drip rate was 1 ml/min) using a dropping funnel while the mixture was being sonicated once more for 30 min at the same temperature. The reaction mixture was then ultrasonically stirred for a further 20 min. The combination was then cooled to room temperature, and MNPs of Pd/Fe_3O_4 NPs were extracted from the solution as a black solid using a magnet and repeated rinses of water and ethyl alcohol.

Aqueous extract of mulberry (*Morus alba* L.) fruit was used as a bio-functional reducing material for the green synthesis of silver, Cu, and Pd metallic NPs. The formation of NPs was monitored visually and characterized by UV–Vis spectroscopy and the morphology and size of the as-prepared NP were evaluated with TEM and dynamic light scattering, respectively. Its synthesis is represented in Fig. 3. The synthesized Ag, Cu, and Pd NPs were found to be spherical and non-regular in shape

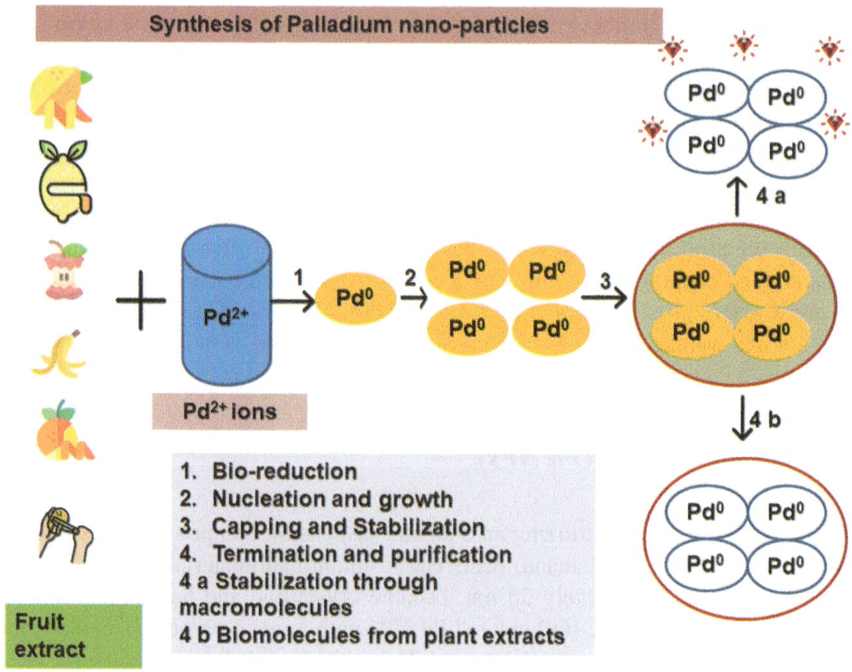

Fig. 3 Green synthesis of Pd NPs from different fruit wastes

with average particle sizes ranging from 80 to 150 nm (Ag NPs), 50 to 200 nm (Cu NPs), and 50 to 100 nm (Pd NPs) [103].

3.6 Siver NPs (Ag NPs)

Ag NPs are said to be created through the reduction of Ag ions to neutral atoms, the aggregation of Ag atoms to form NPs, and the stabilization of the NPs by the active chemicals found in the extract, as per reports just on green synthesis NPs. The production of Ag NPs has utilized waste products from a variety of fruits. Ag NPs were created by treating an aqueous *Annona squamosa* peel extract with $AgNO_3$ (1 mM) at two distinct temperatures, 25 °C and 60 °C [36]. For the synthesis of Ag NPs, fruit peel infusions from Ethiopian cactus pears [8]. 90 mL of 1 mM $AgNO_3$ and 10 mL of Ethiopian cactus pear fruit peel extract were combined under magnetic stirring. A visible color change and UV–visible spectroscopy for the production of Ag NPs were used to track the reaction. To create very stable Ag NPs, [8] used grape seed extract using 1 mL of grape seed extract and 20 mL of a 1×10^{-3} M aqueous solution of $AgNO_3$ was conducted. Likewise, grapes and oranges trash extracts were utilized for Ag NPs [114].

In more detail, 20 mg of the individual lyophilized extracts were combined with 10 mL of aqueous 0.1 M $AgNO_3$, and the reaction mixture was heated to 90 °C for 20 min before being tested for Ag ion reduction using UV–visible spectroscopy. To extract the essence, citrus fruits were crushed and restrained over a nylon mesh with tiny pores. To remove any unwanted contaminants, the recovered substance was centrifuged at 10,000 rpm for 10 min. This juice was used for additional analysis and research. The desired precursor was recognized as a 100 percent extract, and other consolidations of the extract were created utilizing this. Soon, a 50 mL solution of hydrogen tetrachloroaurate trihydrate at 1 mM was brought to a boil and then given a consistent, vigorous stir [15]. Metabolites present in fruits chemically reduce Ag^+ to Ag^0 during the manufacture of Ag NPs using plant or fruit extracts. Ag^0 atoms form small nuclei during the "nucleation" phase, which is followed by the "growth phase," where all these small nuclei are clustered, and lastly, the "capping" phase, where oxidized secondary metabolites surround the surface of the Ag NPs to stabilize the NPs [114]. For the manufacture of Ag NPs, a 1 mM aqueous solution of silver nitrate was produced. Using fruit extract from *A. marmelos* as a capping and reduction agent, co-precipitation is used to create Ag NPs from silver nitrate solution. DLS, UV–VIS spectrophotometer, AFM, and XRD were used to characterize physiologically produced Ag NPs.

3.7 Aluminium NPs (Al NPs)

Unlike citrate-based Al NPs, apple-derived Al NPs contain a lot of pectins. The pectin-rich AX-Al NPs increased the colorimetric detection of Al ions sensitivities. In both simulated samples and genuine data depending on the water supply, the detection limit was around 20 M. It turns out, however, that the apple derive-Al NPs spontaneously gathered following the biochemical test as a result of the liquid degrading [94].

3.8 Stannic NPs (Sn NPs)

Aqueous *Ficus carica* ethanolic extracts and a hydrated solution of Tin (II) chloride ($SnCl_2 \cdot 2H_2O$) were combined with vigorous agitation at 80 °C for 24 h in 2015 to create SnO_2 NPs [95]. Spherical NPs with an average size of 123 nm were produced and these produced SnO_2 NPs were tested to see if they can be employed with a glossy carbon electrode (GCE) to measure Hg^{2+} electrochemically. A linear association between the SnO_2 NPs and GCE was seen in the Hg^{2+} concentration range of 0.001–1.5 mM [95].

4 Application of NPs

Green nanotechnology offers many potential benefits which include application in biomedical sciences, improved food quality, and technology, improvement in nutrient absorption in the soil as well as reduction of agriculture inputs, etc. The application of these NPs and their benefits are discussed briefly.

4.1 Silver NPs (Ag NPs)

Biostatic analysis has shown efficient production and synergistic action of silver NPs as compared to other metals, especially in medical sciences []. Ag NPs have an immense number of applications in various fields. Over the past decade, multiple types of research on plant-derived silver NPs have indicated their role in battling abiotic stresses in plants through antioxidant defense mechanisms, methylglyoxal (MG) detoxification, and reduced oxidative damage [82]. The antibacterial activity of silver NPs extracted from *Phyllantus emblica* against rice bacterial brown stripe is another example of their application in agriculture [78]. AgNPs express significant broad-spectrum antibacterial and anticancer properties in humans [126].

In addition, Ag NPs extracted from *Juniperus procera* have shown cytotoxic and cell proliferation activities [54]. Ag NPs have proven significant in combating different types of tumor development in humans including cervical cancer, lung cancer, breast cancer, hepatic carcinoma, nasopharyngeal carcinoma, glioblastoma, colorectal adenocarcinoma, and prostate carcinoma, both in vitro and in vivo [72, 125]. Silver NPs cause necrosis or cell death in cancer cells by DNA degrading and ROS-producing enzymes [22]. Ag NP-containing NPs also work as biosensors for detecting infections, enzymes, blood sugar, tumor growth, and other pathogenic activities [12, 128]. These exciting aspects portraying the application of Ag NPs in various endeavors inspire us to explore more.

4.2 Gold NPs (Au NPs)

Gold in terms of textural qualities has the highest specific surface among metals this allows gold NPs to have a higher surface area and more dispersion []. Au NPs are being considered for application in multiple disciplines, such as biosensing, anti-cancer medications, and anti-microbial drugs, thanks to their biocompatibility, distinctiveness, and other advantageous qualities. Plant-derived Au NPs have been successful in demonstrating anticancer properties in humans. The pharmacologically important tree *Couroupita guianensis* has been used to production of doping-free completely photosynthetic gold NPs [39].

Two halophytic species; *Atriplex halimus* and *Chenopodium amperosidies* were characterized for green synthesis of gold NPs. Cytotoxicity investigations revealed the prominent effect of Au NPs in the suppression of human breast cancer. Moreover, photocatalytic degradation of methylene blue was exhibited by NPs [46]. Expressivity of the anti-cancer activity of Au NPs can be increased with the integration of organic acids previously present in plant tissues. *Moringa oliefera* with 1 M chloroauric acid was observed to possess good anti-cancer activities in humans [11]. Green Au NPs derived from coconuts have also been examined for their catalytic activity in the treatment of waste water, where these were successful in reducing the hazardous 4-nitrophenol to 4-aminophenol [96]. These reports highlight the significance and potential of the use of green synthesized gold particles in current scenarios.

4.3 Nickel NPs (Ni NPs)

Nickel metal in the form of nickel oxide NPs has been used as a prominent adsorbent of chemical pollutants, particularly in waste water treatments. These particles have been used to absorb dyes, hazardous chemicals as well as heavy metals [117]. Nickel oxide NPs synthesized from *Ocimum sanctum* demonstrated high absorption capacity for dyes and pollutants under different batch conditions [93]. Green synthesized Ni NPs have more catalytic activity than chemically synthesized Ni NPs [93, 117]. NiO NPs have received considerable attention in recent years particularly for electrocatalysis owing to their high chemical stability, superconducting properties, and ability to transmit electrons [117]. These features illustrate the use of Ni NPs and NiO NPs in the field of environmental cleaning and industrial catalysis.

4.4 Copper NPs (Cu NPs)

The chemical synthesis of Copper NPs employs harsh reducing and organic solvents which bind to its surface and thereby, increase their toxicity effects [24]. Environmentally friendly and green nanoparticle production has gained popularity recently as a solution to these issues. Nanobiocomposite copper films constructed via green synthesis from leaves in *Lawsonia inermis* have been used in electrical equipment such as light-emitting diode lamps [24]. Green Cu NPs have also been scrutinized for their anti-bacterial activity against both gram-positive and gram-negative bacteria, *Bacillus subtilis* and *E. coli*, respectively. As compared to silver NPs Cu NPs are more effective against gram-positive bacteria [25]. Cu NPs extracted from the latex of *Calotropisprocera* show anti-cancer activity by apoptotic destruction of tumor/cancer cells [41]. These NPs show NIR-II absorbance and therefore, stability under intense light or heat conditions. This property highlights their potential application in health sciences [121].

4.5 Iron NPs (Fe NPs)

Green iron NPs have major solicitations in the field of environmental pollution control, particularly in reducing harmful dyes such as methylene blue and methyl orange (MO) [107]. Fenton-like catalyst capability of iron NPs in degrading anionic and cationic has been demonstrated. [102], evaluated the heavy metal ion absorption capacity of iron and yeast nanobiocomposite. Chromium (VI) metal absorption increased more than 3 times as compared to conventional bioremediation activities. pH adjustment of waste water from 8.5 to 2 by green tea-derived iron NPs has also been demonstrated [26]. Noticeable bactericidal activity of zerovalent iron NPs against *E. coli* is also useful in medications [106].

4.6 Aluminium Oxide NPs (AlO NPs)

AlO NPs exhibit thermodynamic stability over a wide range of temperatures and possess a hexagonal closed compact structure [40]. This property allows the use of AlO NPs as a great catalyst. Moreover, these NPs are also applicable as biomaterials as well as environmental cleaners [31]. Sumesh et al. (2019), incorporated aluminium oxide NPs extracted from *Muntingia calabura* leaf with a natural composite containing sisal/coir, sisal/banana, and banana/coir to increase the composting activity of the hybrid compost, and the activity of the new hybrid composite was found to increase. AlO NPs have been observed for anti-bacterial activity against gram-negative strains of *E. coli* as well as gram-positive strains of *Staphylococcus aureus* [76]. Aluminium oxide NPs synthesized from leaf extracts of lemon grass have been observed to have bactericidal activity against multi-drug resistant *Pseudomonas aeruginosa* [13].

4.7 Platinum NPs (Pt NPs)

Pt NPs have superior physicochemical properties and great potential in biomedical applications [33]. Its green synthesis was first reported in *Diospyros kaki* and was synthesized in vivo and used as bio-reducing agents [17]. Photo-synthesized Pt NPs derived from *Solanum trilobatum* exhibited anti-bacterial and cytotoxic activity [86]. One of the major advantages of Pt NPs is their size *i.e.* 8 nm, with negligible adverse effects. This size optimization property helps in its application in biomedical sciences. Green synthesized Pt NPs have been used in the treatment of Neuro 2A cell lines without any side effects [75].

5 Advancements

Innovations like "Green Synthesis" in the production of NPs are the demand of the present circumstances. Green synthesis is particularly effective in minimizing the hazardous impacts pertaining to traditionally incorporated NPs. Several advances report the development of metal NPs that are cost-effective, conductive, catalytic, ecofriendly, and preclude the use of hazardous chemicals or reagents. Plant-derived NPs are simply phytofabricated, *i.e.,* metal ions coated with plant extracts under certain controlled conditions [113]. AuNPs coated with seed extracts of *Alpinia katsumadai* were prepared from $AgNO_3$ via the sonication process [42]. These NPs were quasi-spherical with an average diameter of 12.6 mm and showed improved plasmonic surface absorbance. Similarly, eco-friendly AuNPs were synthesized from leaf extracts of *Magnolia kobus* and *Diopyros kaki*, with stable phenolic compounds and showing anti-bacterial activity [112]. However, the catchphrase "Green synthesis of NPs" is not limited to the use of plants only, eco-friendly NPs have been manufactured from micro-organisms such as bacteria, fungi, algae, etc. Bio-NPs have been synthesized from micro-organisms under lab conditions via culture media and a controlled atmosphere [113].

Polysaccharides present in algae species offer better control of the plant morphology NPs as well as reduce metal ions to stable forms. Biogenic NPs have been synthesized from various red and brown algae such as *Chlorella vulgaris, Saragassum sp., Pseudochlorococcum typicum, Chlamydomonas reinhardtii,* etc. [7, 14, 110]. Similarly, extracellular and intercellular bacteria-produced NPs show excellent performance [84]. Another advantage offered by biosynthetic NPs is their ability to be morphologically modified under controlled conditions [68]. Applications of biosynthetic or green synthesized will clearly increase in the future, this will shed light on their long-term applicability and effects on living organisms. Advancements in techniques for manufacturing bio-NPs will help to overcome current obstructions and fully comprehend their true molecular mechanism and dynamics within the bodily tissues and environment, more research is still needed. Metallic green synthesized NPs will surely help in shaping a better nanotechnology-oriented future.

6 Limitation of Green Synthesis

The green synthesis of nanomaterials has already proven its immense potential. However, it has certain limitations such as selection of material, time constraints, conditions for synthesis, product quality, and applications. These pose challenges for large-scale implementation and for adoption at an industrial scale.

6.1 Material Selection

Several plant materials have been exploited for locally available abundant materials for green synthesis. These studies have increased the chance of use of locally available plant materials and difficult to implement them for global-scale production of NPs. For instance, AgNPs were synthesized from coconut which is mainly found in India, Philippines, Sri Lanka, south China and Malaysia [104]. Similarly, the AuNPs synthesis from Fenugreek is broadly distributed on the east coast of the Mediterranean and in China whereas peppermint is native to West Asia and Central Europe [16]. Psoralen is used for iron oxide NPs synthesis however it is restricted to Sri Lanka, Myanmar, and India [83]. The Andean blackberry is used for the synthesis of Cu NPs and is mainly restricted to the Andean region of Central and South America, Colombia, and Ecuador [65]. Hence, material selection using local plants should be explored for the production of large-scale implementation of metallic NPs.

6.2 Time Constraints

Arabica coffea is used for the production of Ag NPs but it requires seven years to become fully mature and prefers to grow at 1500 m altitude [30]. For the peach blossoms, it should be in the flowering period for their collection to limit their availability. Seeds of *Trigonella trifoliate* are used for the production of Au NPs collected during the fruiting time at the time of July to September. Some of the raw materials need further processing because they belong to the second category which further complicates it and increases the cost. Hence, the cost-effectiveness, practicability, and economic feasibility of these materials pose extra challenges. Similarly, carboxymethyl cellulose is utilized to produce Pd NPs [71]. Cellulose is the raw material that is modified by carboxymethylation to be extracted from sago pulp. Therefore, sodium hydroxide, sodium mono-chloroacetate, and related reagents are implemented but make the green synthesis incompatible.

6.3 Synthesis Process

The use of chemical reagents, long reaction times, and excessive energy consumption are the main concerns in the synthesis process. Caroling et al. [23] generated Cu NPs from the fruit extract of guava at 800 °C in the water bath. However, ultrasonic stirring at 80 °C for 120 min can able to generate CuO NPs by the chemical method [80]. Hence, the temperature requirement is very high for green synthesis and it takes a long time posing the requirement of intensive energy may impact adverse effects on the environment. Thus, it deflects from the concept of green synthesis. The long-term extraction process is not appropriate. For example, mango peels are required to be boiled for 12 h before the extraction process [127]. Similarly, mulberry and cherry extracts are needed to be dried at 50 °C for 48 h [100]. It is reported that high surface-to-volume ratio, surface coordination, and 3-D symmetry are highly affected and even the inert metals are oxidized under mild conditions [55]. Additionally, the lack of knowledge regarding the biosynthetic mechanism is another limiting factor in the field of green biosynthesis. For instance, the extract from the peels of the pomegranate can act as an end-capping agent for the $Cu/Cu_2O/CuO/ZnO$ nanomaterial synthesis [35]. The industrial-scale production from the green synthesis still lacks proper guidance on the mass balance as well as the stoichiometric ratio to scale up the process.

6.4 NPs Quality

The shape and size of NPs from different extracts are showing high variability. The high difference in particle size is by using green technology and makes it unfit for large-scale production. The nanoparticle (NZVI-nanoscale Zero-Valent Iron) was generated from the seeds of grapes varied from 63 to 381 nm in size [37]. Similarly, the citron juice and *Rosa canina* fruit were show high variability in size [109]. The defect in the formed product is reduced the practical benefits and limits the application to carry out on very large scale production.

6.5 Applications

The metallic NPs are having strong adsorption, high reducibility, and a large surface area to use in pollutant removal. The BOD and COD removal efficiencies ranged from 12 to 37% by NZVI from the lime, mandarin, orange, and lemon [74]. The low removal rate of NZVI caused raised the concern of iron wastage and inefficient use. The metallic NPs by means of green synthesis have very low efficiency in toxic metal removal. The sewage is generally mixed kinds of heavy metal, therefore, having less efficiency. The application of different nanomaterials formed from the fruit peel is presented in Table 2.

7 Conclusion

Green synthesis has several shortcomings and challenges such as non-uniform particle sizes, low yield, complex extraction procedures, and raw material variability due to regional and seasonal availability are strongly required to be eliminated for more practical production. The improvement of nanoparticle yield, implementation of simple energy economic strategies, and cheaper raw materials are important for the

Table 2 Application of different nanomaterials formed from the fruit peel

Fruit	Scientific name	Dominating compound in peel	Properties/ advantages	References
Coconut (fibers)	*Cocos nucifera*	Phenolics	Larvicidal activity	[104]
Cocoa (husk)	*Theobroma cacao*	Citric acid, p-hydroxybenzoic acid, salicylic acid, linarin, and linoleic acid	Anti-microbial, larvicidal, anti-bacterial activity	[70]
Dragon Fruit	*Selenic ereusundatus*	Pectin, betacyanin, and phyllocactin	Anti-bacterial activity	[99]
Papaya	*Carica papaya*	Phenolics	Anti-bacterial activity	[19]
Peach	*Prunus persica*	Cellulose, lignin, and hemicelluloses	Catalytic activity	[63]
Banana	*Musa paradisiaca*	Tannins, flavonoids, and quinines	Anti-cancer, anti-bacterial activity	[122]
Pomegranate	*Punica granatum*	Phenolics and carboxyl compound	Photocatalytic activity	[59]
Orange	*Citrus* sp.	Phenolics	Anti-bacterial activity	[114]
Grapes	*Vitis vinifera*	Phenolics	Anti-bacterial activity	[114]
Pineapple	*Ananas quamosa*	Coumarins, saponins, terpenoids, carbohydrates, and phenols	Anti-cancer activity	[97]
Passion Fruit	*Passiflora edulis*	Phenolics, and flavonoids	Anti-cancer, anti-bacterial activity	[97]

future in the research field. NPs are functionally modified in several disciplines such as food fortification, pharmaceuticals, personal care, agriculture, and waste water treatment. Thus, it is highly important to establish an eco-friendly and sustainable approach for the generation of NPs with clear morphologies, regulated size, and desired activities. The biological activities are cost-efficient techniques to use naturally produced products and lessen the utilization of NPs synthesis. The waste materials of fruit are abundant, cheaper, bioactive-rich compounds, and biodegradable to supply an excellent raw material in the context of NPs synthesis.

The bio-synthesis of metallic NPs have unique improved properties such as dimensional and optoelectronic properties, efficiency, functional modifications, and enhanced surface-to-volume ratio. The use of material associated with fruit is incredibly significant for the mitigation and remediation of organic waste. Although, more attention is required to biosafety issues, mechanistic aspects, and challenges related to toxicity. Therefore, research areas widen in the future associated with organic fruit material in the extensive bio-compatibility study, industrial-scale synthesis, strategies related to specificity (such as stability, dispersibility, catalytic activity, and adsorption capacity), food processing application, and development of novel strategies for purification, isolation, and functionalization.

References

1. Adebayo AE, Oke AM, Lateef A, Oyatokun AA, Abisoye OD, Adiji IP, Fagbenro DO, Amusan TV, Badmus JA, Asafa TB, Beukes LS (2019) Biosynthesis of silver, gold and silver–gold alloy nanoparticles using *Persea americana* fruit peel aqueous extract for their biomedical properties. Nanotech Environ Eng 4(1):1–5
2. Agarwal H, Kumar SV, Rajeshkumar S (2017) A review on green synthesis of zinc oxide nanoparticles–An eco-friendly approach. Resour Effici Technol 3(4):406–413
3. Ahmad S, Munir S, Zeb N, Ullah A, Khan B, Ali J, Bilal M, Omer M, Alamzeb M, Salman SM, Ali S (2019a) Green nanotechnology: a review on green synthesis of silver nanoparticles—An ecofriendly approach. Inter J Nanomed 14:5087
4. Ahmed S, Ikram S (2015) Silver nanoparticles: one pot green synthesis using *Terminalia arjuna* extract for biological application. J Nanomed Nanotechnol 6(4):1–6
5. Ahmed S, Kaur G, Sharma P, Singh S, Ikram S (2018) Fruit waste (peel) as bio-reductant to synthesize silver nanoparticles with antimicrobial, antioxidant and cytotoxic activities. J Appl Biomed 16(3):221–31
6. Ahmed A, Usman M, Liu QY, Shen YQ, Yu B, Cong HL (2019b) Plant mediated synthesis of copper nanoparticles by using *Camelia sinensis* leaves extract and their applications in dye degradation. Ferroelectrics 549(1):61–69
7. AlNadhari S, Al-Enazi NM, Alshehrei F, Ameen F (2021) A review on biogenic synthesis of metal nanoparticles using marine algae and its applications. Environ Res 194:110672
8. Ali S, Chen X, Shah MA, Ali M, Zareef M, Arslan M, Ahmad S, Jiao T, Li H, Chen Q (2021) The avenue of fruit wastes to worth for synthesis of silver and gold nanoparticles and their antimicrobial application against foodborne pathogens: a review. Food Chem 359:129912
9. Alsammarraie FK, Wang W, Zhou P, Mustapha A, Lin M (2018) Green synthesis of silver nanoparticles using turmeric extracts and investigation of their antibacterial activities. Colloids Surf, B 171:398–405
10. Aminuzzaman M, Ying LP, Goh WS, Watanabe A (2018) Green synthesis of zinc oxide nanoparticles using aqueous extract of *Garcinia mangostana* fruit pericarp and their photocatalytic activity. Bull Mater Sci 41(2):1–10
11. Anand K, Gengan RM, Phulukdaree A, Chuturgoon A (2015) Agroforestry waste Moringaoleifera petals mediated green synthesis of gold nanoparticles and their anti-cancer and catalytic activity. J Indus Eng Chem 21:1105–1111
12. Anderson K, Poulter B, Dudgeon J, Li S-E, Ma X (2017) A highly sensitive nonenzymatic glucose biosensor based on the regulatory effect of glucose on electrochemical behaviors of colloidal silver nanoparticles on MoS2. Sensors 17:1807
13. Ansari MA, Khan HM, Alzohairy MA, Jalal M, Ali SG, Pal R, Musarrat J (2014) Green synthesis of Al_2O_3 nanoparticles and their bactericidal potential against clinical isolates of multi-drug resistant *Pseudomonas aeruginosa*. World J Microbiol Biotechnol 31(1):153–64
14. Arya A, Gupta K, Chundawat TS, Vaya D (2018) Biogenic synthesis of copper and silver nanoparticles using green alga *Botryococcus braunii* and its antimicrobial activity. Bioinorg Chem Appl Article ID 7879403
15. Asiya SI, Pal K, Kralj S, El-Sayyad GS, de Souza FG, Narayanan T (2020) Sustainable preparation of gold nanoparticles via green chemistry approach for biogenic applications. Mater Today Chem 17:100327

16. Aswathy AS, Philip D (2012) Green synthesis of gold nanoparticles using *Trigonella foenum-graecum* and its size-dependent catalytic activity Spectrochim. Acta A Mol Biomol Spectrosc 97:1–5
17. Attar A, Yapaoz MA (2018) Biosynthesis of palladium nanoparticles using Diospyros kaki leaf extract and determination of antibacterial efficacy. Prep Biochem Biotechnol 48:629–634
18. Baig N, Kammakakam I, Falath W (2021) Nanomaterials: a review of synthesis methods, properties, recent progress, and challenges. Mater Adv 2(6):1821–1871
19. Balavijayalakshmi J, Ramalakshmi V (2017) *Carica papaya* peel mediated synthesis of silver nanoparticles and its antibacterial activity against human pathogens. J Appl Res Technol 15(5):413–422
20. Baran MF, Acay H, Keskin C (2020) Determination of antimicrobial and toxic metal removal activities of plant-based synthesized (Capsicum annuum L. Leaves), ecofriendly, gold nanomaterials. Global Chall 4(5):1900104
21. Ben-Othman S, Jõudu I, Bhat R (2020) Bioactives from agri-food wastes: present insights and future challenges. Molecules 25(3):510
22. Bethu MS, Netala VR, Domdi L, Tartte V, Janapala VR (2018) Potential anticancer activity of biogenic silver nanoparticles using leaf extract of Rhynchosiasuaveolens: an insight into the mechanism. Artif cell nanomed 46:104–114
23. Caroling G, Priyadharshini MN, Vinodhini E, Ranjitham AM, Shanthi P (2015) Biosynthesis of copper nanoparticles using aqueous guava extract-characterisation and study of antibacterial effects. Int J Pharm Biol Sci 5(2):25–43
24. Cheirmadurai K, Biswas S, Murali R, Thanikaivelan P (2014) Green synthesis of copper nanoparticles and conducting nanobiocomposites using plant and animal sources. RSC Adv 4(37):19507–19511
25. Chen CW, Hsu CY, Lai SM, Syu WJ, Wang TY, Lai PS (2014) Metal nanobullets for multidrug resistant bacteria and biofilms. Adv Drug Deliv Rev 78:88–104
26. Chrysochoou M, McGuirea M, Dahalb G (2012) Transport characteristics of green-tea nanoscale zero valent iron as a function of soil mineralogy. Chem Eng Trans 28:122–126
27. Coman V, Teleky BE, Mitrea L, Martău GA, Szabo K, Călinoiu LF, Vodnar DC (2020) Bioactive potential of fruit and vegetable wastes. Adv Food Nutr research 91:157–225
28. Deokar GK, Ingale AG (2016) Green synthesis of gold nanoparticles (Elixir of Life) from banana fruit waste extract—an efficient multifunctional agent. RSC Adv 6(78):74620–74629
29. Devi HS, Boda MA, Shah MA, Parveen S, Wani AH (2019) Green synthesis of iron oxide nanoparticles using *Platanus orientalis* leaf extract for antifungal activity. Green Process Synthe 8(1):38–45
30. Dhand V, Soumya L, Bharadwaj S, Chakra S, Bhatt D, Sreedhar B (2016) Green synthesis of silver nanoparticles using *Coffea arabica* seed extract and its antibacterial activity. Mater Sci Eng 58:36–43
31. Duraisamy P (2018) Green synthesis of aluminium oxide nanoparticles by using *Aerva lanta* and *Terminalia chebula* extracts. Int J Res Appl Sci Eng Technol 6:428–433
32. Ebrahiminezhad A, Zare-Hoseinabadi A, Sarmah AK, Taghizadeh S, Ghasemi Y, Berenjian A (2018) Plant-mediated synthesis and applications of iron nanoparticles. Mole Biotechnol 60(2):154–168
33. Fahmy SA, Preis E, Bakowsky U, Azzazy HMES (2020) Palladium nanoparticles fabricated by green chemistry: promising chemotherapeutic, antioxidant and antimicrobial agents. Materials 13(17):3661
34. Fernandes RA, Berretta AA, Torres EC, Buszinski AF, Fernandes GL, Mendes-Gouvêa CC, de Souza-Neto FN, Gorup LF, de Camargo ER, Barbosa DB (2018) Antimicrobial potential and cytotoxicity of silver nanoparticles phytosynthesized by pomegranate peel extract. Antibiotics 7(3):51
35. Fuku X, Kaviyarasu K, Matinise N, Maaza M (2016) Punicalagin green functionalized $Cu/Cu_2O/ZnO/CuO$ nanocomposite for potential electrochemical transducer and catalyst. Nanoscale Res Lett 11(1):1–12

36. Gangapuram BR, Bandi R, Alle M, Dadigala R, Kotu GM, Guttena V (2018) Microwave assisted rapid green synthesis of gold nanoparticles using *Annona squamosa* L peel extract for the efficient catalytic reduction of organic pollutants. J Mole Struct 1167:305–315
37. Gao JF, Li HY, Pan KL, Si CY (2016) Green synthesis of nanoscale zero-valent iron using a grape seed extract as a stabilizing agent and the application for quick decolorization of azo and anthraquinone dyes. Royal Soci Chem 6(27):22526–22537
38. Gayathri V, Kiruba D (2014) Preliminary phytochemical analysis of leaf powder extracts of *Psidium guajava* L. Int J Pharmacogn Phytochem Res 6(2):332–334
39. Geetha R, Ashokkumar T, Tamilselvan S, Govindaraju K, Sadiq M, Singaravelu G (2013) Green synthesis of gold nanoparticles and their anticancer activity. Cancer Nanotech 4:91–98
40. Goutam SP, Avinashi SK, Yadav M, Roy D, Shastri R (2018) Green synthesis and characterization of aluminium oxide nanoparticles using leaf extract of Rosa. Adv Sci Eng Med 10(7–8):719–722
41. Harne S, Sharma A, Dhaygude M, Joglekar S, Kodam K, Hudlikar M (2012) Novel route for rapid biosynthesis of copper nanoparticles using aqueous extract of Calotropisprocera L. latex and their cytotoxicity on tumor cells. Colloids Surf B Biointerfaces 95:284–288
42. He Y, Wei F, Ma Z, Zhang H, Yang Q, Yao B, Huang Z, Li J, Zeng C, Zhang Q (2017) Green synthesis of silver nanoparticles using seed extract of Alpiniakatsumadai, and their antioxidant, cytotoxicity, and antibacterial activities. RSC Adv 7:39842–39851
43. Hemmati S, Yousefi M, Salehi MH, Amiri M, Hekmati M (2020) Palladium nanoparticles immobilized over Strawberry fruit extract coated Fe_3O_4 NPs: a magnetic reusable nanocatalyst for Suzuki–Miyaura coupling reactions. Appl Organomet Chem 34(8):e5653
44. Hoag GE, Collins JB, Holcomb JL, Hoag JR, Nadagouda MN, Varma RS (2009) Degradation of bromothymol blue by greener nanoscale zero-valent iron synthesized using tea polyphenols. J Mater Chem 19(45):8671–8677
45. Horwat D, Zakharov DI, Endrino JL, Soldera F, Anders A, Migot S, Karoum R, Vernoux P, Pierson JF (2011) Chemistry, phase formation, and catalytic activity of thin palladium-containing oxide films synthesized by plasma-assisted physical vapor deposition. Surf Coat Technol 205:S171–S177
46. Hosny M, Fawzy AM, Abdelfatah AM, Fawzy EE, Eltaweil EE (2021) Comparative study on the potentialities of two halophytic species in the green synthesis of gold nanoparticles and their anticancer, antioxidant and catalytic efficiencies. Advan Powder Technol 32(9):3220–3233
47. Huong VTL, Nguyen NT (2021) Green synthesis, characterization and antibacterial activity of silver nanoparticles using *Sapindus mukorossi* fruit pericarp extract. Mater Today: Proc 42:88–93
48. Husen A, Siddiqi KS (2014) Phytosynthesis of nanoparticles: concept, controversy and application. Nanoscale Res Lett 9(229):1–24
49. Husen A (2022) Engineered Nanomaterials for Sustainable Agricultural Production, Soil Improvement and Stress Management. Elsevier Inc., 50 Hampshire St., 5th Floor, Cambridge, MA 02139, USA
50. Husen A, Iqbal M (2019) N*anomaterials*and plant potential: An overview. In: Eds. Husen A, Iqbal M (Eds.) N*anomaterials*and plant potential. Springer International Publishing AG, Gewerbestrasse 11, 6330 Cham, pp. 3–29. https://doi.org/10.1007/978-3-030-05569-1_1
51. Husen A, Jawaid M (2020) Nanomaterialsfor Agriculture and Forestry Applications. Elsevier Inc., 50 Hampshire St., 5th Floor, Cambridge, MA 02139, USA. https://doi.org/10.1016/C2018-0-02349-X
52. Husen A, Siddiqi KS (2022) Advances in Smart Nanomaterials and their Applications. Elsevier Inc., 50 Hampshire St., 5th Floor, Cambridge, MA 02139, USA
53. Ibrahim HMM (2015) Green synthesis and characterization of silver nanoparticles using banana peel extract and their antimicrobial activity against representative microorganisms. J Radiat Res Appl Sci 8(3):265–275
54. Ibrahim E, Kilany M, Ghramh HA, Khan K, Islam SU (2018) Cellular proliferation/cytotoxicity and antimicrobial potentials of green synthesized silver nanoparticles (AgNPs) using *Juniperus procera*. Saudi J Biol Sci 26:1689–1694

55. Ismail E, Khenfouch M, Dhlamini M, Dube S, Maaza M (2017) Green palladium and palladium oxide nanoparticles synthesized via *Aspalathus linearis* natural extract. J Alloys Compd 695:3632–3638
56. Jadoun S, Arif R, Jangid NK, Meena RK (2021) Green synthesis of nanoparticles using plant extracts: a review. Environ Chem Letters 19(1):355–374
57. Jamkhande PG, Ghule NW, Bamer AH, Kalaskar MG (2019) Metal nanoparticles synthesis: an overview on methods of preparation, advantages and disadvantages, and applications. J Drug Dlvr Sci Tech 53:101174
58. Jin-Chul, K, Madhusudhan, A, Husen, A (2021) Smart nanomaterials in biomedical applications. Springer Nature Switzerland AG, Gewerbestrasse 11, 6330 Cham, Switzerland. https://doi.org/10.1007/978-3-030-84262-8
59. Joshi SJ, Geetha SJ, Al-Mamari S, Al-Azkawi A (2018) Green synthesis of silver nanoparticles using pomegranate peel extracts and its application in photocatalytic degradation of methylene blue. Jundishapur J Nat Pharm Prod 13(3):e67846
60. Kaur M, Chopra DS (2018) Green synthesis of iron nanoparticles for biomedical applications. Glob J Nanomed 4(4):68–76
61. Kaur P, Thakur R, Chaudhury A (2016) Biogenesis of copper nanoparticles using peel extract of *Punica granatum* and their antimicrobial activity against opportunistic pathogens. Green Chemistry Lett Rev 9(1):33–38
62. Khani R, Roostaei B, Bagherzade G, Moudi M (2018) Green synthesis of copper nanoparticles by fruit extract of *Ziziphus spina*-christi (L.) Willd.: application for adsorption of triphenylmethane dye and antibacterial assay. J Mole Liq 255:541–549
63. Khodadadi B, Bordbar M, Nasrollahzadeh M. *Achillea millefolium* L. extract mediated green synthesis of waste peach kernel shell supported silver nanoparticles: Application of the nanoparticles for catalytic reduction of a variety of dyes in water. J Colloid Interface Sci 2017 493:85–93
64. Krishnia L, Thakur P, Thakur A (2022) Synthesis of nanoparticles by physical route. In: Thakur, A, Thakur, P, Khurana, SP (eds) Synthesis and applications of nanoparticles 45–59
65. Kumar B, Smita K, Cumbal L, Debut A (2017) Green synthesis of silver nanoparticles using Andean blackberry fruit extract. Saudi J Biol Sci 24(1):45–50
66. Kumar B, Smita K, Debut A, Cumbal L (2018) Utilization of *Persea americana* (Avocado) oil for the synthesis of gold nanoparticles in sunlight and evaluation of antioxidant and photocatalytic activities. Environ Nanotechnol Monit Manag 10:231–237
67. Kumar H, Bhardwaj K, Dhanjal DS, Nepovimova E, Şen F, Regassa H, Singh R, Verma R, Kumar V, Kumar D, Bhatia SK (2020) Fruit extract mediated green synthesis of metallic nanoparticles: a new avenue in pomology applications. Int J Mol Sci 21(22):8458
68. Kuppusamy P, Yusoff MM, Maniam GP, Govindan N (2016) Biosynthesis of metallic nanoparticles using plant derivatives and their new avenues in pharmacological applications-An updated report. Saudi Pharm J 24(4):473–484
69. Lakshmipathy R, Palakshi Reddy B, Sarada NC, Chidambaram K, Khadeer Pasha SK (2015) Watermelon rind-mediated green synthesis of noble palladium nanoparticles: catalytic application. Appl Nanosci 5(2):223–228
70. Lateef A, Azeez MA, Asafa TB, Yekeen TA, Akinboro A, Oladipo IC, Azeez L, Ojo SA, Gueguim-Kana EB, Beukes LS (2016) Cocoa pod husk extract-mediated biosynthesis of silver nanoparticles: its antimicrobial, antioxidant and larvicidal activities. J Nanostruct Chem 6(2):159–169
71. Li G, Li Y, Wang Z, Liu H (2017) Green synthesis of palladium nanoparticles with carboxymethyl cellulose for degradation of azo-dyes. Mater Chem Phy 187:133–140
72. Lin J, Huang Z, Wu H, Zhou W, Jin P, Wei P, Zhang Y, Zheng F, Zhang J, Xu J, Hu Y (2014) Inhibition of autophagy enhances the anticancer activity of silver nanoparticles. Autophagy 10(11):2006–2020
73. Liu YC, Li J, Ahn J, Pu J, Rupa EJ, Huo Y, Yang DC (2020) Biosynthesis of zinc oxide nanoparticles by one-pot green synthesis using fruit extract of *Amomum longiligulare* and its activity as a photocatalyst. Optik 218:165245

74. Machado S, Grosso JP, Nouws HPA, Albergaria JT, Delerue-Matos C (2014) Utilization of food industry wastes for the production of zero-valent iron nanoparticles. Sci Total Environ 496:233–240
75. Manikandan M, Hasan N, Wu HF (2013) Platinum nanoparticles for the photothermal treatment of Neuro 2A cancer cells. Biomaterials 34(23):5833–5842
76. Manikandan V, Jayanthi P, Priyadharsan A, Vijayaprathap E, Anbarasan PM, Velmurugan P (2019) Green synthesis of pH-responsive Al_2O_3 nanoparticles: Application to rapid removal of nitrate ions with enhanced antibacterial activity. J Photochem Photobiol Chem 371:205–215
77. Manikandan V, Velmurugan P, Park JH, Lovanh N, Seo SK, Jayanthi P, Park YJ, Cho M, Oh BT (2016) Synthesis and antimicrobial activity of palladium nanoparticles from *Prunus* × *yedoensis* leaf extract. Mater Lett 185:335–338
78. Masum MMI, Siddiqa MM, Ali KA, Zhang Y, Abdallah Y, Ibrahim E, Qiu W, Yan C, Li B (2019) Biogenic synthesis of silver nanoparticles using *Phyllanthus emblica* fruit extract and its inhibitory action against the pathogen Acidovoraxoryzae strain RS-2 of rice bacterial brown stripe. Front Microbiol 10:820
79. Murgueitio E, Debut A, Landivar J, Cumbal L (2016) Synthesis of iron nanoparticles through extracts of native fruits of Ecuador, as Capuli (*Prunus serotina*) and Mortino (*Vaccinium floribundum*). Biol Med 8(3):1–3
80. Muthuvel A, Jothibas M, Manoharan C (2020) Synthesis of copper oxide nanoparticles by chemical and biogenic methods: photocatalytic degradation and in vitro antioxidant activity. Nanotechnol Environ Eng 5(2):1–19
81. Nabi G, Ain QU, Tahir MB, Nadeem Riaz K, Iqbal T, Rafique M, Hussain S, Raza W, Aslam I, Rizwan M (2022) Green synthesis of TiO_2 nanoparticles using lemon peel extract: their optical and photocatalytic properties. Inter J Environ Anal Chem. 102(2):434–442
82. Nadiyah M, Alabdallah MD, Hasan M (2021) Plant-based green synthesis of silver nanoparticles and its effective role in abiotic stress tolerance in crop plants. Saudi J Biol Sci 28(10):5631–5639
83. Nagajyothi PC, Pandurangan M, Kim DH, Sreekanth TVM, Shim J (2017) Green synthesis of iron oxide nanoparticles and their catalytic and in vitro anticancer activities. J Cluster Sci 28(1):245–257
84. Nagajyothi PC, Prabhakar Vattikuti SV, Devarayapalli KC, Yoo K, Shim J, Sreekanth TV (2020) Green synthesis: photocatalytic degradation of textile dyes using metal and metal oxide nanoparticles-latest trends and advancements. Crit Rev Environ Sci Technol 50(24):2617–2723
85. Naganathan K, Thirunavukkarasu S (2017) Green way genesis of silver nanoparticles using multiple fruit peels waste and its antimicrobial, anti-oxidant and anti-tumor cell line studies. InIOP Confer Ser: Mater Sci Eng 191(1):012009
86. Narendhran S, Manikandan M, Shakila P (2019) Antibacterial, antioxidant properties of Solanumtrilobatum and sodium hydroxidemediated magnesium oxide nanoparticles: a green chemistry approach. Bull Mater Sci 42:1–8
87. Nasiriboroumand M, Montazer M, Barani H (2018) Preparation and characterization of biocompatible silver nanoparticles using pomegranate peel extract. J Photochem Photobiol B: Biol 179:98–104
88. Nava OJ, Soto-Robles CA, Gómez-Gutiérrez CM, Vilchis-Nestor AR, Castro-Beltrán A, Olivas A, Luque PA (2017) Fruit peel extract mediated green synthesis of zinc oxide nanoparticles. J Mole Struct 1147:1–6
89. Naz S, Shams F, Tabassum S, Ul-Haq I, Ashraf M, Zia M (2017) Kinnow peel extract as a reducing and capping agent for the fabrication of silver NPs and their biological applications. Institution of Eng Tech 11(8):1040–1045
90. Nejati M, Rostami M, Mirzaei H, Rahimi-Nasrabadi M, Vosoughifar M, Nasab AS, Ganjali MR (2022) Green methods for the preparation of MgO nanomaterials and their drug delivery, anti-cancer and anti-bacterial potentials: a review. Inorg Chem Commun 136:109107
91. Ogundare SA, Moodley V, Amaku JF, Ogunmoye AO, Atewolara-Odule OC, Olubomehin OO, Awokoya KN, Sanyaolu NO, Ibikunle AA, van Zyl WE (2021) Nanocrystalline cellulose

derived from melon seed shell (*Citrullus colocynthis* L.) for reduction and stabilization of silver nanoparticles: synthesis and catalytic activity. Carbohydr Polym Technol Appl 2:100134

92. Oza G, Reyes-Calderón A, Mewada A, Arriaga LG, Cabrera GB, Luna DE, Iqbal H, Sharma M, Sharma A (2020) Plant-based metal and metal alloy nanoparticle synthesis: a comprehensive mechanistic approach. J Mater Sci 55(4):1309–1330
93. Pandian CJ, Palanivel R, Dhananasekaran S (2015) Green synthesis of nickel nanoparticles using *Ocimum sanctum* and their application in dye and pollutant adsorption. Chinese J Chem Eng 23(8):1307–1315
94. Park H, Kim W, Kim M, Lee G, Lee W, Park J (2021) Eco-friendly and enhanced colorimetric detection of aluminum ions using pectin-rich apple extract-based gold nanoparticles. Spectrochim Acta Part A: Mol Biomol Spectrosco 245:118880
95. Patil SP (2020) *Ficus carica* assisted green synthesis of metal nanoparticles: a mini review. Biotechnol Rep 28:p.e00569
96. Paul K, Bag BG, Samanta K (2014) Green coconut (*Cocos nucifera* Linn) shell extract mediated size controlled green synthesis of polyshaped gold nanoparticles and its application in catalysis. Appl Nanosci 4:769–775
97. Pechyen C, Ponsanti K, Tangnorawich B, Ngernyuang N (2021) Waste fruit peel – Mediated green synthesis of biocompatible gold nanoparticles. J Mater Res Technol 14:2982–2991
98. Phang YK, Aminuzzaman M, Akhtaruzzaman M, Muhammad G, Ogawa S, Watanabe A, Tey LH (2021) Green synthesis and characterization of CuO nanoparticles derived from papaya peel extract for the photocatalytic degradation of palm oil mill effluent (POME). Sustainability 2021 13(2):796
99. Phongtongpasuk S, Poadang S, Yongvanich N (2016) Environmental-friendly method for synthesis of silver nanoparticles from dragon fruit peel extract and their antibacterial activities. Energy Proc 89:239–247
100. Poguberović SS, Krčmar DM, Maletić SP, Kónya Z, Pilipović DDT, Kerkez DV, Rončević SD (2016) Removal of As (III) and Cr (VI) from aqueous solutions using "green" zero-valent iron nanoparticles produced by oak, mulberry and cherry leaf extracts. Ecol Eng 90:42–49
101. Rane AV, Kanny K, Abitha VK, Thomas S (2018) Methods for synthesis of nanoparticles and fabrication of nanocomposites. In: Synthesis of inorganic nanomaterials 121–139
102. Rao A, Bankar A, Kumar AR, Gosavi S, Zinjarde S (2013) Removal of hexavalent chromium ions by Yarrowialipolytica cells modified with phyto-inspired Fe0/Fe_3O_4 nanoparticles. J Contam Hydrol 146:63–73
103. Razavi R, Molaei R, Moradi M, Tajik H, Ezati P, Shafipour Yordshahi A (2020) Biosynthesis of metallic nanoparticles using mulberry fruit (*Morus alba* L.) extract for the preparation of antimicrobial nanocellulose film. Appl Nanosci 10(2):465–476
104. Roopan SM, Madhumitha G, Rahuman AA, Kamaraj C, Bharathi A, Surendra TV (2013) Low-cost and eco-friendly phytosynthesis of silver nanoparticles using *Cocos nucifera* coir extract and its larvicidal activity. Ind Crop Prod 43:631–635
105. Sadhasivam S, Vinayagam V, Balasubramaniyan M (2020) Recent advancement in biogenic synthesis of iron nanoparticles. J Mole Struct 1217:128372
106. Saif S, Tahir A, Chen Y (2016) Green synthesis of iron nanoparticles and their environmental applications and implications. Nanomaterials 6:209
107. Shahwan T, Sirriah SA, Nairat M, Boyacı E, Eroğlu AE, Scott TB, Hallam KR (2011) Green synthesis of iron nanoparticles and their application as a Fenton-like catalyst for the degradation of aqueous cationic and anionic dyes. Chem Eng J 172(1):258–266
108. Sharma R, Tripathi A (2022) Green synthesis of nanoparticles and its key applications in various sectors. Mater Today: Proc 48(5):1626–1632
109. Shende S, Ingle AP, Gade A, Rai, M (2015) Green synthesis of copper nanoparticles by Citrus medica Linn. (Idilimbu) juice and its antimicrobial activity. World J Microbiol Biotechnol 31(6):865–873
110. Siddiqi KS, Husen A (2016) Fabrication of metal and metal oxide nanoparticles by algae and their toxic effects. Nanoscale Res Lett 11:1–11

111. Singh P, Kim YJ, Yang ZD, DC (2016) Biological synthesis of nanoparticles from plants and microorganisms. Trends Biotechnol 34(7):588–599
112. Song JY, Jang HK, Kim BS (2009) Biological synthesis of gold nanoparticles using *Magnolia kobus* and *Diopyros kaki* leaf extracts. Process Biochem 44:1133–1138
113. Soni V, Raizada P, Singh P, Cuong HN, S R, Saini A, Saini RV, Le QV, Nadda AK, Le TT, Nguyen VH, (2021) Sustainable and green trends in using plant extracts for the synthesis of biogenic metal nanoparticles toward environmental and pharmaceutical advances: a review. Environ Res 202:111622
114. Soto M, Quezada-Cervantes CT, Hernández-Iturriaga M, Luna-Bárcenas G, Vazquez-Duhalt R, Mendoza S (2019) Fruit peels waste for the green synthesis of silver nanoparticles with antimicrobial activity against foodborne pathogens. LWT 103:293–300
115. Subramaniyam V, Subashchandrabose SR, Thavamani P, Megharaj M, Chen Z, Naidu R (2015) Chlorococcum sp. MM11—a novel phyco-nanofactory for the synthesis of iron nanoparticles. J Appl Phycol 27(5): 1861–1869
116. Suhag R, Kumar R, Dhiman A, Sharma A, Prabhakar PK, Gopalakrishnan K, Kumar R, Singh A (2022) Fruit peel bioactives, valorisation into nanoparticles and potential applications: A review. Crit Rev Food Sci Nutr 1–20
117. Thema FT, Manikandan E, Gurib-Fakim A, Maaza M (2016) Single phase Bunsenite NiO nanoparticles green synthesis by *Agathosma betulina* natural extract. J Alloys Comp 657:655–661
118. Timoszyk A (2018) A review of the biological synthesis of gold nanoparticles using fruit extracts: scientific potential and application. Bullet Mater Sci 41(6):1–11
119. Ting ASY, Chin JE (2020) Biogenic synthesis of iron nanoparticles from apple peel extracts for decolorization of malachite green dye. Water Air Soil Poll 231(6):278
120. Turunc E, Binzet R, Gumus I, Binzet G, Arslan H (2017) Green synthesis of silver and palladium nanoparticles using *Lithodora hispidula* (Sm.) Griseb. (Boraginaceae) and application to the electrocatalytic reduction of hydrogen peroxide. Mater Chem Phy 202:310–319
121. Veerasamy R, Xin TZ, Gunasagaran S, Xiang TFW, Yang EFC, Jeyakumar N, Dhanaraj SA (2011) Biosynthesis of silver nanoparticles using mangosteesn leaf extract and evaluation of their antimicrobial activities. J Saudi ChemSoc 15:113–120
122. Vijayakumar S, Vaseeharan B, Malaikozhundan B, Gopi N, Ekambaram P, Pachaiappan R, Velusamy P, Murugan K, Benelli G, Kumar RS, Suriyanarayanamoorthy M (2017) Therapeutic effects of gold nanoparticles synthesized using *Musa paradisiaca* peel extract against multiple antibiotic resistant *Enterococcus faecalis* biofilms and human lung cancer cells (A549). Microb Pathog 102:173–183
123. Vinnarasi J, Raj AAA, Augustin M, Revathi G (2021) Green synthesis of zinc oxide nanoparticles using hydro methanolic extract of *Flueggea leucopyrus* Wild Fruit. Acta Sci Pharma Sci (ISSN: 2581–5423), 5(12)
124. Vishnupriya B, Nandhini GE, Anbarasi G (2020) Biosynthesis of zinc oxide nanoparticles using Hylocereus undatus fruit peel extract against clinical pathogens. Mater Today: Proc
125. Wang ZX, Chen CY, Wang Y, Li FX, Huang J, Luo ZW, Rao SS, Tan YJ, Liu YW, Yin H, Wang YY (2019) Ångstrom-scale silver particles as a promising agent for low-toxicity broad-spectrum potent anticancer therapy. Adv Funct Mater 29(23):1808556
126. Xu L, Wang YY, Huang J, Chen CY, Wang ZX, Xie H (2020) Silver nanoparticles: synthesis, medical applications and biosafety. Theranostics 10(20):8996–9031

127. Yirsaw BD, Megharaj M, Chen Z, Naidu R (2016) Reduction of hexavalent chromium by green synthesized nano zero valent iron and process optimization using response surface methodology. Environm Technol Inno 5:136–147
128. Zeng F, Xu D, Zhan C, Liang C, Zhao W, Zhang J, Feng H, Ma X (2018) Surfactant-free synthesis of graphene oxide coated silver nanoparticles for SERS bio-sensing and intracellular drug delivery. ACS Appl Nano Mater 1(6):2748–2753

Green Synthesis of Metal-Oxide Nanoparticles from Fruits and Their Waste Materials for Diverse Applications

Anam Khan, Reena Vishvakarma, Poonam Sharma, Swati Sharma, and Archana Vimal

Abstract Nanotechnology is the state-of-the-art technology providing new horizons of ideas and scope to unlimited possibilities in almost all genres of day-to-day life ranging from diagnostic, therapeutic, agricultural, chemical to microelectronics, sensors, etc. In this connection, one of the main concerns regarding the metal-oxide nanoparticles is their synthesis, application in a safe mood to protect the overall toxic impact on the food web. To resolve this concern, a greener and cleaner method of producing the nanoparticles is being looked upon as a favorable alternative more commonly referred to as green chemistry or green synthesis. The metal-oxide nanoparticles synthesized from biological sources have demonstrated fulfilling properties and shown antibacterial, antiviral, antifungal, drug delivery, catalytic activity, etc., response. In this chapter, the role of fruits and their wastes in the green synthesis of metal-oxide nanoparticles is discussed. This could efficiently reduce the cost and is safer for the environment thus allowing us to implore more on their benefits without impending much harm to the environment.

Keywords Nanotechnology · Flavonoids · Anti-microbial · Polyphenols · Drug delivery · Biosensor

List of Abbreviations

ATP	Adenosine triphosphate
COD	Chemical oxygen demand
DO	Dissolved oxygen
DPW	Date pulp waste
FESEM	Field-emission scanning electron microscopy

A. Khan · R. Vishvakarma · P. Sharma · A. Vimal (✉)
Department of Bioengineering, Integral University, Lucknow, UP, India
e-mail: vimal.archana@gmail.com

S. Sharma
Department of Biosciences, Integral University, Lucknow, UP, India

A. Husen (ed.), *Nanomaterials from Agricultural and Horticultural Products*,
Smart Nanomaterials Technology,
https://doi.org/10.1007/978-981-99-3435-5_5

FTIR	Fourier transform infrared
HRTEM	High-resolution transmission electron microscopy
IONPs	Iron oxide nanoparticles
MOF	*Moringa oleifera* fruit
MONPs	Metal oxide nanoparticles
MPTMS	3-Mercaptopropyltrimethoxysilane
MRI	Magnetic resonance imaging
NMs	Nanomaterials
PDT	Photodynamic treatment
PTT	Photo thermal therapy
SEM	Scanning electron microscopy
SPIONs	Superparamagnetic iron oxide nanoparticles
TDS	Total dissolved solids
TEM	Transmission electron microscopy
XRD	X-ray diffraction

1 Introduction

The synthesis, characterization, and applications of nanomaterials are the main topics of the interdisciplinary field of study known as nanotechnology. Nanomaterials (NMs) are nano objects with at least one-dimensional size in the 1–100 nm range and particular features such as size, shape, porosity, and so on. The key factors contributing to the popularity of nanomaterials are their distinctive optical, magnetic, chemical, and mechanical properties that set them apart from their bulk counterparts.

Metal oxide nanoparticles are nanoscale particles that comprise both intentionally made and naturally occurring particles emerging from natural and human processes. A significant family of materials, transition metal oxide nanoparticles have desirable magnetic, electrical, and optical characteristics (Gebru and Sendeku 2019). Due to their highly porous structure with good thermal/chemical stability and huge surface areas (>100 $m^2\ g^{-1}$), the MONPs (Metal oxide nanoparticles) are employed as nanoparticles and have several advantages such as being effective, simply synthesizable, changeable, and catalytic. Numerous applications, including those involving optics, electronics, hyperthermia therapy, catalysts, MRI (Magnetic Resonance Imaging), magnetically targeted medication delivery, and cell / nucleic acid separations, have made use of MONPs as valuable nanomaterials with distinctive features [27].

The great mass-to-volume ratio and small dimensions of edges on MONPs' surfaces are thought to be responsible for their unusual chemical and physical characteristics. For instance, changes in cell characteristics have been seen in NPs of metal oxides such as ZnO, Al_2O_3, MgO, AgO, TiO_2, etc., owing to structural changes associated with the size. As the size of nanoparticles drops, a greater number of interface and surface atoms along with nearby structural perturbations cause strain or stress. The magnetic, conductive, chemical, and electrical characteristics of a nanoparticle can change depending on its precise size. In Fe_2O_3 nanoparticles, size-dependency has been found; while 12 nm nanoparticles display superparamagnetic behavior without hysteresis, 55 nm particles exhibit ferromagnetic activity. While causing the transition to superparamagnetic, the overall magnetic polarization will decrease with a reduction in particle size [8].

The nanoparticle manufacturing methods adhere to the two standard methodologies, including "Top-Down" and "Bottom-Up," and are classified into three subcategories: physical, chemical, and biological. The bottom-up strategy, as opposed to the top-down method, results in the production of nanoparticles via the "nucleation" mechanism. Although chemical and physical procedures are widely used to produce nanoparticles, they are not only expensive but also harmful to the environment and living things. The extensive use of manufactured goods requires high energy use, the use of potentially harmful organic solvents, the creation of novel disposal intermediaries, the risk of biological hazards, and environmental deterioration.

A lot of attention has been placed on "greener synthesis," a bottom-up strategy in materials science and technology for creating environmentally friendly goods, to deal with the drawbacks of the methodologies discussed [27]. The creation of green nanoparticles requires the use of milder synthesis conditions, such as sol–gel techniques, biomolecules such amino acids, proteins, and carbohydrates, as well as live animals, plants, and cells. Many have become interested in green synthesis since it is an economical and environmentally responsible method of creating nanoparticles that minimize or negate the use of hazardous reaction precursors. It is a new way of reducing nanoparticle toxicity that is often linked with traditional chemical and physical production processes. It employs non-hazardous chemicals, straightforward methods, and moderate reaction conditions. Green synthesis approaches involve the use of biological organisms such as bacteria, fungus, algae, and plants, among others, as stabilizing/reducing agents, or both in certain situations [5, 6, 16-18, 20, 46]. Biosynthesis is the process of creating nanoparticles using bacteria and plants [54]. In addition to the actual food, food preparation waste and any food that is not eaten are abundant sources of phytochemicals with a variety of biological functions, A sizable amount of food—nearly, one-third of the food produced globally for human consumption—is wasted every day, leading to substantial environmental and economic issues once they are disposed of. Concern over environmental sustainability is raised by the pressure on global food security and the scarcity of natural resources. Fruit and fruit wastes are a rich and affordable source of advantageous phytochemicals,yet, recovering these compounds and using them as functional additions in various products represents a difficult subject for researchers [26]. Thus, using biowaste materials like fruit waste to create highly efficient, biocompatible, cost-effective, and environmentally friendly metal oxide nanoparticles might help waste valorization and lead to environmental sustainability [15].

2 Synthesis of Metal Oxide Nanoparticles

Metal oxide nanoparticle production techniques are often classified as physical, chemical, and biological. A brief description of the methods is as follows (Fig. 1). The various metal oxide nanoparticles synthesized by different methods is summerized in Table 1.

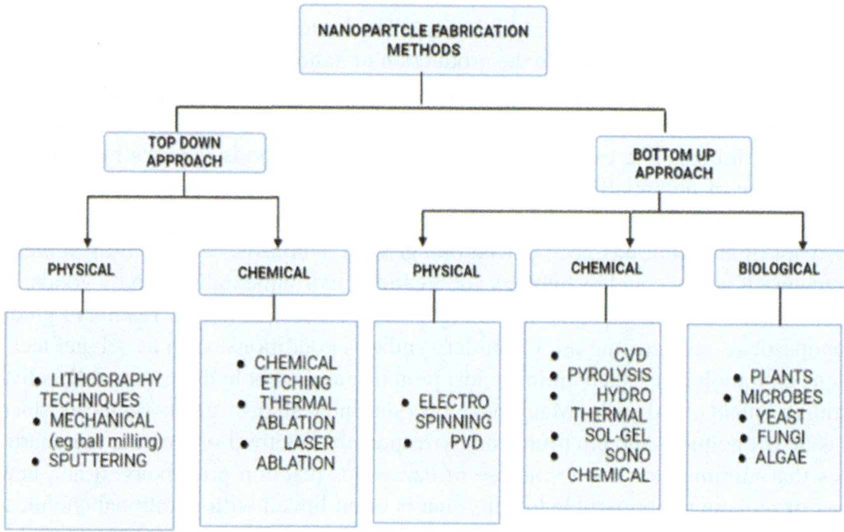

Fig. 1 Different methods of nanoparticle synthesis

Table 1 List of metal oxide NP's obtained using different methods

S. no	Method used	Metal oxide NP's	References
Solution based synthesis			
1	Sonochemical	TiO_2, ZnO, CeO_2, MoO_3, V_2O_5	[47]
2	Coprecipitation	ZnO, MnO_2, MgO, SnO_2, Cu-doped ZnO	[47]
3	Solvo-thermal	Nb_2O_5, MgO, TiO_2, $MnFe_2O_4$, Fe_3O_4	[47]
4	Sol–gel	TiO_2, ZnO, MgO, CuO, ZrO_2, and Nb_2O_5	[47]
5	Microwave	MnO, Fe_3O_4, CeO_2, CaO, $BaTiO_3$, ZnO, Mn_2O_3, MgO	[47]
6	Micro-emulsion	Fe_2O_3, NiO, CeO_2, TiO_2, ZnO, CuO	[47]

(continued)

Table 1 (continued)

S. no	Method used	Metal oxide NP's	References
Vapor state synthesis			
1	Laser ablation	ZnO, NiO, SnO_2, ZrO_2, iron-oxide, Al_2O_3, Au-SnO_2, Cu/Cu_2O	[47]
2	Chemical vapor based method	ZnO, magnetite, Cu_2O, MgO, CaO, SnO_2, SrO, CoO, and Co_3O_4	[47]
3	Combustion method	ZnO, FeO, CuO, MgO, CdO, Co_3O_4 or Ag supported on MgO surface, Co_3O_4 on CuO nanowire arrays	[47]
4	Template mediated method	MoO_2, α-Fe_3O_4, Co_3O_4, Fe_2O_3, mesoporous $NiMn_2Ox$	[47]
Biological synthesis			
	Plant	ZnO, MgO, TiO, Al_2O_3, CuO, CoO, MgO, Ag_2O	[3]

2.1 Physical Methods

2.1.1 Sonochemical Method

A flow of escalating ultrasonic waves is used in this method to dissolve metal salts, breaking the chemical connections between the constituents. The passage of the ultrasonic waves through the solution causes cycles of compression and relaxation. Acoustic cavitation is the development, expansion, and implosive collapse of bubbles in a liquid as a consequence. In addition, small bubbles rapidly burst as a result of the pressure differential, causing shock waves to occur inside the gaseous phase of the falling bubbles. When millions of bubbles burst at once, a tremendous quantity of energy is released into the solution. This technique has been used to create a variety of nanomaterials, including metals, alloys, metal oxides, metal sulfides, metal nitrides, metal-polymer composites, metal chalcogenides, and metal carbides, and others. Benefits of sonochemical methods include better phase purity of metal oxide nanoparticles, homogenous size dispersion, a larger surface area, and a faster rate of reaction [47].

2.1.2 Laser Ablation Method

By laser irradiating the targets, colloidal solutions generated from bulk materials dissolved in aqueous or non-aqueous solvents are used to produce nanoparticles. This method has been used to synthesize ternary metal oxides such as Au-SnO_2 and Cu/Cu_2O as well as ZnO, NiO, SnO_2, ZrO_2, iron oxide, Al_2O_3, and many more. The type of liquid media and laser fluence are two factors that can be modified to affect the size of the nanoparticles. With an increase in the size of the nanoparticles, the thickness of the molten layer rises, which in turn causes an increase in laser fluence. Some drawbacks of laser ablation include the need for capping, a lack of long-term stability in solution, and a propensity for nanoparticle aggregation [47].

2.1.3 Chemical Vapor Deposition (CVD)

In chemical vapor deposition (CVD), substrates are heated to high temperatures and exposed to precursor materials while they are gaseous. The precursors interacted with or disintegrated on the surface of the substrate to produce the nanomaterial. In a hot-wall reactor, under optimum chemical conditions, pure metal or metal–organic salts are added after being heated in a flow reactor to the vapor phase. An inert gas like Ar is commonly used to convey the gaseous reactants to the reaction zone, where crystal nucleation and formation take place. The product is then moved to a much cooler location, where the temperature difference allows it to shift from a gaseous form to a solid state, allowing it to be used. These techniques are frequently used to produce uniform and impurity-free metal oxide nanoparticles and films, such as ZnO nanowires and films as well as defect-free magnetite, Cu_2O, MgO, CoO, and Co_3O_4 nanoparticles. An additional advantage of this method is reproducibility [47].

2.2 *Chemical Methods*

2.2.1 Co-Precipitation Method

Salts such as chlorides and nitrates are dissolved in water (or another solvent) along with a base that precipitates the oxo-hydroxides [24]. There is a transient burst of nucleation followed by a growth phase when a threshold concentration of species in solution is reached. Co-precipitation is frequently employed to produce magnetic nanoparticles such as magnetite [47].

The most used synthetic method for creating iron oxide nanoparticles is co-precipitation. At moderate temperatures (100 °C), the base is added to aqueous solutions containing ferric ($Fe^{3+)}$ and ferrous (Fe^{2+}) salts as precursors [24]. The benefits of this method are minimal cost, reliable reaction parameters, including synthesis at low temperatures, the capability to perform immediate production in water, simplified processing, convenience in scaling up, and adaptability in modifying inner and surface characteristics [47]. Co-precipitation reactions are often used, although the mechanism of the reaction is still unknown because it is poorly understood how intermediates are generated. The products may contain intermediates to varying degrees, which frequently makes it challenging to produce repeatable synthesis [24].

2.2.2 Solvothermal Method

Using an autoclave or boiling, metal complexes are degraded. To control particle size growth and avoid agglomeration, the reaction media are often supplemented with an appropriate surfactant ingredient [12]. The method is known as hydrothermal synthesis when water serves as the solvent. The final particle formation is influenced by thermodynamic characteristics as well as the type, concentration, $_3$and content of the reactants). The basicity and hydrolysis ratio of the reacting media, together with the steric or electrostatic stabilization of reactive molecules, all have an impact on the particle size, shape, composition, and crystal structure. MgO, TiO_2, $MnFe_2O_4$, $CoFe_2O_4$, and Fe_3O_4 nanoparticles have been produced using this method [47].

2.3 *Template/Surface-Mediated Synthesis*

Its foundation is the creation of required nanomaterial inside the pores or channels of nanoporous templates. Different nanomaterial morphologies, including rods, fibrils, and tubules, can be created based on the characteristics of the templates [47]. There are two fundamental types of tools used in the template process: soft templates (surfactants) and hard templates (porous solids such as carbon or silica). Self-assembling systems have been created using precursor nanoparticles that are surface- and template-mediated [12]. Highly monodisperse nanostructures can be produced with enhanced activity and stable form. Examples include mesoporous $NiMn_2Ox$, Co_3O_4, and Fe_2O_3. Another is mesoporous MoO_2 nanoparticles with improved electrochemical properties. Mesoporous zeolites, porous alumina, track-etch membranes, and other nanoparticle structures can be used as templates for these synthesis processes [47].

2.4 *Biological Synthesis (Green Synthesis)*

The production and design of nanoparticles utilizing nontoxic chemicals, renewable resources, ecologically friendly solvents, and lastly degradable waste products are only a few of the 12 green chemistry principles that are the foundation of the green synthesis strategy for nanoparticles. According to a green chemistry perspective, three crucial phases in the production of nanoparticles are the use of a safe solvent, a nontoxic reducing agent, and ecologically friendly stabilizing agents [3]. It is possible to illustrate the main benefits of "greener synthesis," such as decreasing waste and pollution and utilizing conventional solvent systems, organic systems, and renewable raw materials. Therefore, via direct regulation, control, and repair processes, nanomaterials' greener production results in increased efficiency and environmental adaptability [27]. The chemical process is replaced in these syntheses by an enzymatic reaction, a biological synthesis also requires less energy than its physicochemical equivalents. The primary disadvantage of biological synthesis is the difficulty to produce nanoparticles in the correct size and/or form and low yield. The process is often slow, taking many hours or perhaps a few days. Additionally, produced nanoparticles may decompose after a given amount of time. The various advantages offered by green chemistry of nanoparticles synthesis is summerized in Fig. 2. However, this approach continues to be quite alluring when it comes to the creation of prospective antibacterial drugs because of its biocompatibility [47].

Green synthesis of nanomaterials uses a bottom-up strategy similar to chemical reduction, substituting extracts from various biological sources such as plant parts, microbes, waste products, etc. for a costly chemical-reducing agent. Various biological resources, such as plants and plant products, algae, fungus, yeast, bacteria, and viruses, are present in nature and can be used to create metal oxides. It has been demonstrated that both single- and multicellular organisms may create extracellular or intracellular inorganic nanomaterials. In comparison to microorganisms, using plant parts is more beneficial and simpler since it does not require any specialized or difficult processes like isolation, culture preparation, and culture maintenance. Nanoparticle synthesis by plants is quicker than by microbes, more affordable, and simple to scale up for the production of huge amounts of nanoparticles (Gebru and Sendeku 2019) (Fig. 3).

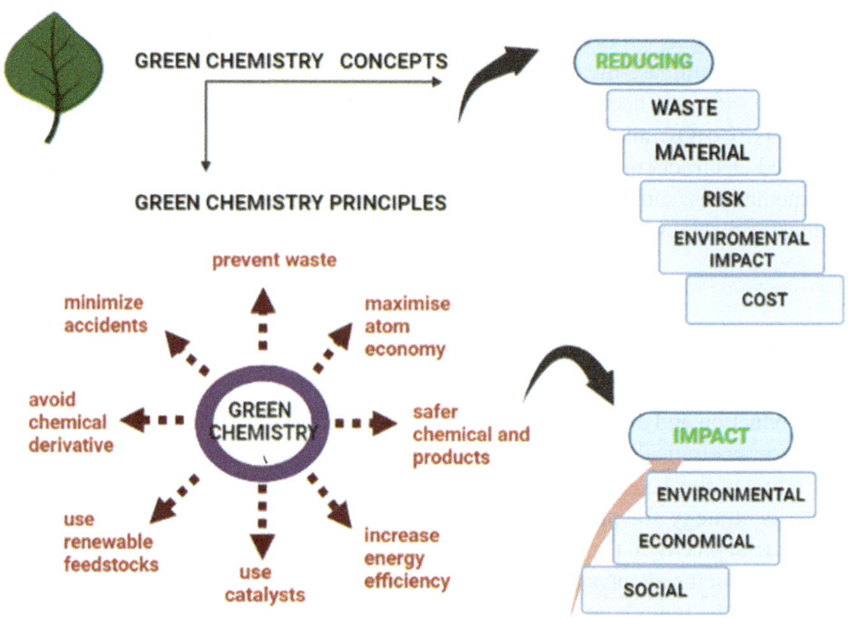

Fig. 2 Efficacy of green chemistry in nanomaterial synthesis

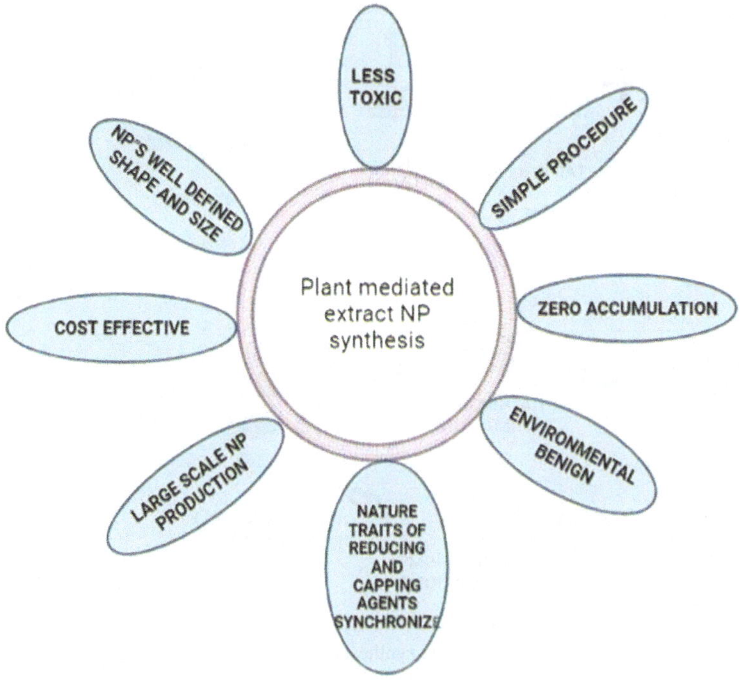

Fig. 3 Advantages of plant extract mediated nanoparticle synthesis

2.4.1 Plant-Mediated Synthesis

Phytomining is the cultivation of plants having a hyperaccumulation effect on soils high in metal ions to absorb metal ions. The metal ions diffuse into certain plant organelle cavities, where primary or secondary metabolites inside those organelles transform the metal ions into nanoparticles. This approach is sluggish, however, in addition to the fact that the plants must be grown over a lengthy period to allow for metal ion penetration into the plant system for the creation of nanoparticles. Additionally, it is difficult to recover the metals in their original form. In vitro plant-based, MONPs synthesis is growing in favor of research as a result of the aforementioned difficulties. In vitro biogenic synthesis uses either the entire plant (biomass) or extracts from certain plant components (roots, stem, leaves, flowers, and fruit peels). Nanoparticle stabilizing agents are present in the extracts. These include proteins, polysaccharides, organic acids, terpenoids, flavonoids, phenolic acids, and others. Additionally, the plant components may be used as templates to guide the development of the nanostructures. To enhance the surface area, the required section is cut off from the plant, cleaned to eliminate contaminants, dried, and ground or chopped. The plucked section is either soaked at room temperature or heated until the necessary amount of extract is produced to extract the phytochemicals.

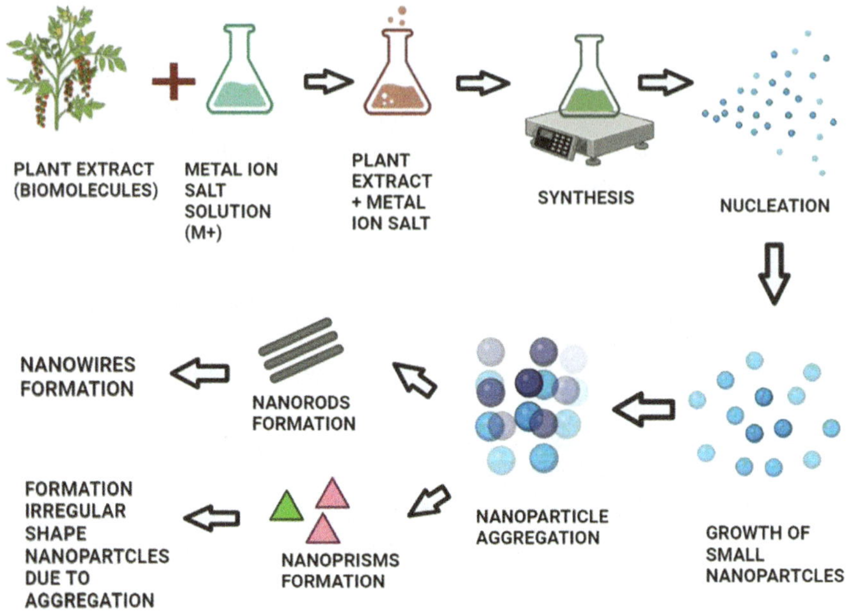

Fig. 4 Process of metal oxide nanoparticle synthesis through green chemistry

The metal ions is then added, which facilitates the complexation of the metal ions with the accessible functional groups. Metal oxide nanoparticle production is encouraged by the alkaline pH and is stabilized by phytochemicals in the extract (Fig. 4). Due to its qualities of interest, including its antibacterial action and photocatalytic activity, zinc oxide is the most often produced metal oxide nanoparticle via biological means [54].

2.4.2 Algae-Mediated Synthesis

Alkaloids, polyketides, cyclic peptides, polysaccharides, proteins, phlorotannins, diterpenoids, stereos, quinones, lipids, and glycerol are just a few of the active metabolites found in macroalgae that may be exploited to create nanoparticles. After being cleaned under running water to get rid of any dirt, salt, or other foreign objects, the gathered seaweeds were then immersed in deionized water. The cleaned algae or seaweed was pulverized after drying and being immersed in deionized water for many hours. Paper filtering was applied to the obtained extract. To produce the metal oxide nanoparticles, a solution containing a metal ion source salt was added last [13].

2.4.3 Bacterial Controlled Synthesis of MNOPs

Iron-reducing bacteria like magnetotactic bacteria are very adept at biomineralization, another term for bacterial controlled synthesis. The magnetotactic bacterium is a member of the broad group of fastidious prokaryotes, which also includes the vibrio, cocci spirilla, and bacilli-shaped vibrio bacteria. These species create vesicles with phospholipid membranes that are attached to linear arrays of magnetite nanocrystals (magnetosome). The multistep chemical process for producing iron oxide nanoparticles involves the creation of phospholipid membrane-bound structures known as magnetosome vesicles, the introduction of iron into the magnetosome vesicles, and lastly the crystallization of iron oxide nanoparticles (magnetite) [54].

2.4.4 Fungal-Mediated Synthesis

Myco-nanotechnology is the production of nanoparticles using fungal biomass and the metabolites they produce. The cell wall of fungal mycelia facilitates the absorption of metal ions to form metal nanoparticles and offers a large surface area for interaction. Additionally, it has been discovered that fungi offer greater polydispersity, stable structures, and variety in nanoparticle dimensions. Furthermore, since they produce more enzymes and metabolites than other types of fungi, filamentous fungi are thought to be superior at producing nanoparticles. They may be grown on both basic and sophisticated mediums, and they are simpler to manage [28]. In addition to synthesizing more proteins under similar synthesis circumstances, fungi grow quicker than bacteria, which contributes to the faster creation of nanoparticles. Additionally, the majority of the metal oxide nanoparticles produced by fungus are monodispersed nanoparticles with clearly defined morphologies [54].

3 Active Compounds Present in Fruit and Fruit Waste

Fruits and vegetables are essential dietary items that must be consumed every day for humans to meet their nutritional needs. Huge amounts of waste are produced in the food business, and this trash is classified according to its biological and chemical oxygen demands. These fruits and fruit wastes include a variety of beneficial components, including phenolic acids, carotenoids, and flavonoids. The polarity, solubility, molecular size, bioavailability, and metabolic routes of these phytochemicals differ. The various active compounds are as follows.

3.1 Polyphenols

Polyphenols are the most significant secondary metabolites found in fruits. They consist of an aromatic ring containing -OH groups. Molecules with this structure can be either simple or polymeric. These substances have antioxidant, anti-inflammatory, and antibacterial effects. Oleuropein and hydroxytyrosol are polyphenols that may both neutralize reactive oxygen species and guard against neurotoxin-induced cell death. The phenolic pigment curcumin has been shown to have neuroprotective properties because of its significant antioxidant activity.

It also reduces protein oxidation products and inflammation by decreasing the activity of microglia and lipoxygenase enzymes. The primary components of phenolic acids, stilbenes, lignans, and tannins are non-flavonoid polyphenols [38]. Oleuropein, a secoiridoid found in olive leaves and olives, is what gives fruit and olive oil their bitter flavor [25].

3.2 Flavonoids

The most prevalent polyphenolic components in plant food are flavonoids, which have anti-free radicals and anti-ROS properties. Additionally, they have been found to have antibacterial, antiviral, anti-inflammatory, anti-allergenic, _5antithrombotic, and vasodilatory properties. Their antioxidant and chelating characteristics, which also include antimutagenic and antitumoral qualities, contribute to the health benefits of their ingestion by halting the spread of germs or tumor metastases. An example of a flavonoid that has cardioprotective antioxidant properties is flavon-3-ol. Many fruits and vegetables have red, violet, or blue coloring because of a class of flavonoids called anthocyanins. When photosynthesis is occurring, these chemicals act as photo-protectors, protecting against the damage that free radicals might cause. Because they lower oxidative stress by lowering the production of free radicals and lipid peroxidation, anthocyanins are beneficial for the neural system. By preventing the production of inflammatory mediators, they also have anti-inflammatory effects [38].

3.3 Phenolic Acids

The primary source of antioxidants in plant diets is phenolic acids. They can be classified as hydroxybenzoic acid derivatives and hydroxycinnamic acid derivatives based on the presence of a C=O group attached to a benzene ring. These substances also have a role in the prevention of oxidative stress-related tissue damage brought on by long-term illnesses and anticancer properties. Examples are chlorogenic, gallic, and caffeine.

3.4 Stilbenes

Due to their numerous benefits to the human body ranging from protection against bacterial infection to neural, and cardiac systems, stilbenes have become more important in recent years. Resveratrol, for instance, slows the aging process, preserving cellular health, it also has anti-inflammatory characteristics. Due to its antioxidant characteristics, this substance also decreases blood pressure by producing NO, which dilates blood vessels.

3.5 Carotenoids

Many fruits and vegetables contain carotenoids, which give them their yellow, orange, and red hues. Through their antioxidant capacity for ROS, fruits high in these chemicals are said to strengthen immunity and lower the risk of several illnesses, including cancer, type 2 diabetes, and cardiovascular issues. Two of the most common carotenoids are lycopene and lutein [38]. α and β carotene are found in apricots, carrots, pumpkins, and sweet potatoes, lycopene in tomatoes [53]. Our bodies store lutein and zeaxanthin, two carotenoids, in the retina and lens of our eyes. According to several studies, a high intake of lutein and zeaxanthin, especially from xanthophyll-rich foods like spinach, broccoli, and eggs, is associated with a considerable decline in age-related cataracts (over 20%) and macular degeneration (over 40%).

3.6 Phytosterols

Triterpenes known as phytosterols (plant sterols) play a crucial structural role in the structure of plant membranes. Free phytosterols maintain phospholipid bilayers in plant cell membrane.1 or 2 C=C bonds, normally one in the sterol nucleus and occasionally a second in the alkyl side chain is present in the majority of phytosterols, which have 28 or 29 carbons [32]. Sitosterol, stigmasterol, and campesterol are the most prevalent phytosterols in the diet. These have a similar structural make-up to cholesterol and compete with it for intestine absorption, decreasing blood cholesterol levels [38].

3.7 Vitamins

Because humans are unable to produce vitamins, they must be consumed through food. They are vital to human nutrition and health and have complicated biochemistry.

Plants are a good source of vitamins B, C, and K1. Less common, a-carotene and b-cryptoxanthin are mostly found in fruits like pumpkins. Ascorbate is the most prevalent and pervasive cellular antioxidant (vitamin C). Due to its outstanding capacity to scavenge reactive oxygen species, ascorbate plays a significant part in the antioxidant defense network of plants (ROS) [4].

Fatty acid, protein, and carbohydrate metabolism are all impacted by vitamins. Additionally, they take part in the creation of several substances that the body needs to operate properly. Vitamins A, D, E, and K, which are fat-soluble vitamins, have antioxidant characteristics that can help lower the risk of neurological or cardiovascular illness. Numerous water-soluble vitamins, including thiamine, riboflavin, niacin, and vitamin C, are abundant in fruits and vegetables, especially their leftover bits. The B complex vitamins, which include thiamine, riboflavin, and niacin, are crucial cofactors in biochemical processes and are necessary for healthy skin, normal body growth and development, neuron and heart function, and the production of red blood cells. B vitamins play a direct role in the metabolism of energy [38]. The different phytochemicals obtained from a variety of fruits, their role in metal oxide nanopartcle synthesis is presented below (Table 2).

Table 2 Phytochemicals obtained from different fruits for the synthesis metal oxide nanoparticles and their applications

S. No	Fruit varieties	Phytochemicals	Role	Type of metallic NPs	Phytochemicals as capping agents	Reference
1	Banana, pomegra-nate	Trans β carotene, β-sitosterol, gallagic acid	Protection against cancer	CuO	Phenols, amines, sterols, fatty acids, hydroxyl, terpenoids, protein carboxyl,	[21]
2	Amla, pomegranate, guava, citron	p-Coumaric acid, vitamin-c, catechin, guavin-B, β-Bisabolene	Protection against free radical damage	Au	Flavonoids, vitamins, lycopene, glycosides, amino acids, alkaloids	[21]
3	Guava, Citron	Ferulic acid, 1,6-bis β-o-galloyl-β -D-glucose, Citral-B, Limonene	Antibacterial Activity	Ag	Carbohydrates, Proteins, Flavonoids	[21]
4	Lemon, Grape, Pineapple	Punicic acid, β-pinene, Stilbenoid, bromelain	Protect Skin	ZnO	Phenols, Flavonoids, Anthocyanin	[21]

4 Synthesis of Metal Oxide Nanoparticles from Fruit and Fruit Wastes

4.1 Aluminium Oxide Nanoparticles

Alumina having the atomic number Al_2O_3 is one of the least dense and insoluble metals in all chemical reagents (density 2.7 g/cm^3). Additionally, Al_2O_3 is inert at ambient temperature and has been known to include materials like aluminum hydroxide oxide (AlO(OH)) and aluminum trioxide for over a century. The Al_2O_3 is finely polished and exhibits good corrosion resistance [27]. Al_2O_3 is the most thermodynamically stable of the Al_2O_3 allotropes. Due to its stability, several different processes have historically been used to create Al_2O_3 NPs, including hydrothermal reactions, laser ablation, pyrolysis, sol–gel, etc. Nanotechnologists have focused on phyto-mediated synthesis of Al_2O_3 NPs to produce ecologically benign, nontoxic, cost-effective, and quick techniques because these methods are not environmentally friendly [3].

Sutradhar et al. created Al_2O_3 NPs using triphala extract as the only source of energy. These phenolic components are used as a reducing agent against aluminium nitrate. The results showed that the triphala extract produced oval-shaped nanoparticles with an average size of 200–400 nm. In contrast to the traditional and microbiological methods, this technique amply demonstrated the quick rate of alumina nanoparticle synthesis using microwave irradiation utilizing plant extract [3].

About 60–70% of the fruit waste that is processed during the abstraction of citrus juices is turned into wastes: peels, seeds, and membranes are left over. Grapefruit peel extract was used to reduce aluminium nitrate salt to produce Al_2O_3 NPs. As a precursor for this biosynthesis, 2 molar of aluminium nitrate, weight of 375.13 g/mol, were collected and dissolved in distilled water. Al_2O_3 nanoparticles are formed when brownish-yellow precipitates develop [7].

Utilizing leftover jackfruit rind, green aluminium oxide nanoparticles have been created (*Artocarpus heterophyllus*). These nanoparticles are employed in the photodegradation of Congo red and methyl red. Aluminum nitrate was utilized to create aluminium oxide nanoparticles with a molecular weight of 375.23 g/mol by combining it with distilled water at a 2 molar concentration, to which the extract from the previous step was added in a 3:2 ratio. Add 5 mL of sodium hydroxide (10%), followed by 40 min of sonication. After that, it was rotated for six hours at 1290 rpm. The precipitate is yellow–brown. TEM, XRD, and EDS analysis reveal the production of aluminium oxide nanoparticles [33].

4.2 Copper Oxide Nanoparticles

Copper has the atomic number 29 with the symbol (Cu), and an electron configuration of [Ar] $3d^{10}4s^1$. It is paramagnetic owing to the unpaired electron in the d orbital.

High thermal and electrical conductivity characteristics are present in copper metal. Four oxygen atoms and copper atoms are coordinated to form an almost square-shaped planar structure [27]. CuO NPs produced by biological synthesis have been proven to be economical, ecologically safe, and come in a range of sizes based on the source. The UV absorption band, which relates to the Cu metal interband transition of the core electrons and the CuO band edge transition, and the spherical form of green-produced CuO NPs, respectively, are located between 265 and 285 nm (strong band) and roughly 670 nm (weak band) [3].

The fruit extract from *Myristica fragrans*' pericarp was used to create copper oxide (CuONPs) nanoparticles at a reasonable cost and with no environmental impact. The fruit extract (10 mL), added to a solution of copper acetate was stirred and then microwave irradiated in the microwave for 5 min. The solution's hue changes from blue to green to eventually a greenish-brown suspension when heated in a microwave. Centrifugation was used to recover the produced CuO nanoparticles, which were then cleaned with deionized water. High-Resolution Transmission Electron Microscopy (HRTEM) and Field-Emission Scanning Electron Microscopy (FESEM) were used to determine the size and morphology of the particles. The copper nanoparticles exhibited a Surface Plasmon Resonance band at 360 nm. The size of the synthetic copper oxide was found to be between 10 and 50 nm. EDS spectrum was used to establish the element's existence. By using FTIR (Fourier Transform Infrared) spectroscopy, it was demonstrated that phytochemicals were involved in the stability and reduction of the nanoparticles [44].

4.3 Cerium Oxide Nanoparticles

Cerium (Ce), a chemical element with the atomic number 58 and the first member of the lanthanide group, has desirable chemical characteristics due to the close closeness of its 4f electrons and outer orbitals. Different oxidation modes of cerium (Ce), including Ce^{+4}, and Ce^{+3}, exist. The form of Ce^{4+} transforms into Ce^{3+} when the particle size lowers [27].

The red pomelo (*Citrus maxima*) is a common fruit crop throughout the Orient, including India and also the biggest citrus fruit. Pectin, a biopolymer that can be recycled and is safe to use, was taken out of the peels of Indian red pomelo fruit and utilized to create cerium oxide nanoparticles (CeO_2-NPs.The relative and intrinsic viscosities of the Indian red pomelo were determined from the apparent viscosity of the pectin solution (IRP–P). The pectin solution was progressively added to 50 mL of a 0.5 M cerium nitrate solution while being vigorously stirred for 30 min at 60 °C, further liquid ammonia was added until the pH level reached 10. One hour was given for the solution to stir. The solution was initially light yellow in hue, but when the ammonia content rose, it became yellow. To clean it of nitrate, ammonia, and organic contaminants, the resultant yellow precipitate was centrifuged and washed repeatedly with acetone and then with water. The washed precipitate was dried for 12 h at 60 °C, then annealed for 4 h at 400 °C and submitted to pectin characterization

(IRP-P). CeO_2-NPs' photoluminescence tests showed that the wide emission was made up of seven distinct bands. Pectin's participation in the synthesis and stability of CeO_2-NPs was confirmed by FTIR analysis. A prominent Raman active mode peak was visible in FT-Raman spectra at 461.8 cm^1, which was caused by the Ce–O vibration's symmetrical stretching mode. The produced CeO_2-NPs were polydispersed, spherical in form, with a cubic fluorite structure, and an average particle size of around 40 nm, according to DLS, FESEM, EDX, and XRD analyses. These CeO_2-NPs demonstrated significant antioxidant and non-cytotoxic properties as well as a broad range of antibacterial activities [36].

4.4 Iron Oxide Nanoparticles

A magnetic core and its chemical combinations, which exhibit magnetic characteristics in the presence of an external magnetic field, make up the most basic structure of nanoparticles. The materials exhibit six different forms of magnetic behavior such as diamagnetic, paramagnetic, ferromagnetic, super-magnetic, etc., depending on how they react to the external magnetic field. Superparamagnetic nanoparticles (Fe_3O_4; magnetite) are the most frequently utilized after the removal of the external magnetic field because they are nontoxic, simple to make, have high biocompatibility, and have no residual magnetism. They are one of the types of magnetic iron oxide nanoparticles. Fe_2O_3-NPs is another magnetic nanoparticle that is present in a variety of minerals, including hematite (–Fe_2O_3), maghemite (–Fe_2O_3), –Fe_2O_3, magnetite (–Fe_3O_4), and wüstite (–Fe–O) [2]. The biological source contributes to the reduction or stabilization of -Fe_2O_3 NPs and Fe_3O_4 NPs during their production extract [3].

Despite being a waste product, the *Punica granatum* (pomegranate) fruit peel contains 30–40% of the fruit's protein. *P. granatum* exhibits strong anticancer properties in the treatment of human bladder cancer cells, cervical cancer HeLa cells, prostate cancer cells, breast cancer cells, and thyroid cancer cells. Iron salts (Fe^{2+}/Fe^{3+}), sodium hydroxide, and *P. granatum* fruit peel extract all functioned as stabilizers, iron suppliers, and reducing agents, respectively. By using X-ray powder diffraction, Fourier-transform infrared spectroscopy, and scanning electron microscopy, the extract and synthetic IONPs were assessed. After the addition of Fe^{+3}, Fe^{+2}, and NaOH solution, the aqueous extract suspension of *P. granatum* fruit peel changed from brown to black, suggesting the effective formation of stable IONPs [52].

In India, the edible plant *Moringa oleifera* is frequently used as a flocculant to purify water as well as a medicine. Using aqueous *Moringa oleifera* fruit (MOF) extracts, iron (Fe) NPs were synthesized. By adding 0.5 M of Fe-salt drop-by-drop to the aqueous plant extract in a sonicated reactor, biogenic Fe NPs were created. Iron was present, as shown by the formation of NPs with a black hue. The majority of the Fe NPs, which had crystallite sizes of 35–40 nm, were Fe_2O_3 and FeOOH, according to

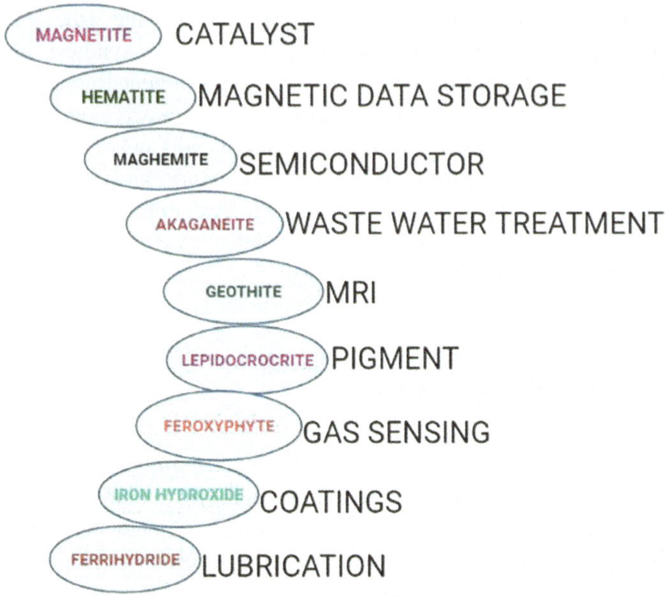

Fig. 5 Different forms of iron oxide nanoparticles and their applications

the XRD examination. Non-spherical nanoparticles could be seen in the SEM (Scanning Electron Microscopy) and TEM (Transmission Electron Microscopy) pictures of the Fe NPs [19].

Iron oxide nanoparticles were created using dried fruit extract from *Ficus carica*, often known as the common fig. By decreasing the iron precursor salt, biomaterials in the dried fruit extract of the common fig created iron nanoparticles, which subsequently served as capping and stabilizing agents. The high phenolic component concentration led to the production of nanoparticles with diameters less than 20 nm, which then underwent oxidation. The nanoparticles' 9 4 nm sizes and metallic core-oxide shell structure were seen in TEM pictures. The nanoparticles had a monodisperse distribution and spherical forms. Signals from EDX, XRD (X-ray diffraction), and FTIR analyses demonstrated the production of iron oxyhydroxide/oxide [10]. This metal oxide nanoparticles synthesized in different forms for a wide range of applications (Fig. 5).

4.5 *Zinc Oxide Nanoparticles*

ZnO nanoparticles (ZnO-NPs) have emerged in recent years as one of the most significant metal oxide nanoparticles due to their distinctive form and range of applications, high biocompatibility, favorable economics, and low cytotoxicity. Additionally, this material may be employed in a range of applications due to its advantageous physical characteristics, such as high stability and high melting point [27]. When compared to ZnO NPs that are generated normally, green ZnO NPs are shown to be highly efficient in photocatalytic, antibacterial, and anticancer activities extract [3].

The green synthesis method was used to create zinc oxide nanoparticles (ZnO NPs) from orange peels. To examine antibacterial activity, gram-negative *Pseudomonas aeruginosa*, gram-positive Bacillus subtilis, and gram-positive Staphylococcus aureus are utilized. The generation of ZnO NPs was verified by XRD, FTIR-ATR, and UV–Vis measurements. Additionally, it was confirmed by XRD, UV–Vis data, optical band gap, and HRTEM that chemically produced ZnO NPs are larger than those made using green synthesis techniques. At concentrations of 50 and 100 ppm, the antibacterial action of ZnO manufactured chemically over that prepared environmentally friendly showed a modest advantage, whereas *Pseudomonas aeruginosa* and *B. subtilis* showed no differences at 150 ppm [30].

Date pulp waste (DPW) is a useful bio-reductant that has been used in the green synthesis of ZnO-NPs. For the DPW-mediated ZnO-NPs (DP-ZnO-NPs) production, a low-temperature, environmentally friendly technique was used. Microscopic examinations verified the spherical and non-agglomerative character of the particles, which had a mean diameter of 30 nm. The chemical composition and product purity of NPs were determined by EDX and XPS analysis. Studies using UV and photoluminescence revealed that DP-ZnO-NPs are fluorescent and had surface plasmonic resonance at 381 nm. A DSC/TG investigation revealed thermal stability up to 700 °C with a 10-weight-percent loss. NPs were used in the photocatalytic degradation of potentially harmful methylene blue and eosin yellow dyes, which demonstrated a quick breakdown rate and 90% degradation efficiency. By measuring the zone of inhibition using the disc-diffusion method, DP-ZnO-NPs showed substantial antibacterial activity on a variety of harmful bacteria. Thus, the DP-ZnO-NPs in their produced form are suited for the treatment of industrial wastewater [39].

It has also been possible to create discarded pineapple peels into zinc oxide nanoparticles. Waste pericarp contains a variety of phytochemicals that act as a capping and reducing agent. The artificial nanoparticles exhibit good size and purity. By measuring the turbidity and conductivity, synthetic nanoparticles are employed to remediate the textile effluent. This research shows that dissolved oxygen (DO), total dissolved solids (TDS), chemical oxygen demand (COD), and total suspended solids play significant roles in the water purification process [31].

4.6 Cobalt Oxide Nanoparticles

A ferromagnetic element with a glossy grey tint is called cobalt. The oxidation states of this transition metal are CoO_2, Co_2O_3, CoO (OH), CoO, and Co_3O_4. Co^{2+} ions can occupy tetrahedral sites in the cubic spinel structure of the Co_3O_4 nanoparticles, whereas Co^{3+} ions are also linked to octagonal sites. Due to their low cost and robust electrical power, Co_3O_4-NPs are used as antibacterial, antifungal, electrochromic, heterogeneous catalysts, and energy storage sensors. Co_3O_4-NPs are also thought of as photocatalysts due to their optical band gap, which is in the region of 1.48–2.19 eV. Due to the abundance of oxidation states with low, medium, and high Co_3O_4 rotational angles, it is also advantageous in spintronics [27].

Co-precipitation was used to create GCoO-NPs (green cobalt oxide nanoparticles) using the Jumbo Muscadine grape (*Vitis rotundifolia*). Using a mixer grinder, the Jumbo Muscadine pulp was crushed into a fine powder after the seeds and covers were taken off. 40 mL of pulp extract was added drop by drop while 0.5 M of cobalt chloride ($CoCl_2.6H_2O$) was added to a beaker and stirred at 500 rpm for 60 min at 75 °C. As a catalyst, 0.01 M NaOH solution was gradually added until the mixture's pH reached 9; a white precipitate was produced and allowed to settle. The resulting residue was air-dried for three to four days after being washed with water and ethanol concurrently. The leftover material was further crushed and calcined for four hours at 600 °C. Finally, a powder with a dark brown hue was produced, and more characterization was done. Future uses of the produced GCoO-NPs can cure microorganisms, which also have outstanding photocatalytic activity [43].

Cobalt (II) acetate tetrahydrate and *Terminalia chebula* fruit were simply thermalized to create carbon-supported cobalt oxide nanoparticles (Co3O4@C NPs). The phytoconstituents of *T. chebula* were carbonized and supported the Co_3O_4 NPs, according to the ATR-FTIR study. The XRD patterns revealed that the synthesized materials had carbon support and were extremely crystalline. The particle size is estimated from microscopic photographs to be between 15 and 25 nm, with a deformed spherical shape. Three significant elements, including Co, O, and C, were present, according to the EDS data. Co_3O_4 NPs have a computed specific surface area of 22 m^2/g. In comparison to Ag/AgCl, the Co_3O_4@C NPs had a maximum specific capacitance of around 642 F/g at 1 A/g of discharging current density. Co_3O_4@C NPs' high specific capacitance may be due to a combination of pseudocapacitance and double-layer capacitance [11].

4.7 Nickel Oxide Nanoparticles

Ni^{+2} has an ion radius of 0.69 A. Ni^{+2} is the most frequent type of nickel oxidation, although Ni^{+3} and Ni^{+1} are less frequent. The chemical formula for nickel oxide (II) NiO, a P-type semiconductor with a large band gap (3.6–4.0 eV) and a molar mass of 74.69 g mol^{-1}, can serve as an electron receiver. NiO-NPs are 8–10 nm in size and morphology, with cubic or spherical forms and low, porous densities, which can be attributed to the tremendous amount of heat produced during the combustion process. The maximal absorption of NiO-NPs is in the range of less energy in the UV region than in bulk, according to an analysis of the optical reflection spectrum. NiO-NPs should have photocatalytic activity because electrons flow from the valence band to the band that transmits UV light [27].

By employing date palm fruit extract as a reducing or stabilizing agent with deposition on the surface of graphene oxide sheets, Alshatwi and Athinarayanan have effectively manufactured NiO NPs. By modifying Hummers' method and adding nickel nitrate to a solution of graphene oxide in distilled water, NiO nanoparticles were made using a green method. After adding fruit extract to convert the graphene oxide to graphene, the mixture was heated at 85 C for an hour while being continuously stirred. Their research demonstrated the evenly anchored NiO NPs with an average size in the range of 20–30 nm and a spherical shape produced by biological synthesis extract [3].

Limonia accidissima natural fruit juice has been successfully used to create NiO nanoparticles (NPs). *Limonia acidissima* fruit juice was made in various amounts, and nickel nitrate hexahydrate was used to carry out the synthesis. With continual stirring, $Ni(NO_3)_2 \cdot 6H_2O$ was dissolved in varying amounts of fruit juice. The agitated solution was held in a muffle furnace that had been warmed for 15 min. at 500 10 °C before being calcined for 3 h. at 500 °C. The cubic structure of NiO, with typical crystallite sizes of 13, 21, 20, 12, and 10 nm, is confirmed by the X-ray diffraction pattern. The Ni–O bond's stretching mode is indicated by the band at 430 cm^{-1} in the FT-IR. The material's band gap was determined to be 3.4 eV. According to SEM and TEM studies, the morphologies are spherical and include particles that are between 20 and 25 nm in size. Under UV light, one photocatalytic activity photodegrades MB dye with a degradation efficiency of 99.6% for 90 min. Additionally, it exhibits strong antioxidant activity in tests using the hydroxyl, ferrous ion, and DPPH radicals. The produced NPs also exhibit strong anti-angiogenic action. Finally, with a detection limit of 11 M concentration levels, NiO demonstrates outstanding electrochemical applicability toward the sensing of dopamine [22].

4.8 *Titanium Oxide Nanoparticles*

TiO_2 NPs are among the most often produced metal oxide NPs due to their reputable and varied properties derived from their physical, chemical stability, optical, and electrical activity [27]. Water, organic acids, or diluted alkaline solutions do not cause TiO_2-NPs to dissolve; only hot sulfuric and hydrofluoric acids do. Three different forms of TiO_2-NPs, anatase (tetragonal), rutile (cubic), and brookite, can crystallize. The three most important physical properties that influence how much TiO_2-NPs are used in industrial applications are densities, melting point, and optical refraction [3].

TiO_2 nanoparticles have been produced environmentally friendly utilizing orange fruit peel extract. Because citric acid is the major component of orange peel, it serves as a reducing agent in the manufacture of TiO_2. The sample's average crystallite size in the XRD study was 19 nm, and it had a tetragonal structure. The particle size analyzer determined the average particle size to be 24 nm. In the FT-IR spectrum, vibrational bonds between TiO_2 were seen. The sample's weight loss was calculated using TG–DTA curves as 2.5%, with support from the DTA curve [41].

The biogenesis of rutile TiO_2 nanoparticles (TiO_2 NPs) has also been accomplished by employing a fresh, practical method that makes use of fruit peel extract from Annona squamosa. 100 mL of distilled water was agitated with $TiO(OH)_2$ for two hours. *A. squamosa's* aqueous extract was combined with 80 mL of 5 mM TiO_2

at room temperature and stirred for 6 h. The ideal temperature for TiO_2 nanoparticle production is 60 °C. Therefore, 6 h is shown to be the most efficient for producing TiO_2 nanoparticles. Polydisperse nanoparticles with spherical forms could be seen in the TEM pictures. The distribution of the particles is in the 23–2 nm size range [42].

Additionally, rosaceous fruit peel extract, which served as a reducing and capping agent, was used to create titanium dioxide nanoparticles. These rosaceous fruits included *Prunusdomestica* L. (Plum), *Prunus Persia* L. (Peach), and *Actinidia deliciosa* (Kiwi). It was evident from the XRD-diffraction peak that TiO_2 nanoparticles occur in Anatase form and represent TiO_2's nanocrystalline structural structure. Titanium dioxide produced from plum peels had a cylindrical form, 200 nm overall size, and particle sizes of 47.1 and 63.21 nm. In another manner, 54.17 and 85.13 nm-sized and cylindrical-shaped biosynthesized TiO_2 was found to be produced by Kiwi peels. Another TiO_2 nanoparticle (NP) with a consistent size of 200 nm and a cylindrical shape was made from peach peels. Titanium dioxide NPs have both antibacterial and antioxidant properties as confirmed by various assays such as DPPH free radical scavenging, hydrogen peroxide scavenging, and nitric oxide scavenging [2].

4.9 Silver Oxide Nanoparticles

A p-type semiconductor with a band gap of roughly 3.3–4.0 eV is a silver oxide (Ag_2O). The Ag_2O-NPs are widely used in photocatalysts, antimicrobials, and electronic industries. Ag_2O-NPs for disease pathogens are thought to be hazardous. This Ag_2O-NPs feature results from the impact of raising the surface area, which increases the material's reactivity and the subsequent matter from quantum chemistry and physics in the nano mode. The Ag_2O-NPs' microbial feature increases to more than 99% when reduced to small dimensions at the nano scale, making them useful for treating infections and wounds [27].

The biggest fruit produced by a tree is known as the Jackfruit, or *Artocarpus heterophyllus*. It is a member of the Moraceae mulberry family. To create silver nanoparticles in an environmentally responsible manner, jackfruit fruit rind extract was employed. To create silver oxide nanoparticles (Ag_2O NPs), jackfruit rind extract (10 mL) was mixed with 90 mL of a solution of 1 mM silver nitrate in 250 ml Erlenmeyer flasks. The reaction was then carried out at room temperature. When the reaction mixture's color changed from light yellow to dark brown, it was determined that silver oxide nanoparticles were present. With excellent production, the greatest peak absorbance (measured by UV–Vis) was obtained for 1.0 mM silver ions. 180 min later, the greatest yield was noted. Ag_2ONPs made from jackfruit rind extract have distinct bands at 3408, 2922, 1613, 1383, 1020, and 610 cm^{-1} in their FT-IR spectra. The bands of the carbonyl and hydroxyl groups shifted when the FT-IR spectra of the extract and Ag_2O NPs were compared. This can be seen in the IR spectra of the silver solution, which displays the distinctive peaks of the principal biomolecules from the extract that were coated on the Ag_2O NPs surface. Ag_2O NPs (11.2–24.5 nm) are seen in the HRTEM picture, and they are mostly spherical. The SAED pattern's brilliant circular rings attest to the crystallinity of the jackfruit-derived Ag_2O NPs. Actual silver (I) oxide nanoparticles were present, according to X-ray diffraction of Ag_2O NPs. The XRD spectra supported the typical Ag_2O spectra. The monodisperse, spherical Ag_2O NPs had a diameter between 10 and 30 nm. When tested against a plant pathogenic fungus, Ag_2O NPs' antifungal activity showed moderate to good effectiveness with few infections. Based on these results, it was determined that this approach may produce Ag_2O NPs on an industrial scale from a routinely discarded waste jackfruit rind [50].

4.10 Magnesium Oxide Nanoparticles

Magnesium (Mg) is an essential inorganic mineral for plants. The odorless, non-toxic white powder known as magnesium oxide nanoparticles (MgO-NPs) has a high melting point and a high hardness. Because of its alkaline characteristics and reactive oxygen species (ROS), magnesium oxide has high antibacterial action. It has been established that the MgO-NPs' antibacterial action is produced by creating superoxide on their surface and by raising pH by hydrating MgO with water. According to studies, MgO-NPs harm cell membranes, which leads to content leaking into the cell and ultimately results in bacterial cells dying light [27].

As a natural ligation agent, *Nephelium lappaceum* L. peels were successfully employed in the manufacture of magnesium oxide nanoparticles. The crystallinity and spherical shape of the biosynthesized nanoparticles were demonstrated by XRD and SEM. From XRD and SEM investigation, it was determined that the particles were between 60 and 70 nm in size. Magnesium oxide powders as-produced were measured by PSA to have a particle size of about 100 nm. Using XRD, SEM–EDX, and PSA analyses, the successful production of magnesium oxide nanoparticles was verified [48].

Aqueous *C. aurantium* fruit peel extract was used in the green route approach to creating MgO nanoparticles. Then, 10 ml of *C. aurantium* peel extract was added to the magnetic stirrer, which was then heated to 80 °C for two hours, along with 0.1 M of Mg $(NO_3)2.6H_2O$ and 50 ml of distilled water. The extract was then centrifuged for 10 min at 10,000 rpm. The separated magnesium complex was next dehydrated for 8 h at 40 °C in an oven, followed by a 450 °C calcination in a muffle furnace to produce MgO nanoparticles. Utilizing the UV–visible spectrum, the optical absorption peak at 290 nm was discovered. According to XRD and FE-SEM results, the MgO nanoparticles' average particle size is between 50 and 60 nm. EDAX analysis was used to determine the elemental composition of MgO nanoparticles, which shows the purity of plant-mediated MgO nanoparticles. The antibacterial efficacy against Staphylococcus aureus, Staphylococcus epidermis, *Pseudomonas aeruginosa, Klebsiella pneumoniae, Candida albicans*, and *Aspergillus niger* was examined using the agar well technique. When compared to commercial medications, evaluation of antimicrobial efficacies has shown superior antibacterial activity. Because of their remarkable antibacterial properties, the produced MgO nanoparticles have been approved for use in biomedical applications [51].

4.11 Calcium Oxide Nanoparticles

CaO has drawn special interest since it is thought to be safe for both people and animals to consume. Applications in drug delivery systems are well suited for CaO-NPs with nanostructures. Additionally, they can be exploited as possible drug delivery systems, photodynamic treatment (PDT), and photo-thermal therapy (PTT) because of their special structural and optical features. CaO-NPs made from plants are often more widely utilized than techniques using microorganisms because they can be quickly improved, have fewer biological risks, and do not need the development stage of cell culture [27].

Red dragon fruit peel extract (*Hylocereus polyrhizus*) was used as the physiologically reductive agent in the green production of calcium oxide (CaO) nanoparticles. Tropical plants known as red dragon fruit (*H. Polyrhizus*) are common in Indonesia. Red dragon peels contain a lot of phenolic and flavonoid chemicals (*H. Polyrhizus*). CaO nanoparticles were created by mixing an aqueous extract of *H. Polyrhizus* with calcium metal precursor ($CaCl_2 2H_2O$) 0.2 M. (1:1 of volume ratio). NH_4OH was also employed in this procedure as a pH controller, with a 24-h reaction time at room temperature. After that, a centrifuge operation lasting 15 min at 10,000 rpm was performed. Following that, its precipitation and separated supernatants were calcined at 700 °C for three hours. The existence of the Ca-O bond in the biosynthesized sample was shown by a maximum absorbance at 450 nm in the UV–Vis spectrum. Using Fourier-transform infrared spectroscopy (FTIR), further analysis revealed the usual wave numbers that were seen at 505.35 and 540.07 cm^{-1}. Physical investigation using a scanning electron microscope (SEM) demonstrated the biosynthesized CaO were rod-shaped (fibber) morphology, and Energy Dispersive X-ray (EDX) validated the sample's contents. Calcium (29.06%) and oxygen were present (43%). According to X-ray diffraction (XRD) analysis, the average size of the CaO produced through biosynthesis was 18.98 nm. Using an antifungal experiment against *Candida albicans* at varied concentrations of 4500, 5900, and 6600 g/mL using the turbidimetry technique, anti-microorganism activity of the biosynthesized CaO was detected [40]. The various metal oxide NP synthesized by utilizing different fruits has been shown in Table 3.

Table 3 Metal oxide NPs synthesized using different parts of the fruits

S No	Fruit	Part used	NPs	Reaction conditions	Shape	Size	References
1	North Arcot	Whole fruit	CuO	50 °C/2 h	Sphere	3–68 nm	[21]
2	Christ's thorn Jujube	Pulp	CuO	80 °C/ND	Sphere	4–19 nm	[21]
3	Caperberry	Whole fruit	CuO	60 °C/24 h	Sphere	20–35 nm	[21]
4	Citron	Juice	CuO	60–100 °C/ND	ND	9–59 nm	[21]
5	Strawberry	Whole fruit	CuO	RT/1 h	Sphere	11–29 nm	[21]
6	Guava	Whole fruit	CuO	80 °C/2 h	Flakes	16–29 nm	[21]
7	Pomegranate	Peel	CuO	80 °C/10 min	Sphere	16–19 nm	[21]
8	Pomelo	Juice	ZnO	400 °C/5–10 min	Agglomerated	10–20 nm	[21]
9	Pineapple	Juice	ZnO	240 °C/5 min	ND	31–56 nm	[21]
10	Pomegranate	Peel	ZnO	80 °C ND	Sphere and hexagonal	31–80 nm	[21]

5 Surface Modification and Functionalization of Metal Oxide Nanoparticles

Metal oxide nanoparticles (NPs) have attracted a lot of attention lately. The nanoparticles exhibit distinctive characteristics when contrasted with the identical chemical composition on a bulk scale. Metal oxide NPs' great inclination to aggregate, which is a result of their high surface area to volume ratio and consequently high surface energy, is the main problem with their use. Surface modification of the NPs has lowered the surface energy of nanoparticles as well as their propensity to combine. Ideal surface modification is the enhancement of nanoparticle surface properties without changing their bulk qualities [1]. The second component of modification is the capacity to make nanoparticles compatible with another phase. The modification can avoid compatibility and homogeneity problems between the two phases. The third purpose for alteration is to allow nanoparticles to self-organize [34]. By utilizing physicochemical interactions between the metal oxide NPs and the modifiers, the surface of the NPs may be changed.

5.1 Physical Techniques

One of the simplest methods for improving the surface stability of NPs is to coat the surface of the nanoparticles with an ionic or polymeric surfactant. Both hydrophobic groups (their tails) and hydrophilic groups are present in the chemical structure of surfactants (their heads). The foundation of the surface modification procedure is the electrostatic or chemical bond-based adsorption of a hydrophilic group onto the NP's surface. Surfactants can reduce particle interactions and reduce the impact of interfacial forces. Surface tension and the rate of particle aggregation are reduced by the surfactant's adsorption on the surfaces of suspended particles. The particles become more stable as a result. In the physical modification strategy, the modifiers have weak hydrogen bonds or van der Waals forces on the surface of the NPs. The NPs that have undergone physical modification are thus thermally and solvolytically unstable. This technique has been used to modify the surfaces of NPs such as TiO_2, Fe_2O_3, Al_2O_3, and Al_2O_3-CuO utilizing surfactants [1].

5.2 Chemical Techniques

An efficient way to develop the surface properties of nanomaterials that can result in a stable nanosystem for additional applications is by chemical surface modification. These methods rely on covalent bonds between the modifier and the surface of the NPs. The development of the NPs dispersion in diverse media has made use of a range of coupling agents, including thiols, amines, organophosphorus compounds, carboxylic acids, polymers, and silanes. For chloroauric ($[AuCl_4]$) absorption, Roto et al. created $Fe_3O_4@SiO_2$ core–shell nanoparticles that were then treated with a thiol group (3-mercaptopropyltrimethoxysilane (MPTMS) [1].

The main compounds used for modifying metal oxide nanoparticles are phosphonates or silanes.

5.2.1 Silanes

The capacity of silane modifiers to form a strong chemical connection between organic and inorganic materials is their most crucial quality. The silane modifiers ($X(CH_2)_nSiR_3$), on the other hand, have two different functional groups. Because of its reactivity or compatibility with organic materials, the organic functional group (X) was chosen. Methoxy, ethoxy, and other hydrolyzable groups (R) are intermediates in the synthesis of silanol groups for attaching to inorganic or NPs surfaces [1]. Alkoxy and chlorosilanes engage in a condensation process with the OH groups on the metal oxide surface. Since the reactivity of chlorosilanes R_2SiCl_2 or $RSiCl_3$ is so strong, neither water nor a catalyst is required. Due to the diand trichlorosilanes' strong reactivity, they may react with neighboring M-OH groups as well as a single M-OH at the surface, leading to multidentate attachments [34].

5.2.2 Phosphonates

M–O–P bonds are produced when phosphonate groups modify the surface of a metal oxide. Precursors including P-OH, P–OR (R=alkyl), and P-O can introduce phosphonate groups. Ti–O–P bonds are extremely stable against hydrolysis, in contrast to the silane alteration of titania particles that results in unstable Ti–O–Si bonds. The fact that phosphonates only react with surface OH groups and do not go through homo-condensations, i.e., do not create P-O-P bonds, is another significant distinction between them and silanes. The anchoring can be monodentate, bidentate, or tridentate, which is one of the key characteristics of the attachment of a phosphonate group on a metal oxide surface [34].

5.3 Chemical Functionalization of Nanoparticles

Functional groups can be introduced using one of two methods. The first approach (referred to as "method 1") entails introducing the entire functional ligand in a single phase. Bifunctional organic compounds are needed for this, where one functionality (X) is employed to bind to the surface of the nanoparticle and the second group (Z) is the group that functionalizes the nanoparticle. Thiols are the capping ligands that enable the introduction of different functions on metal nanoparticles. Thiol ligands most frequently introduce the functionalities COOH, NH_2, or OH. Once adhered to the particle surface, these groups are capable of undergoing further chemical processes. The functional groups may also make the nanoparticles more compatible with their surroundings. An illustration is the creation of water-dispersible nanoparticles by the grafting of carboxylate-terminated alkanethiols.

A bifunctional molecule X–Y is initially reacted in "method 2," where the group Y serves as a coupling site and can be changed, in a subsequent step, into the ultimate functionality Z. Whenever feasible, Method 1 should be used because it has one lesser step. The primary reason to avoid using the first technique is that the functional group Z is incompatible with the preparation process, as a step-by-step method may cause steric hindrance. By gradually altering an aminopropylsilyl ligand, Ding et al. added an atom transfer radical polymerization catalyst to Fe_3O_4 nanoparticles. The aminopropyl ligand's NH_2 group was first used to introduce an olefinic group, which was then used in a subsequent reaction with a triamine to create a highly chelating ligand for the coordination of copper bromide [34].

6 Factors Influencing Synthesis of Metal Oxide Nanoparticles

A few other essential factors that affect the synthesis of nanoparticles are pH, temperature, the concentration of the extracts utilized, the concentration of the raw materials used, size, and—most importantly—the protocols used for the synthesis process. Below is a discussion of a few of the factors:

6.1 *Method/Technique*

Nanoparticles may be created using a variety of approaches, from physical ones involving mechanical processes to chemical or biological ones involving various organic or inorganic compounds and living things. Each method has unique advantages and disadvantages. However, as opposed to conventional procedures, biological approaches for the production of nanoparticles employ nontoxic and ecologically safe components along with green technology, making them more palatable and ecofriendly [21].

6.2 *pH*

The size of NPs is directly impacted by the pH shift. This is because changing the pH affects the secondary metabolites' charge, which affects how well they can adsorb metal ions. As opposed to acidic settings, very alkaline circumstances typically result in the production of smaller, distributed MONPs. The reason for this is that more functional groups are available for attaching to the metal ions and stabilizing the NP during the nucleation and development phases at higher pH levels. Therefore, at higher pH levels, the resulting NPs are less aggregated [54].

6.3 *Temperature*

The type of MO-NPs that is generated depends on the temperature during synthesis or incubation. Chemical processes need temperatures lower than 350 °C, while physical processes need the greatest temperature (>350 °C). Most of the time, ambient temperature or temperatures lower than 100 °C are needed to synthesize nanoparticles utilizing green technology [21].

In comparison to lower reaction temperatures, greater temperatures often increase the yield, speed up the reaction pace, and generate more crystalline nanoparticles [54]. During the synthesis of silver nanoparticles, it was observed that the production of tiny silver nanoparticles was made possible by the quick decrease in the rate of Ag^+ ions caused by an increase in reaction temperature and the homogenous nucleation of silver nuclei that followed. It has been discovered that when the reaction mixture's temperature rises, both the stability and the pace of nanoparticle formation decrease. Additionally, the size of the silver nanoparticles produced at higher temperatures is smaller [23].

6.4 Concentration

The shape of the NPs is influenced by the concentration of the plant extract or metal salt. Ghidan et al. produced less stable nanoplates at low extract concentrations and very stable nanoflowers at high concentrations, Rajiv et al. produced spherical or hexagonal ZnO NPs using *Parthenium hysterophorus* leaf extract. Various organisms have different levels of phytochemicals. Additionally, the varied regions of the body have different concentrations of phytochemicals in the same organism, therefore varied morphologies and sizes will be obtained when the same NPs are created with leaves, roots, or the same plant's bark [54].

6.5 Time

The amount of time the reaction medium is incubated has a significant impact on the type and quality of nanoparticles produced utilizing green technology. Similar to how the properties of naturally occurring nanoparticles changed with time, the synthesis method, exposure to light, storage conditions, and other factors all had a significant impact. Particles may aggregate as a result of long-term storage, they may contract or expand over that period, they may have a shelf life, and so forth, all of which have an impact on their potential [21].

6.6 Environment

The nature of the produced nanoparticles is significantly influenced by their surroundings. The manufactured nanoparticles develop a covering in a biological system that increases their thickness and size. In addition, the chemical and physical makeup of the produced nanoparticles is influenced by their surroundings. Some reports demonstrate how the environment affects the nature of produced nanoparticles. When the environment of the zinc sulfide nanoparticles shifted from a wet state to a dry state, the crystalline nature of the particles quickly altered. Similar to this, the amount of peroxide in the fluid in which cerium nitrate nanoparticles are floating changes their chemical make-up [21].

7 Merit and Demerit of Fruit and Fruit Waste Metal Oxide Nanoparticle Synthesis

Fruit extracts are advantageous for synthesizing nanoparticles since they are less expensive, economical, energy-efficient, safe, and environmentally friendly, don't harm people's health, and generate less waste. In the realm of nanotechnology, these environmentally friendly nanoparticles are being assessed for a variety of applications. A further benefit of using fruit extracts for nanoparticle synthesis over other biological methods is that they don't take as long and don't demand the upkeep of cultures [21]. Pharmacological compounds on the surfaces of the NPs from medicinal plants boost their effectiveness against antibacterial actions. Therefore, using plant extracts to create nanoparticles might have a significant influence in the years to come.

To boost the NPs' biocompatibility and effectiveness, additional functional groups can be attached to the strong capping layer that is created during the biosynthesis of MONPs. Contrary to physiochemical synthesis, biologically produced metal oxides skip stages like functionalizing the nanoparticle as it has already been done so by the phytochemicals, shortening the synthesis process [54].

It has been demonstrated that commercial controls are less efficient than biologically generated MONPs at inhibiting bacterial activity. According to reports, Cu_2ONPs made from *Ziziphus spinachristi* L. extract have antibacterial action against *Staphylococcus aureus* but not *Escherichia coli*. In contrast to *E. coli* and *Klebsiella pneumoniae*, Cu_2ONPs made from fruit extract of *Capparis spinosa* were reported in another investigation to have robust antibacterial action against *Bacillus cereus* and *S. aureus*. Also, strawberry-derived Cu_2ONPs have been reported to show impressive concentration-dependent DPPH radical scavenging ability [21].

Following are the main concerns with the production and synthesis of green NPs:

- A lack of comprehensive expertise to create green NPs with plant entities.
- A definite size and form should be developed for green NPs using the logical approach.
- NPs should be uniform, it should be noted. NPs produced by plants have a wider range of sizes, shapes, and structures.
- The fundamental issue to be solved is salt conversion to ion. The greatest amount of salt to ion conversion should occur during plant-mediated synthesis.
- It is important to clarify the specific function of plant components in NPs. These compounds function as stabilizing and reducing agents.
- The homogeneity of the manufactured NPs, given that different biological resources were used to synthesize them.
- Implementing technology transfer procedures is necessary to move NP fabrication from the laboratory to the industrial level.
- It should have been possible to produce NPs in industrial quantities using methods that are benign to the environment and that prioritize ease of synthesis, resource utilization, particle creation (monodispersity, homogeneity, repeatability), waste management, and toxicity [37].

- Even though one of the benefits of biogenic synthesis is that it is easy to scale up, it is unclear if the amount of biomass now accessible is sufficient for long-term NP synthesis [54].
- To manufacture NPs devoid of toxicity is a dream. Thermally created Co_3O_4 nanoparticles damage DNA, provoke inflammatory reactions, and put human lymphocytes under oxidative stress. The induction of apoptosis is brought on by oxidative stress, which is a significant contributor to toxicity. Sun et al. showed substantial toxicity on cell viability after subjecting the A549, H1650, and CNE-2Z cell lines to chemically produced CuO nanoparticles [45].

8 Applications of Metal Oxide Nanoparticles

8.1 Antimicrobial

Due to the declining effectiveness of conventional antibiotics and the rise of bacteria with antibiotic resistance, attention has been drawn more and more to the potential use of nanoparticles as efficient antibacterial agents. The bactericidal effects of nanoparticles have been hypothesized to be due to their size and high surface-to-volume ratio, which make it simple for them to interact strongly with the bacterial membrane, rather than only being the result of the release of metal ions [3]. To understand the bactericidal characteristics of metal oxide nanoparticles, conclusions on potential modes of action have been reached. Some of these elements that affect the bacteria both internally and outside include reactive oxygen species (ROS), electrostatic interaction, accumulation, ions provided, and direct contact with NPs [49].

8.1.1 Formation of Reactive Oxygen Species

Another way that NPs kill bacteria is by producing reactive oxygen species (ROS) or oxygen free radicals like hydrogen peroxide (H_2O_2) or superoxide anions. ROS is indirectly produced by the NPs themselves. As a result of the severe oxidative stress and damage that ROS cause to the cell's macromolecules, lipid peroxidation, protein alteration, enzyme inhibition, and nucleic acid damage eventually occur. This severe oxidative stress will cause cell lysis and may cause the bacterial membrane to form holes or pits [14].

8.1.2 Electrostatic Attraction and Accumulation

Metal-based NPs can modify the potential and integrity of cell membranes by electrostatically adhering to the bacterial cell wall and/or releasing metallic ions. Through membrane disruption and an increase in oxidative stress, these interactions cause damage to the bacterial proteins. A large volume of cytosolic water is released as a result of the cell membrane being broken. Through the bacteria's proton efflux pumps and electron transport, cells attempt to make up for this loss. These transmembrane systems suffer serious harm as a result of the increased demand for these ions. Overall, this ion-membrane equilibrium is out of balance, which disrupts energy transfer, impairs respiration, and ultimately leads to cell death [37]. This method has also been connected to the disruption of the bacterium's DNA and adenosine triphosphate (ATP) replication, which results in the bacterium's destruction [49].

8.1.3 Loss of Homeostasis by Metal Ions

There will be a disturbance in the metabolic processes when the bacteria have an excess of metals or metal ions. By creating cross-links between and within the DNA strands, metal ions bind to DNA and change its helical structure. The metal ions promote the permeabilization of the outer membrane and balance the charges in LPS. By producing significant numbers of hydroxyl radicals and diffusing through bacterial cells, the ions of metal oxides may also contribute to the breakdown of bacterial cells.

8.1.4 Protein and Enzyme Dysfunction

The oxidation of the side chains of amino acids, which produce carbonyls attached to proteins, is catalyzed by the metal ions. Protein oxidative damage is shown by the degree of carboxylation that exists inside the protein molecule. In the case of enzymes, this carboxylation of proteins will result in a loss of catalytic activity, which in turn causes the breakdown of proteins.

8.1.5 Transduction Signal Inhibition

In bacteria, phosphotyrosine is a crucial part of the signal transduction pathway. The phosphotyrosine residues are dephosphorylated by NPs, which prevents signal transmission and eventually prevents bacterial growth.

8.2 Biomedical Applications

Metal and metal oxide nanoparticles have a distinctive form, a large surface area, biocompatibility, fascinating redox and catalytic activity, and exceptional mechanical stability. Because of these traits and unique plasmonic properties, metal and metal oxide nanoparticles have attracted a lot of attention in the field of biomedical applications, including bioimaging, biosensing, neurochemical monitoring and diagnostics, etc. [3].

The goal of targeted drug delivery, a significant biomedical application, is to deliver anticancer medications to the precise location of the tumor without harming neighboring healthy cells. Magnetic hyperthermia therapy is a crucial biological use of nanoparticles. Tumors are heated to a temperature over 42 °C during magnetic hyperthermia therapy to kill malignant cells. This method has an advantage over chemotherapy in that it just affects the tumor and spares the surrounding healthy tissue. Using light-sensitive materials, photoablation treatment can eradicate unhealthy tissue, including malignant tumors. A crucial biological application for sensing a range of biomolecules is the development of biosensors [9].

8.3 *Targeted Drug Delivery*

Due to the ineffectiveness of traditional treatment, cancer is regarded as the top cause of death in the majority of industrialized nations. Conventional chemotherapy is ineffective because of non-specific drug distribution, drug resistance at the location of the tumor, quick clearance of anti-cancer drugs, low efficiency, and occasionally pre-existing anti-cancer drugs. Therefore, medication delivery plays a very intriguing part in biomedical areas to combat issues like side effects and to entertain the greatest advantage of the medicine intended. The circulatory system is necessary for chemotherapy because it carries anticancer medications to the tumor. The therapy has drawbacks, such as the drug's toxicity and non-specificity, which causes it to attack both healthy and malignant cells and organs.

Magnetic nanoparticles are employed in targeted medication delivery to deliver the medication to a specified place. An external magnetic field is employed to direct the drug/nanoparticle combination to the particular tumor location after administration [29]. One of the most common and utilized nanoparticles for targeted medication administration is superparamagnetic iron oxide nanoparticles (SPIONPs). The loaded drug's release, as well as the transmission of cell death through temperature-induced apoptosis, can both be triggered by magnetic properties. For clinical usage, Feridex, a SPIONP-based formulation, and ferumoxytol for iron replacement treatment are both licensed. Additionally, SPIONPs have been described as a stable and secure nanodelivery device for chemotherapy, genes, and proteins [3].

Iron oxide nanoparticles (NPs) are magnetic, making them ideal for site-specific medication administration, diagnostics, and magnetic separation of biological products and cells. Additionally, ZnO NPs are desirable candidates for the delivery of cancer drugs due to their biodegradable properties. It was discovered that the doxorubicin-ZnO nanocomplex functions as an effective drug delivery system against hepatocarcinoma cells, improving treatment effectiveness by raising the concentration of doxorubicin in the intercellular space [35].

8.4 Magnetic Hyperthermia

A treatment method called hyperthermia uses heat to kill malignant cells and tissue. To eliminate malignant cells without harming healthy cells, the sick or contaminated region is heated to 41–46 °C. Compared to healthy cells, cancerous cells are more temperature sensitive. The hyperthermic effect, which occurs when malignant cells are heated to 41–46 °C, causes cell apoptosis. If the cells are heated above 46–48 °C—a process known as thermoablation—necrosis will take place. Local, regional, and whole-body hyperthermia can all be treated in three distinct ways. Heat is given to a small region in local hyperthermia therapy utilizing a variety of methods, including radio frequency, microwave, and ultrasound. Typically, regional hyperthermia is utilized to treat vast tissue regions. An organ or limb is heated externally during this procedure. Treatment for metastatic cancer that has spread throughout the body frequently involves whole-body hyperthermia [29].

Due to its outstanding capacity to produce heat energy in alternating magnetic fields at the targeted location, superparamagnetic iron oxide nanoparticles (SPIONs) have been extensively researched for application in hyperthermia therapy of cancer. Strong correlations exist between the tight particle size distribution, superparamagnetic behavior, and crystal structure, which all contribute to the capacity to function as "heaters" [9].

8.5 Sensors/Biosensors

A biosensor is a type of analytical tool used to examine biological materials. An electrical signal is produced by converting a chemical, biological, or biochemical reaction. The following three elements are crucial to a biosensor. (i) The transducer, which can be electrochemical, optical, electronic, piezoelectric, pyroelectric, or gravimetric, (ii) The bioelement or bioreceptor, which is typically made up of enzymes, nucleic acids, antibodies, cells, or tissues, and (iii) The electronic unit, which includes the amplifier, processor, and display. The target analyte or substrate of interest is recognized by the bioreceptor, and the transducer converts the resultant signal into an electrical signal that is easier to quantify. Once they have been given a bioresponsive shell, nanoparticles can be employed as bioreceptors. Biosensors are used in a variety of sectors, including food, pharmaceutical, environmental, and medical ones [29].

Nanobiosensors for the detection of tiny molecules can be enzyme-based, genosensors, immunosensors, cytosensors, and biosensors. H_2O_2, glucose, or dopamine may all be detected using small molecule electroanalysis. For instance, –Fe_2O_3 NPs in the shape of cubes were employed as a glucose biosensor material for non-enzyme catalytic oxidation with excellent sensitivity and quick reaction. They were generated from hydrophobic iron-containing fluids under hydrothermal conditions. The working electrode surface of the enzyme-based biosensors has a thin coating of the immobilized enzyme on it. For instance, the shape of the NPs utilized affects how the lactase enzyme, which detects dopamine, is immobilized. A biosensor with a 0.4 mol L1 detection limit for lactic acid was developed using lactate dehydrogenase immobilized on silica sol–gel modified gold electrodes.

The immunosensors are based on particular antigen–antibody identification, whereas single strained DNA fragments are fixed on the electrode surface in genosensors. Fast electrochemical biosensors are often used in traditional DNA detection methods due to their high specificity, mobility, and affordability. Due to their superior electron transfer kinetics, robust adsorption capacity, and increased biosensing properties, MO nanostructures have been frequently employed to immobilize enzymes. Antibodies, receptors, glycans, and other molecules that are overexpressed on a target cell's cell membrane are recognized by the cytosensors.

Nanobiosensors are sensitive enough to pick up even the smallest hydrogen bond breaks since they are quickly moved and distorted in response to pressures as low as 10 pN. It is clear that the main properties of MONPs, including their high sensitivity and selectivity, rapid response and recovery times, reversibility, and integration in different scales, make these substances suitable for monitoring infection diseases, drug pharmacokinetics, detecting biomarkers (cancer and disease), etc. [35].

8.6 Environmental Applications

8.6.1 Water Treatment

Nanotechnology has the potential to improve water and wastewater treatment by raising treatment efficiency. Because they display a range of favorable physicochemical traits, several different types of nanomaterials are being explored as possible components for water filtration. The use of nanoparticles for disinfection, particularly those based on fullerene, ZnO, and photocatalytic nanoparticles, is receiving a lot of attention (like TiO_2). Nano adsorbents are effective in removing organic molecules and metal ions, and functionalization can increase their selectivity for particular pollutants. Nanoscale metal oxides, such as titanium dioxides, iron oxides, zinc oxides, alumina, etc., have been studied as low-cost, efficient adsorbents for water treatment due to their size and adsorption efficacy.

8.6.2 Removal of Heavy Metals

Numerous research has looked at the use of nanoparticles for the adsorption of heavy metals due to the simplicity with which their surface functionality may be changed and the high surface area to volume ratio of their nanoparticles, which boosts their adsorption capacity, efficiency, and reusability. The removal efficiency of Cd(II) was only approximately 55% when the adsorbent dosage was 20 mg/L, according to Srivastava et al., but when the dose was increased to 200 mg/L, the removal efficiency of 92% during an hour of contact time was noted. Ehrampoush et al. investigated the elimination of Cd(II) using tangerine peel extract-derived synthetic iron oxide nanoparticles. These particles were spherical and 50 nm in size on average. Peel extract with Fe_2O_3 nanoparticles efficiently eliminates cadmium ions. At pH 4 and 0.4 g/100 mL of adsorbent, about 90% of the cadmium ions were successfully removed from the peel extract of Fe_2O_3 nanoparticles over a 90 min contact time.

8.6.3 Photocatalytic Applications

Photocatalysis is the process of accelerating chemical processes (oxidation/reduction) by using ultraviolet (UV) or visible light to activate a catalyst, often a semiconductor oxide. To initiate or hasten a chemical change, light, and a catalyst are required. Wastewater contains a lot of organic contaminants, which are particularly bad for both aquatic and terrestrial ecosystems. Examples of these pollutants include textile dyes, pesticides, and pharmaceutical waste. They can create substantial bioaccumulation in the environment, which will have a lengthy half-life and the smallest amount can cause biological variation Due to their strong polarity and resistance to environmental degradation processes, they are toxic.

The jackfruit (*Artocarpus heterophyllus*) leaf template was used to synthesize ZnO nanoparticles with a hexagonal wurtzite structure and a 15–25 nm particle size range. Jayaseelan et al. also confirmed this work. They exhibited remarkable photocatalytic degradation efficiency (>80%, 0.24 g/L, 1 h) against Rose Bengal dye, the main water pollutant released by the textile industries. The photodegradation of methylene blue (MB) dye by ZnO nanoflowers generated by bacteria (*Bacillus licheniformis*) achieves a remarkable 83% with a three times efficient recovery of the ZnO nanoflowers, according to Tripathi et al. [13].

9 Conclusion and Future Aspects

This review summarises the development, direction, and present applications of research on green synthesis, which produces metal and metal oxide nanoparticles using fruit and fruit waste. The usage of these green nanoparticles in a variety of fields, including electronics, photonics, optics, medical, home and waste management, has generated a great deal of attention. The nanoparticles show great promise for widespread use in medical-related disciplines, including drug delivery, cancer therapies, antibacterial agents, bandages, pharmaceuticals, and consumer goods. Consequently, this green chemistry strategy employing artificially produced nanoparticles made from plant extracts offers a novel chance to boost design and development study. With cutting-edge methods, we can create metal and metal oxide nanoparticles with the appropriate characteristics for biosensors, applications in electrochemistry, catalysis, antimicrobials, and cancer diagnostics and treatment.

References

1. Ahangaran F, Navarchian AH (2020) Recent advances in chemical surface modification of metal oxide nanoparticles with silane coupling agents: a review. Adv Coll Interface Sci 286:102298
2. Ajmal N, Saraswat K, Bakht MA, Riadi Y, Ahsan MJ, Noushad M (2019) Cost-effective and eco-friendly synthesis of titanium dioxide (TiO_2) nanoparticles using fruit's peel agro-waste extracts: characterization, in vitro antibacterial, antioxidant activities. Green Chem Lett Rev 12(3):244–254
3. Annu AA, Ahmed S (2018) Green synthesis of metal, metal oxide nanoparticles, and their various applications. Handb Ecomater 2018:1–45
4. Asensi-Fabado MA, Munné-Bosch S (2010) Vitamins in plants: occurrence, biosynthesis and antioxidant function. Trends Plant Sci 15(10):582–592
5. Bachheti RK, Abate L, Deepti, Bachheti A, Madhusudhan A, Husen H (2021) Algae-, fungi-, and yeast-mediated biological synthesis of nanoparticles and their various biomedical applications. In: Kharisov B, Kharissova O (eds) Handbook of Greener Synthesis of Nanomaterials and Compounds, Volume 1: Fundamental Principles and Methods. Elsevier Inc. 50 Hampshire St., 5th Floor, Cambridge, MA 02139, USA, pp. 701–734
6. Bachheti RK, Gupta V, Husen A, Joshi A, Konwarh R (2019) Green Synthesis of Iron Oxide Nanoparticles: Cutting Edge Technology and Multifaceted Applications. pp. 239–259. In: Husen A, Iqbal M (eds) N*anomaterials* and Plant Potential. Springer International Publishing AG, Gewerbestrasse 11, 6330 Cham pp. 239–259. (https://doi.org/10.1007/978-3-030-055 69-1_9) ISBN 978-3-030-05569-1
7. Bokhary KA, Maqsood F, Amina M, Aldarwesh A, Mofty HK, Al-Yousef HM (2022) Grapefruit extract-mediated fabrication of photosensitive aluminum oxide nanoparticle and their antioxidant and anti-inflammatory potential. Nanomaterials 12(11):1885
8. Chavali MS, Nikolova MP (2019) Metal oxide nanoparticles and their applications in nanotechnology. SN Appl Sci 1(6):1–30
9. Crippa F, Rodriguez-Lorenzo L, Hua X, Goris B, Bals S, Garitaonandia JS, Petri-Fink A (2019) Phase transformation of superparamagnetic iron oxide nanoparticles via thermal annealing: implications for hyperthermia applications. ACS Appl Nano Mater 2(7):4462–4470
10. Demirezen DA, Yıldız YŞ, Yılmaz Ş, Yılmaz DD (2019) Green synthesis and characterization of iron oxide nanoparticles using *Ficus carica* (common fig) dried fruit extract. J Biosci Bioeng 127(2):241–245
11. Edison TNJI, Atchudan R, Sethuraman MG, Lee YR (2016) Supercapacitor performance of carbon supported Co_3O_4 nanoparticles synthesized using *Terminalia chebula* fruit. J Taiwan Inst Chem Eng 68:489–495
12. Fernández-Garcia, M., & Rodgriguez, J. (2007). *Metal oxide nanoparticles* (No. BNL-79479-2007-BC). Brookhaven National Lab.(BNL), Upton, NY (United States)
13. Gebre SH, Sendeku MG (2019) New frontiers in the biosynthesis of metal oxide nanoparticles and their environmental applications: an overview. SN Appl Sci 1(8):1–28
14. Gold K, Slay B, Knackstedt M, Gaharwar AK (2018) Antimicrobial activity of metal and metal-oxide based nanoparticles. Adv Ther 1(3):1700033
15. Gowda BJ, Ahmed MG, Chinnam S, Paul K, Ashrafuzzaman M, Chavali M, Gupta PK (2022) Current trends in bio-waste mediated metal/metal oxide nanoparticles for drug delivery. J Drug Deliv Sci Technol 103305
16. Husen A (2019a) Natural product-based fabrication of zinc oxide nanoparticles and their application. In: Husen A, Iqbal M (eds) N*anomaterials* and plant potential. Springer International Publishing AG, Gewerbestrasse 11, 6330 Cham pp. 193–291. https://doi.org/10.1007/978-3-030-05569-1_7
17. Husen A (2019b) Medicinal plant-product based fabrication nanoparticles (Au and Ag) and their anticancer effect. In: Kintzios SE, Barberaki M, Flampouri (eds) Plants that Fight Cancer – Second Edition. Taylor & Francis/CRC Press, pp. 133–147

18. Husen A, Iqbal M (2019) Nanomaterials and plant potential: an overview. In: Husen A, Iqbal M (eds) N*anomaterials* and plant potential. Springer International Publishing AG, Gewerbestrasse 11, 6330 Cham pp. 3–29. https://doi.org/10.1007/978-3-030-05569-1_1
19. Jegadeesan GB, Srimathi K, Srinivas NS, Manishkanna S, Vignesh D (2019) Green synthesis of iron oxide nanoparticles using *Terminalia bellirica* and *Moringa oleifera* fruit and leaf extracts: antioxidant, antibacterial and thermoacoustic properties. Biocatal Agric Biotechnol 21:101354
20. Joshi A, Sharma A, Bachheti RK, Husen A, Mishra VK (2019) Plant-mediated synthesis of copper oxide nanoparticles and their biological applications. In: Husen A, Iqbal M (Eds) N*anomaterials* and Plant Potential. Springer International Publishing AG, Gewerbestrasse 11, 6330 Cham pp. 221–237. (https://doi.org/10.1007/978-3-030-05569-1_8) ISBN 978-3-030-05569-1
21. Kumar H, Bhardwaj K, Dhanjal DS, Nepovimova E, Şen F, Regassa H, Kuča K (2020) Fruit extract mediated green synthesis of metallic nanoparticles: a new avenue in pomology applications. Int J Mol Sci 21(22):8458
22. Kumar MS, Soundarya TL, Nagaraju G, Raghu GK, Rekha ND, Alharthi FA, Nirmala B (2021) Multifunctional applications of Nickel oxide (NiO) nanoparticles synthesized by facile green combustion method using *Limonia acidissima* natural fruit juice. Inorg Chim Acta 515:120059
23. Lade BD, Shanware AS (2020) Phytonanofabrication: methodology and factors affecting biosynthesis of nanoparticles. In: *Smart Nanosystems for Biomedicine, Optoelectronics and Catalysis*. IntechOpen.
24. LaGrow AP, Besenhard MO, Hodzic A, Sergides A, Bogart LK, Gavriilidis A, Thanh NTK (2019) Unravelling the growth mechanism of the co-precipitation of iron oxide nanoparticles with the aid of synchrotron X-Ray diffraction in solution. Nanoscale 11(14):6620–6628
25. Lucarini M, Durazzo A, Bernini R, Campo M, Vita C, Souto EB, Romani A (2021) Fruit wastes as a valuable source of value-added compounds: a collaborative perspective. Molecules 26(21):6338
26. Mahmoudi G, Sufimahmoudi E, Sajadi SM (2020) Bioactive metal oxide nanoparticles from some common fruit wastes and *Euphorbia condylocarpa* plant. Food Sci Nutr 8(10):5521–5531
27. Marouzi S, Sabouri Z, Darroudi M (2021) Greener synthesis and medical applications of metal oxide nanoparticles. Ceram Int 47(14):19632–19650
28. Mathur P, Saini S, Paul E, Sharma C, Mehtani P (2021) Endophytic fungi mediated synthesis of iron nanoparticles: characterization and application in methylene blue decolorization. Curr Res Green Sustain Chem 4:100053
29. McNamara K, Tofail SA (2017) Nanoparticles in biomedical applications. Adv Phys: X 2(1):54–88
30. Menazea AA, Ismail AM, Samy A (2021) Novel green synthesis of zinc oxide nanoparticles using orange waste and its thermal and antibacterial activity. J Inorg Organomet Polym Mater 31(11):4250–4259
31. Mirgane NA, Shivankar VS, Kotwal SB, Wadhawa GC, Sonawale MC (2021) Waste pericarp of ananas comosus in green synthesis zinc oxide nanoparticles and their application in waste water treatment. Mater Today: Proc 37:886–889
32. Moreau RA, Whitaker BD, Hicks KB (2002) Phytosterols, phytostanols, and their conjugates in foods: structural diversity, quantitative analysis, and health-promoting uses. Prog Lipid Res 41(6):457–500
33. Nayak SS, Wadhawa GC, Shivankar VS, Inamadar RF, Sonawale C (2020) Green synthesis of nanostructured aluminum: antibacterial activity and dye degradation. Eur. J. Mol. Clin. Med 7:2640–3265
34. Neouze MA, Schubert U (2008) Surface modification and functionalization of metal and metal oxide nanoparticles by organic ligands. MonShefte Für Chem-Chem Mon 139(3):183–195
35. Nikolova MP, Chavali MS (2020) Metal oxide nanoparticles as biomedical materials. Biomimetics 5(2):27
36. Patil SN, Paradeshi JS, Chaudhari PB, Mishra SJ, Chaudhari BL (2016) Bio-therapeutic potential and cytotoxicity assessment of pectin-mediated synthesized nanostructured cerium oxide. Appl Biochem Biotechnol 180(4):638–654

37. Patil S, Chandrasekaran R (2020) Biogenic nanoparticles: a comprehensive perspective in synthesis, characterization, application and its challenges. J Genet Eng Biotechnol 18(1):1–23
38. Pereira JA, Berenguer CV, Andrade CF, Câmara JS (2022) Unveiling the bioactive potential of fresh fruit and vegetable waste in human health from a consumer perspective. Appl Sci 12(5):2747
39. Rambabu K, Bharath G, Banat F, Show PL (2021) Green synthesis of zinc oxide nanoparticles using *Phoenix dactylifera* waste as bioreductant for effective dye degradation and antibacterial performance in wastewater treatment. J Hazard Mater 402:123560
40. Ramli M, Rossani RB, Nadia Y, Darmawan TB, Ismail YS (2019). Nanoparticle fabrication of calcium oxide (CaO) mediated by the extract of red dragon fruit peels (*Hylocereus Polyrhizus*) and its application as inorganic–anti-microorganism materials. In: IOP Conference Series: Materials Science and Engineering (Vol. 509, No. 1, p. 012090). IOP Publishing
41. Rao KG, Ashok CH, Rao KV, Chakra CS, Rajendar V (2015) Synthesis of TiO_2 nanoparticles from orange fruit waste. Synthesis 2(1):1
42. Roopan SM, Bharathi A, Prabhakarn A, Rahuman AA, Velayutham K, Rajakumar G, Madhumitha G (2012) Efficient phyto-synthesis and structural characterization of rutile TiO_2 nanoparticles using *Annona squamosa* peel extract. Spectrochim Acta Part A Mol Biomol Spectrosc 98:86–90
43. Samuel MS, Selvarajan E, Mathimani T, Santhanam N, Phuong TN, Brindhadevi K, Pugazhendhi A (2020) Green synthesis of cobalt-oxide nanoparticle using jumbo Muscadine (*Vitis rotundifolia*): Characterization and photo-catalytic activity of acid Blue-74. J Photochem Photobiol, B 211:112011
44. Sasidharan D, Namitha TR, Johnson SP, Jose V, Mathew P (2020) Synthesis of silver and copper oxide nanoparticles using *Myristica fragrans* fruit extract: Antimicrobial and catalytic applications. Sustain Chem Pharm 16:100255
45. Seabra AB, Durán N (2015) Nanotoxicology of metal oxide nanoparticles. Metals 5(2):934–975
46. Siddiqi KS, Husen A, Rao RAK (2018) A review on biosynthesis of silver nanoparticles and their biocidal properties. J Nanobiotechnol 16(14):1–28
47. Stankic S, Suman S, Haque F, Vidic J (2016) Pure and multi metal oxide nanoparticles: synthesis, antibacterial and cytotoxic properties. J. Nanobiotechnol 14(1):1–20
48. Suresh J, Yuvakkumar R, Sundrarajan M, & Hong SI (2014) Green synthesis of magnesium oxide nanoparticles. In: Advanced Materials Research (Vol. 952, pp. 141–144). Trans Tech Publications Ltd
49. Vega-Jiménez AL, Vázquez-Olmos AR, Acosta-Gío E, Álvarez-Pérez MA (2019). In vitro antimicrobial activity evaluation of metal oxide nanoparticles. Nanoemulsions Prop Fabr Appl 1–18
50. Velu M, Pyong-In Y, Palanivel V, Palaniyappan J, Sung-Chul H, Seong-Ho J, Subpiramaniyam S (2017) Production, optimization and characterization of silver oxide nanoparticles using Artocarpus heterophyllus rind extract and their antifungal activity. Afr J Biotech 16(36):1819–1825
51. Vijayakumar S, Punitha VN, Parameswari N (2022) Phytonanosynthesis of MgO nanoparticles: green synthesis, characterization and antimicrobial evaluation. Arab J Sci Eng 47(6):6729–6734
52. Yusefi M, Shameli K, Ali RR, Pang SW, Teow SY (2020) Evaluating anticancer activity of plant-mediated synthesized iron oxide nanoparticles using *Punica granatum* fruit peel extract. J Mol Struct 1204:127539
53. Zakynthinos G, Varzakas T (2016) Carotenoids: from plants to food industry. Curr Res Nutr Food Sci 4(1):38–51
54. Zikalala N, Matshetshe K, Parani S, Oluwafemi OS (2018) Biosynthesis protocols for colloidal metal oxide nanoparticles. Nano-Struct & Nano-Objects 16:288–299

Palm Waste Utilisation for Nanoparticles Synthesis and Their Various Application

Radwa A. El-Salamony

Abstract After the fresh fruit bunches are harvested from oil palm plants, a tremendous amount of garbage is created each year. This waste includes empty fruit bunches, mesocarp fibre, shell fractions or palm kernel shells, oil palm fronds, oil palm trunks, and palm leaves. The ecology, plants, animals, and people are all negatively impacted by these wastes in different ways. Therefore, the current chapter aims to highlight the utilisation of palm tree waste extracts for producing green metallic NPs, carbon materials, lignin, nanofluids, and ester as well as their applications. These results support the environmentally acceptable use of waste palms as emulsifying agents, heat transfer fluid, antimicrobial, and wastewater treatment biocatalysts.

Keywords Palm waste · Green synthesis · Metal-oxides NPs · Carbon materials · Lignin · Nanofluids · Natural Ester · Antimicrobial

1 Introduction

Green technology is seen as a powerful driver of economic expansion and a tool for halting environmental deterioration, particularly in developing nations. A tropical and subtropical tree, the date palm is a member of the Arecaceae family. It is abundantly distributed throughout the Middle East region [9], and it is known as the "tree of life" because it offers materials for construction, transportation, and healthcare. Phoenix dactylifera is a widely distributed tree with significant economic, nutritional, and therapeutic value; its many sections are utilised to create several types of environmentally beneficial nanoparticles. It contains a lot of phytochemicals, including phenolics, sterols, carotenoids, anthocyanins, and flavonoids, as well as carbs [61]. Most tropical climates with high rainfall cultivate oil palm (*Elaeis guineensis*) as a plantation crop, mostly for processing palm fruits to obtain edible and technical oils

R. A. El-Salamony (✉)
Reactor Engineering Lab, Process Development Department, Egyptian Petroleum Research Institute (EPR), Cairo 11727, Egypt
e-mail: radwa2005@hotmail.com

A. Husen (ed.), *Nanomaterials from Agricultural and Horticultural Products*,
Smart Nanomaterials Technology,
https://doi.org/10.1007/978-981-99-3435-5_6

[23]. Around 76.54 million tonnes of the total oils and fats produced globally in 2021 were palm oil [62].

It is concerning because oil palm-related garbage poses significant disposal challenges [59]. In order to resolve the usage of oil palm wastes, technological, cost-effective, energy balance, and ecofriendly are required in a balanced ratio. The maxim "wastes to wealth"—which is geared toward the "zero-waste industry"—has led to the usage and manufacturing of valuable materials from "green wastes" produced by oil palm industries [44].

With numerous applications in the fields of agriculture, biomedicine, catalysis, cosmetics, energy, electronics, mechanics, optics, pharmaceutics, sensors, and textile, nanotechnology has emerged as a cutting-edge scientific study [3, 26, 29, 30, 35, 39, 41–43, 50, 51]. These plants are abundant sources of phytochemicals that have functional groups with aromatic hydroxyl, aldehyde, and carboxyl [8, 11, 24, 25]. Metal nanoparticles have been successfully synthesised using a variety of plant resources [27, 28]. Therefore, the current chapter intends to highlight the use of palm tree waste extracts for synthesising green metallic NPs, carbon materials, lignin, nanofluids, and ester obtained through various palm waste extracts as well as their applications. These findings boost the ecofriendly use of waste palm oil in water treatment, heat transfer applications, and emulsifying agents.

2 Palm Waste Consisting

Around forty-three countries around the world already cultivate oil palm, and if growth rates do not slow down, more will be planted. A tremendous amount of trash, including empty fruit bunches (EFB), mesocarp fibre (MF), shell fractions or palm kernel shells (PKS), oil palm fronds (OPF), oil palm trunks (OPT), and palm leaves, are created each year after the harvest of fresh fruit bunches (FFB) [4] (Fig. 1).

3 Advantages of Green Synthesis

The utilisation of green chemistry in their production is the major justification for taking into account naturally occurring plant-based extracts for the creation of NPs. The main advantage of employing green chemistry to create NPs is that: an ecofriendly reducing agent. The plant extract contains a variety of substances, including alkaloids, amines, amides, flavanones, terpenoids, proteins, phenolics, and pigments, which help reduce and stabilise metal ions during the manufacture of green NPs [37, 40]. The process is simple, highly reproducible, and stable; therefore, it could be used for large-scale production with more efficiency in cost investment. An example of a plant extract is that derived from date palm (Fig. 2).

Fig. 1 Illustration of oil palm biomass consisting of fruit bunches, leaves, and trunks (adopted from Zakaria et al. [62])

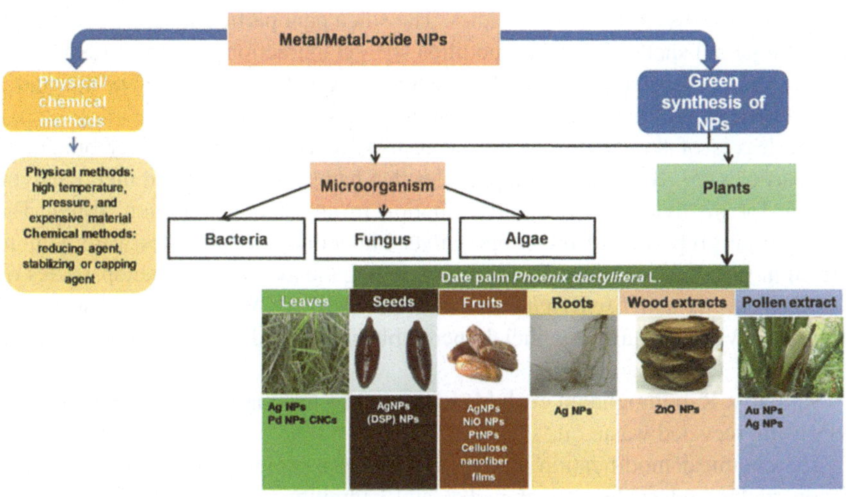

Fig. 2 Methods of synthesis of NPs using physical/chemical methods or the green methods like that from the date palm tree (adopted from Baazaoui and Sghaier-Hammami [9])

4 Preparation of Metal Oxide NPs

The ionic liquid was studied for colloidal gold stabilisation and flavonol extraction from oil palm leaves (OPL) [32, 33]. The total degradation of MB was 98.30 and 97.70% using AuNPs synthesised on day 1 and day 60, respectively. This suggested that AuNPs are still well dispersed in colloidal solutions. Palm oil leaf extracts were used as a stabilising agent and ultrasonic radiation as a reducing agent during the effective one-pot green synthesis of spherical gold nanoparticles (AuNPs) [60]. The produced AuNPs are stabilised by the amine groups (–NO) and hydroxyl group (–HO) in the POLE. For employing 2 and 5% of POLE, the average particle size ranged from 92.37 to 112.3 nm, respectively. As well, silver nanoparticles were synthesised by using an extract solution of palm leaves [11].

Likewise, Sadrolhosseini et al. [48] discussed the use of the laser ablation approach for the synthesis of Au NPs distributed in palm oil at various temperatures. Whereas, long hydrocarbon chains in palm oil hinder the aggregation of NPs, and polar ester bonds can cap the NPs.

The palm oil mill effluent (POME) is produced during the manufacture of palm oil [12] and used in the synthesis of spherical gold nanoparticles without adding an external surfactant, capping agent, or template [18].

Numerous industries, such as medicine delivery, lightweight aggregates, and energy storage, use silica nanoparticles. The silica nanoparticles were created from the palm kernel shell ash using a modified sol–gel extraction process (PKSA) [31]. The isolated silica nanoparticles have a unit size between 50 and 98 nm and a relatively large specific surface area (438 m^2g^{-1}).

The leaves of the date palm tree were used to create copper-silver bimetallic nanoparticles [5]. Bimetallic nanoparticles that had been created had their catalytic activity for methylene blue dye degradation in an aqueous solution tested. The antibacterial properties of the copper-silver bimetallic nanoparticles were demonstrated in well diffusion investigations employing the as-produced nanoparticles on *Bacillus subtilis* and *Escherichia coli* bacteria (Fig. 3). Waste palm oil was used to create environmentally beneficial, incredibly stable, and biodegradable CuO NPs [34].

Similarly, Ramimoghdam et al. [46] suggested a technique for synthesising ZnO NPs using recycled waste oil.

The chemical modification of oil palm bagasse biomass (OPB) with alumina nanoparticles and evaluation of nickel and cadmium uptake [23]. At pH 6, the maximum removal yields of 87% for Cd^{+2} and 81% for Ni^{+2} were attained.

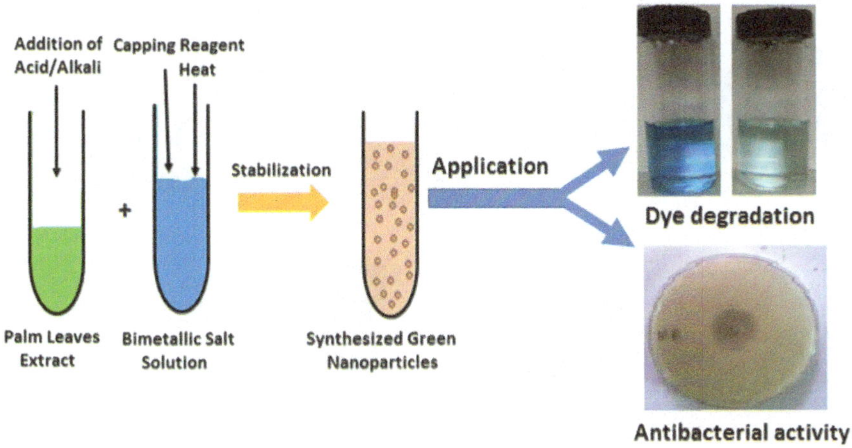

Fig. 3 Preparation and application of copper-silver Bimetallic nanoparticles (adopted from Al-Haddad et al. [5])

5 Preparation of Carbon Materials

Carbon nanomaterials have a crucial role in nanotechnology. It is the most common and feasible material used in various applications, from energy generation to purification of water pollution.

5.1 Preparation of Carbon Dots (CDs)

Newly developed carbon nanostructures called carbon dots (CDs) have uses in a variety of industries, including bioimaging, sensing, photovoltaics, and catalysis [52]. Smaller than 10 nm carbon nanoparticles, also known as carbon quantum dots (CQDs), make up the majority of carbon dots (CDs) [57]. Despite the intriguing characteristics of CQDs, certain research revealed that unmodified, naked CQDs have poor spectral efficiency and photosensitisation [22]. With the use of triflic acid and palm shell powder, nitrogen and sulphur co-doped CQDs were successfully created [54]. The as-prepared CQDs showed high photoluminescence, good dispersibility, and a quick, sensitive, and selective reaction to nitrophenols over other organic pollutants like phenol, p-Chlorophenol, and monocrotophos. Using discarded oil palm frond as an alternative precursor for NCQDs, a simple in-situ hydrothermal process has been used to successfully create NCQDs/TiO_2 nanocomposite (Fig. 4) [22]. FESEM and TEM images of pure TiO_2 and NCQDs/$TiO_{2\text{–}4}$ nanocomposites were represented in Fig. 5. The photocatalytic activity of NCQDs/$TiO_{2\text{–}4}$ towards MB degradation was 5.8-fold faster than that of pure TiO_2 under visible light.

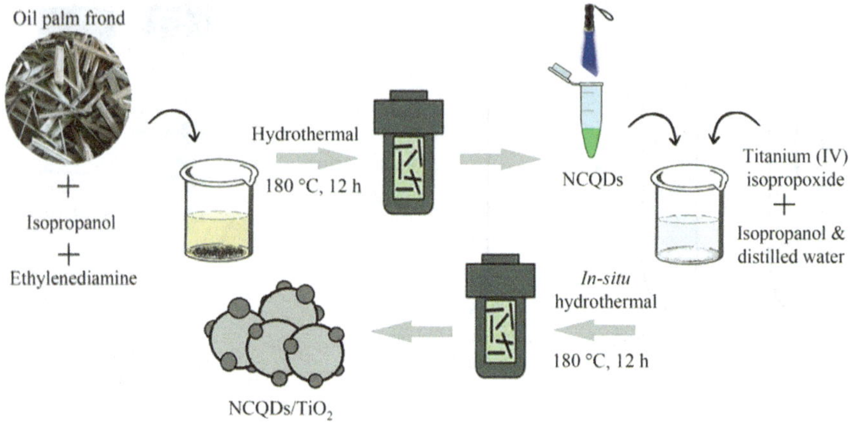

Fig. 4 Illustration of the synthesis process of NCQDs and NCQDs/TiO_2 nanocomposite (adopted from Heng et al. [22])

5.2 *Preparation of Activated Carbon (AC)*

Ag nanoparticles were supported on activated carbon that was made from palm shell agro-waste by a one-pot synthesis [56]. The as-prepared AgNPs-PSAC was used as an efficient catalyst for the conversion of nitrophenols from aqueous solutions to aminophenols, offering a natural remedy for nitrophenol remediation. The reduction was finished quickly—in just 5 min—and followed pseudo-first-order kinetics. Also; Agnp-PSAC demonstrated four cycles of recyclability with no change in activity. SEM images of the prepared AC from oil palm fibres were represented in Fig. 6.

5.3 *Carbon Nanofibres (CNFs)*

Due to their exceptional mechanical and electrical capabilities, CNFs fall under a new category of remarkable nanostructured materials [38]. CNFs have been derived from palm kernel shells [20], and oil palm trunks [15], which can be widely adopted for water purification.

5.4 *Graphene Oxide (GO)*

For both graphene oxide (GO) and reduced graphene oxides rGO preparation, oil palm byproducts such as palm kernel shells (PKS), oil palm leaf (OPL), and empty fruit bunch (EFB) are highly affordable carbon sources as promising materials for supercapacitor applications [44]. The samples of reduced graphene oxide exhibit

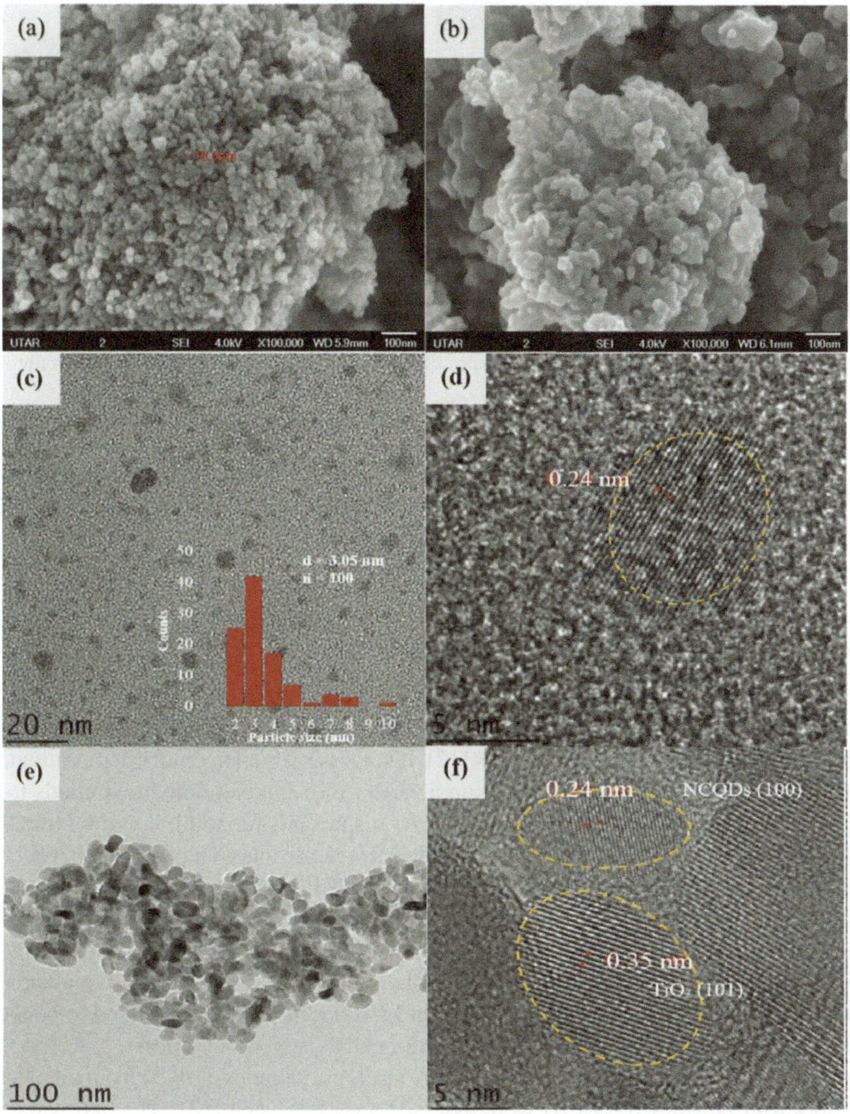

Fig. 5 FESEM images of **a** pure TiO_2 and **b** NCQDs/TiO_2-4 nanocomposite. **c** TEM image and **d** HRTEM image of NCQDs. **e** TEM image and **f** HRTEM image of NCQDs/TiO_2-4 nanocomposite (adopted from Heng et al. [22])

a rise in specific capacitance in the following sequence: commercially acquired graphite (rGOCG) < (rGOPKS) < (rGOOPL) < (rGOEFB). Waste frying palm oil for the manufacture of graphene oxide at various temperatures was proven by Robaiah

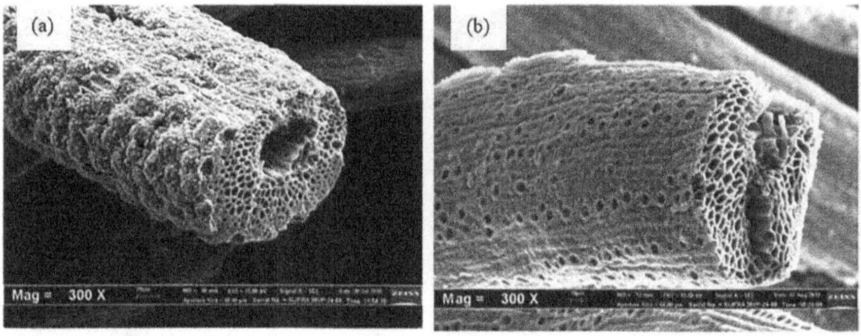

Fig. 6 SEM images for **a** char from oil palm fibres and **b** activated carbon from oil palm fibres (adopted from Foo and Hameed [16])

et al. [47]. For the manufacture of GO nano tablets, Azam et al. [7] used palm-based waste chicken frying oil. Using oil palm fibre and fruit waste, Tahir et al. [58] developed pure GO on copper sheets by CVD.

6 Preparation of Nano Lignin

In the manufacture of several polymer blends, lignin and its derivatives were employed as a source of carbon fibres [17]. It serves as a binder, emulsifier, pesticide dispersing agent, and other functions in a variety of chemically modified goods for which it is utilised as a raw material [10]. As dispersants for solid–liquid mixtures, they can stabilise liquid–liquid mixtures and lower the interfacial tension between oil and water [21]. Oil palm empty fruit bunch (OPEFB) biomass was used to create a non-toxic plant-based emulsifying agent (Fig. 7) [49]; using a chemical homogenizer, soda lignin samples, and several nanosized soda lignin samples were effectively created. For various dosages of emulsifying agents in a water–oil emulsion, the nanosized soda lignin homogenizer at a 12,400 rpm sample was found to be the top performer in terms of stability in creaming and visual observation.

In the green synthesis of AgNPs, the various lignin samples derived from oil palm empty fruit bunches (OPEFB) were assessed and contrasted [63]. The antioxidant activity of the silver nanoparticles was related to the lignin utilised in their manufacture. Additionally, the lignin-mediated silver nanoparticle and control groups did not significantly differ in the disc diffusion antimicrobial susceptibility test against Escherichia coli (Fig. 8).

Fig. 7 Extraction of nano-sized lignin from Oil Palm Empty Fruit Bunch waste, and its applications (adopted from Sekeri et al. [49])

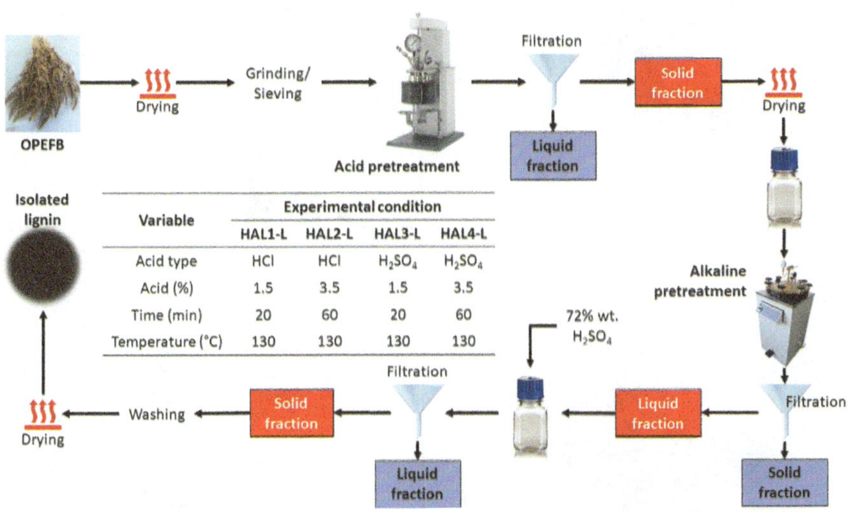

Variable	Experimental condition			
	HAL1-L	HAL2-L	HAL3-L	HAL4-L
Acid type	HCl	HCl	H_2SO_4	H_2SO_4
Acid (%)	1.5	3.5	1.5	3.5
Time (min)	20	60	20	60
Temperature (°C)	130	130	130	130

Fig. 8 Extraction sequence of lignin from oil palm empty fruit bunches (adopted from Zevallos Torres et al. [63])

7 Preparation of Nanofluids

The upcoming generation of heat transfer fluids is predicted to be nanofluids. These fluids are a combination of different nano additions with improved qualities not present in the traditional fluids [13]. CuO nanoparticles were dispersed into used palm oil to create stable nanofluids [34]. These nanofluids are shown better stability improvements (up to 6 months) without aggregation (Fig. 9). CuO nanofluid with a 0.7% nanoparticle concentration was found to have up to 190% more thermal conductivity than pure palm oil.

A novel type of nanofluid has been successfully created, consisting of Polyaniline Nanofibres (PANI), Copper Oxide (CuO) nanoparticles, and CuO-PANI nanocomposites dispersed in refined, bleached, and deodorised palm olein (RBDL) base fluids (Fig. 10) [53]. According to the thermal conductivity measurement, nanofluid containing 10 Wt% CuO-PANI nanocomposites, with a 31.34% improvement, attained the most exceptional thermal conductivity.

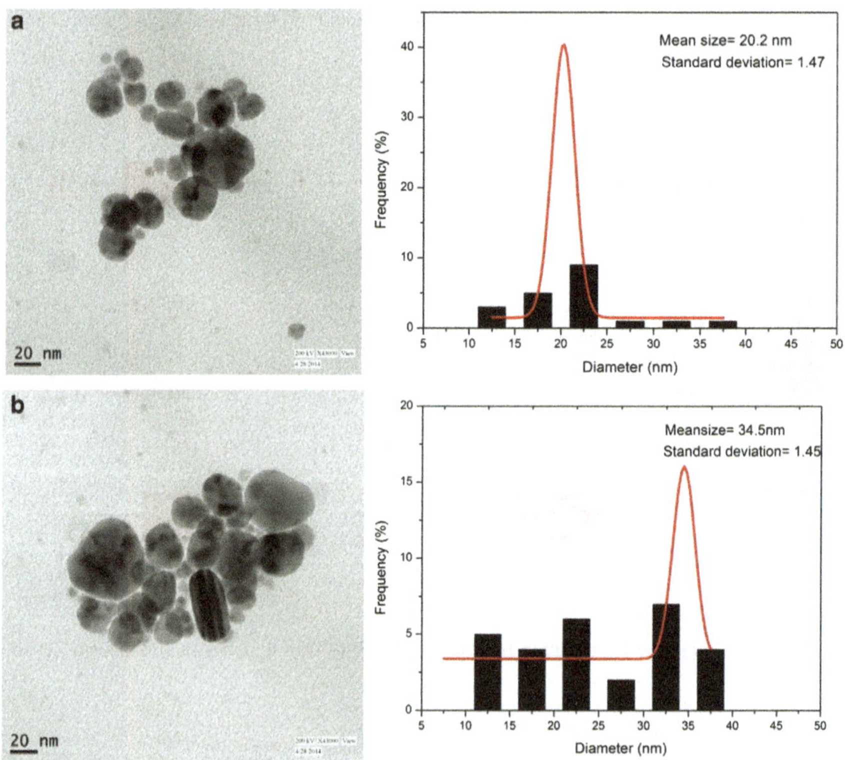

Fig. 9 TEM of waste palm oil-based CuO nanofluid after stored for **a** 3 months and **b** 6 months. Right-hand side images represent the corresponding size distributions (adopted from Javed et al. [34])

Fig. 10 Sedimentation photograph of CuO/RBDL, PANI/RBDL, and CuO-PANI/RBDL nanofluids (adopted from Sofiah et al. [53])

8 Production of Natural Ester

A form of dielectric insulating fluid known as a "natural ester" is made from seed oil and has some physicochemical characteristics that are superior to those of basic seed oil, such as low viscosity and low pour point. Natural ester fluid has been presented as an alternative by many researchers working on acceptable substitute insulating oil for transformers recently [19]. For example, when Abdelmalik [2] modified the electrical insulation oil made from palm kernel oil, it reduced the oil's viscosity making it suitable for cooling. Abdelmalik's [1], work on chemically altering palm kernel oil ester produced a breakdown voltage that was higher than that of mineral oil. In comparison to the base alkyl ester, the study found that the Al_2O_3 and TiO_2 nanofluid samples had lower dielectric losses and better breakdown voltage.

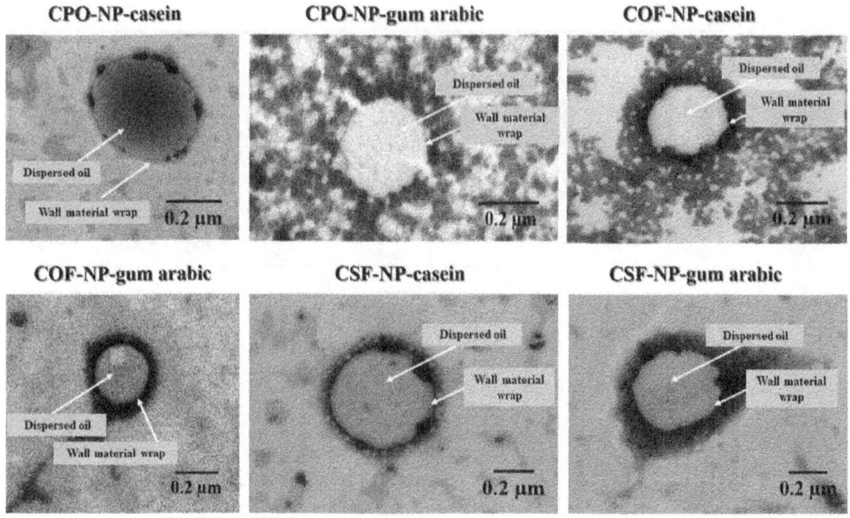

Fig. 11 Transmission Electron Microscopy (TEM) images of crude palm oil nanoparticles of casein (CPO-NP-casein); (CPONP- gum arabic); (COF-NP-casein); (COF-NP-gum arabic); crude palm stearin fraction nanoparticles of casein (CSF-NP-casein); and (CSF-NP-gum arabic), (adopted from Nuryawan et al. [45] , Duarte Ferreira Ribeiro et al. [14])

9 Preparation of Nano-oil

Crude palm oil nanoparticles (CPO-NP), crude palm olein fraction nanoparticles (COF-NP), and crude palm stearin fraction nanoparticles are new bioactive nanoparticles generated from crude palm oil (CSF-NP) [14]. The nanoparticles were created by homogenising the ingredients and encasing them in either biodegradable casein or gum Arabic (Fig. 11). The bioactive nanoparticles were able to maintain the carotenoids and are being promoted as a green alternative to artificial colorants and antioxidants in food. This study showed that CPO-NP, COF-NP, and CSF-NP have the potential to be used in long-lasting refrigerated food matrices and can act as a green, novel product to replace synthetic colorants and antioxidants in meals, particularly in slightly acidic, semi-liquid, and pasty foods.

10 Others

For the layers of the oil palm trunk (OPT) veneer and empty fruit bunch fibre mat, phenol–formaldehyde (PF) resins were produced and loaded with varying amounts of oil palm ash (OPA) nanoparticles [45] (Fig. 12). The resulting hybrid plywood demonstrated how considerably the inclusion of OPA nanoparticles influenced the plywood panels' physical, mechanical, and thermal capabilities. The plywood panels with the maximum OPA nanoparticles loading in PF resin showed significant improvements in dimension from water absorption and thickness swelling studies (Table 1).

A polyester used for microbiological preservation is polyhydroxyalkanoate (PHA). Various kinds of microorganisms naturally produce it. PHA's similarities to some commercially available polymers' characteristics make it a potential renewable replacement for some petrochemical plastics [55]. Palm oil is a multipurpose oil that is currently utilised to make oleochemicals as well as food oils. The palm oil

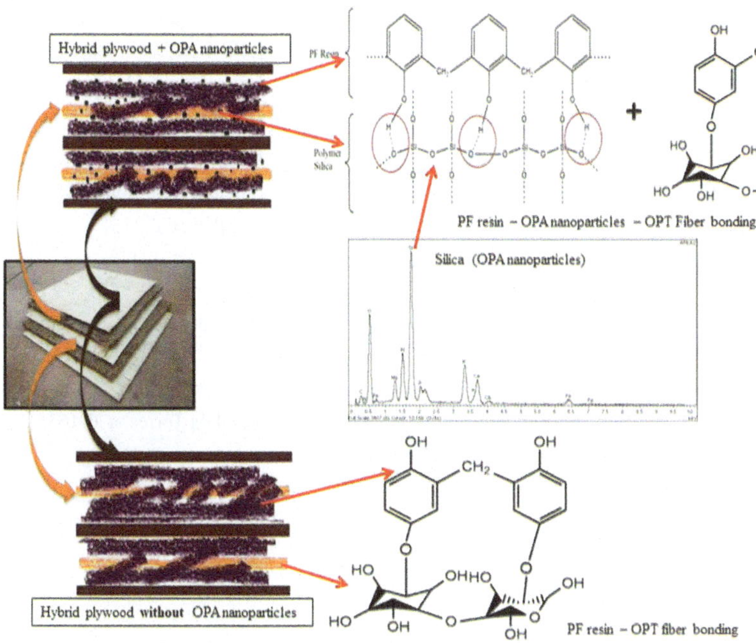

Fig. 12 Schematic diagram showing the influence of the nanoparticles on the bonding between the layers (adopted from Nuryawan et al. [45])

Table 1 Materials prepared from different palm oil waste

Palm waste	Materials	References
Oil palm leaves (OPL)	AuNPs	Irfan et al. [32, 33] Usman et al. [60]
	AgNPs	Chand [11]
	Cu–Ag bi-metallic NPs	Al-Haddad et al. [5]
	GO	Nasir et al. [44]
	Oleochemicals	Sudesh et al. [55]

(continued)

Table 1 (continued)

Palm waste	Materials	References
Palm oil waste	AuNPs	Sadrolhosseini et al. [48]
	CuO NPs	Javed et al. [34]
	ZnO NPs	Ramimoghdam et al. [46]
	CuNF	Javed et al. [34], Daniali et al. [13]
	Crude palm oil-NPs (CPO-NP)	Duarte Ferreira Ribeiro et al. [14]
Palm oil mill effluent (POME)	AuNPs	Chou et al. [12], Gan et al. [18]
Palm kernel shell ash (PKSA)	SiO_2 NPs	Imoisili et al. [31]
	CNFs	Hassan et al. [20]
Oil palm frond	CQDs	Heng et al. [22]
Palm shell waste	AC	Sudhakar and Soni [56]
Oil palm trunk	CNFs	Fadillah et al. [15]
	Phenol-formaldehyde (PF) resins	Nuryawan et al. [45]
Waste frying palm oil	GO	Robaiah et al. [47], Azam et al. [7], Tahir et al. [58]
Oil palm empty fruit bunch (OPEFB) biomass	Emulsifying agent	Sekeri et al. [49]
	AgNPs	Zevallos Torres et al. [63]
	phenol–formaldehyde (PF) resins	Nuryawan et al. [45]
Seed oil	Nature ester	George [19], Abdelmalik [2]
Date palm seeds	Antimicrobial NPs	Awad et al. [6]

industry produces a lot of by-products made of triglycerides and fatty acids that can be used by microorganisms [55].

The nanoparticles from date palm seeds can be utilised to kill microbes or stop their growth. The effective concentration of date palm seed nanoparticles in the solution may be around 1.5 M [6]. By soaking the seeds in water, drying them, and then grinding them, date palm seed powder can be made. The seeds can be ground in a heavy-duty grinder and accordingly, the powder can have a particle size from 1 to 2 mm.

11 Conclusion

This chapter resolved that there are millions of tonnes of palm garbage created each year without any benefits. By using various techniques, researchers have employed these wastes to synthesise various nanoparticles, including metal oxide, carbon nanoparticles, nano lignin, nanofluid, and nano-oil. These NPs are employed to alleviate environmental issues and serve as antimicrobials, food colorants and antioxidants, emulsifiers, electrical capacitors, cooling fluids, and antimicrobials. Researchers will need to utilise these NPs in a variety of eco-friendly and affordable products in the future.

References

1. Abdelmalik AA (2014) Chemically modified palm kernel oil ester: a possible sustainable alternative insulating fluid. Sustain Mater Technol 1:42–51
2. Abdelmalik AA (2012) The feasibility of using a vegetable oil-based fluid as electrical insulating oil
3. Abate L, Bachheti A, Bachheti RK, Husen A, Pandey DP, Megra MB (2022) Coir fiber-based cellulose, nanocellulose, and their cutting-edge applications. In: Jawaid M (ed) Coir fiber and its composites. Elsevier Inc., Cambridge, MA, USA, pp 309–325
4. Ahmad A, Buang A, Bhat AH (2016) Renewable and sustainable bioenergy production from microalgal co-cultivation with palm oil mill effluent (POME): a review. Renew Sustain Energy Rev 65:214–234
5. Al-Haddad J, Alzaabi F, Pal P, Rambabu K, Banat F (2020) Green synthesis of bimetallic copper–silver nanoparticles and their application in catalytic and antibacterial activities. Clean Technol Environ Policy 22:269–277
6. Awad MAG, Al Olayan EM, Yehia HM, Ortashi KMO, Ali HSM, Elkhadragy MF (2017) Method of preparing date palm seed nanoparticles. US Patents
7. Azam MA, Abd Mudtalib NESA, Seman RNAR (2018) Synthesis of graphene nanoplatelets from palm-based waste chicken frying oil carbon feedstock by using catalytic chemical vapour deposition. Mater Today Commun 15:81–87
8. Bachheti RK, Gupta V, Husen A, Joshi A, Konwarh R (2019) Green synthesis of iron oxide nanoparticles: cutting edge technology and multifaceted applications. In: Husen A, Iqbal M (eds) Nanomaterials and plant potential. Springer International Publishing AG, Cham, pp 239–259. https://doi.org/10.1007/978-3-030-05569-1_9
9. Baazaoui N, Sghaier-Hammami B (2021) Green synthesis of nanoparticles from date palm (Phoenix dactylifera L.). In: The date palm genome, vol 2. Springer, pp 51–69
10. Calvo-Flores FG, Dobado JA (2010) Lignin as renewable raw material. Chemsuschem 3:1227–1235
11. Chand D (2019) Green synthesis and characterization of silver nanoparticles using extract of palm tree bark, palm leaf and ghaf leaf. Orient J Chem 35:547
12. Chou KW, Norli I, Anees A (2010) Evaluation of the effect of temperature, NaOH concentration and time on solubilization of palm oil mill effluent (POME) using response surface methodology (RSM). Biores Technol 101:8616–8622
13. Daniali O, Toghraie D, And SE-PASM, 2020 U (2020) Thermo-hydraulic and economic optimization of Iranol refinery oil heat exchanger with Copper oxide nanoparticles using MOMBO. Phys A Stat Mech Appl 540:123010
14. Duarte Ferreira Ribeiro C, Barbosa Schappo F, da Silva Sales I, Santos Assunção L, Murowaniecki Otero D, Teixeira Magalhães-Guedes K, Aparecida Souza Machado B, Mara

Block J, Izabel Druzian J, Larroza Nunes I (2022) Novel bioactive nanoparticles from crude palm oil and its fractions as foodstuff ingredients. Food Chem 373
15. Fadillah A, Hartoyo APP, Solikhin A (2020) Business model development of oil palm empty fruit bunch and trunk carbon nanofibers–based water purifier. In: IOP conference series: earth and environmental science. IOP Publishing, p 12018
16. Foo KY, Hameed BH (2011) Microwave-assisted preparation of oil palm fiber activated carbon for methylene blue adsorption. Chem Eng J 166:792–795
17. Frank E, Hermanutz F, Buchmeiser MR (2012) Carbon fibers: precursors, manufacturing, and properties. Macromol Mater Eng 297:493–501
18. Gan PP, Ng SH, Huang Y, Li SFY (2012) Green synthesis of gold nanoparticles using palm oil mill effluent (POME): a low-cost and eco-friendly viable approach. Biores Technol 113:132–135
19. George M, On PM-2016 IC, 2016 U (no date) Dielectric performance of solid dielectric immersed in vegetable oil with antioxidant
20. Hassan M, Abou-Zeid R, Hassan E, Berglund L, Aitomäki Y, Oksman K (2017) Membranes based on cellulose nanofibers and activated carbon for removal of Escherichia coli bacteria from water. Polymers 9:335
21. Hay WT, Fanta GF, Felker FC, Peterson SC, Skory CD, Hojilla-Evangelista MP, Biresaw G, Selling GW (2019) Emulsification properties of amylose-fatty sodium salt inclusion complexes. Food Hydrocolloids 90:490–499
22. Heng ZW, Chong WC, Pang YL, Sim LC, Koo CH (2021) Photocatalytic degradation of organic pollutants using green oil palm frond-derived carbon quantum dots/titanium dioxide as multifunctional photocatalysts under visible light radiation. Chin J Chem Eng
23. Herrera-Barros A, Tejada-Tovar C, Villabona-Ortíz A, Gonzalez-Delgado AD, Benitez-Monroy J (2020) Cd (II) and Ni (II) uptake by novel biosorbent prepared from oil palm residual biomass and Al_2O_3 nanoparticles. Sustain Chem Pharm 15:100216
24. Husen A (2019a) Natural product-based fabrication of zinc oxide nanoparticles and their application. In: Husen A, Iqbal M (eds) Nanomaterials and plant potential. Springer International Publishing AG, Cham, pp 193–291. https://doi.org/10.1007/978-3-030-05569-1_7
25. Husen A (2019b) Medicinal plant-product based fabrication nanoparticles (Au and Ag) and their anticancer effect. In: Kintzios SE, Barberaki M (eds) Flampouri Plants that fight cancer, 2nd edn. Taylor & Francis/CRC Press, pp 133–147
26. Husen A (2022) Engineered nanomaterials for sustainable agricultural production, soil improvement and stress management. Elsevier Inc., Cambridge, MA, USA
27. Husen A Iqbal M (2019a) N*anomaterials*and plant potential: an overview. In: Husen A, Iqbal M (eds) Nanomaterials and plant potential. Springer International Publishing AG, Cham, pp 3–29. https://doi.org/10.1007/978-3-030-05569-1_1
28. Husen A, Jawaid M (2020) Nanomaterials for agriculture and forestry applications. Elsevier Inc., Cambridge, MA, USA. https://doi.org/10.1016/C2018-0-02349-X
29. Husen A, Siddiqi KS (2014) Phytosynthesis of nanoparticles: concept, controversy and application. Nanosc Res Lett 9(229):1–24
30. Husen A, Siddiqi KS (2022) Advances in smart nanomaterials and their applications. Elsevier Inc., Cambridge, USA
31. Imoisili PE, Ukoba KO, Jen T-C (2020) Green technology extraction and characterisation of silica nanoparticles from palm kernel shell ash via sol–gel. J Market Res 9:307–313
32. Irfan M, Moniruzzaman M, Ahmad T, Mandal PC, Bhattacharjee S, Abdullah B (2017) Ionic liquid based extraction of flavonoids from Elaeis guineensis leaves and their applications for gold nanoparticles synthesis. J Mol Liq 241:270–278
33. Irfan M, Moniruzzaman M, Ahmad T, Osman OY, Mandal PC, Bhattacharjee S, Hussain M (2018) Stability, interparticle interactions and catalytic performance of gold nanoparticles synthesized through ionic liquid mediated oil palm leaves extract. J Environ Chem Eng 6:5024–5031
34. Javed M, Shaik AH, Khan TA, Imran M, Aziz A, Ansari AR, Chandan MR (2018) Synthesis of stable waste palm oil based CuO nanofluid for heat transfer applications. Heat Mass Transf 54:3739–3745

35. Jin-Chul K, Madhusudhan A, Husen A (2021) Smart nanomaterials in biomedical applications. Springer Nature Switzerland AG, Cham, Switzerland. https://doi.org/10.1007/978-3-030-84262-8
36. Joshi A, Sharma A, Bachheti RK, Husen A, Mishra VK (2019) Plant-mediated synthesis of copper oxide nanoparticles and their biological applications. In: Husen A, Iqbal M (eds) Nanomaterials and plant potential. Springer International Publishing AG, Cham, pp 221–237. https://doi.org/10.1007/978-3-030-05569-1_8
37. Khanna P, Kaur A, Goyal D (2019) Algae-based metallic nanoparticles: synthesis, characterization and applications. J Microbiol Methods 163:105656
38. Khanna V, Bakshi BR, Lee LJ (2008) Carbon nanofiber production: life cycle energy consumption and environmental impact. J Ind Ecol 12:394–410
39. Kumar H, Bhardwaj K, Dhanjal DS, Nepovimova E, Şen F, Regassa H, Singh R, Verma R, Kumar V, Kumar D (2020) Fruit extract mediated green synthesis of metallic nanoparticles: a new avenue in pomology applications. Int J Mol Sci 21:8458
40. Kumar H, Bhardwaj K, Kuča K, Kalia A, Nepovimova E, Verma R, Kumar D (2020) Flower-based green synthesis of metallic nanoparticles: applications beyond fragrance. Nanomaterials 10:766
41. Kumar A, Choudhary A, Kaur H, Guha S, Mehta S, Husen A (2022) Potential applications of engineered nanoparticles in plant disease management: a critical update. Chemosphere 295:133798. https://doi.org/10.1016/j.chemosphere.2022.133798
42. Kumar A, Choudhary A, Kaur H, Mehta S, Husen A (2021) Smart nanomaterial and nanocomposite with advanced agrochemical activities. Nanosc Res Lett 16(156):1–26. https://doi.org/10.1186/s11671-021-03612-0
43. Mondal R, Dam P, Chakraborty J, Paret ML, Katı A, Altuntas S, Sarkar R, Ghorai S, Mandal AK, Husen A (2022) Potential of nanobiosensor in sustainable agriculture: the state-of-art. Heliyon. https://doi.org/10.2139/ssrn.4134873
44. Nasir S, Hussein MZ, Yusof NA, Zainal Z (2017) Oil palm waste-based precursors as a renewable and economical carbon sources for the preparation of reduced graphene oxide from graphene oxide. Nanomaterials 7:182
45. Nuryawan A, Abdullah CK, Mohamad Hazwan C, Olaiya NG, Bashir Yahya E, Risnasari I, Masruchin N, Baharudin MS, Khalid H, S Abdul Khalil HP, Sultan Salahuddin Abdul Aziz Shah P, usahawan P, Alam S (2020) Enhancement of oil palm waste nanoparticles on the properties and characterization of hybrid plywood biocomposites. mdpi.com 12
46. Ramimoghadam D, Bin HMZ, Taufiq-Yap YH (2013) Synthesis and characterization of ZnO nanostructures using palm olein as biotemplate. Chem Cent J 7:1–10
47. Robaiah M, Rusop M, Abdullah S, Khusaimi Z, Azhan H, Asli NA (2017) Synthesis graphene layer at different waste cooking palm oil temperatures. In: AIP conference proceedings. AIP Publishing LLC, p 30008
48. Sadrolhosseini AR, Abdul Rashid S, Zakaria A (2017) Synthesis of gold nanoparticles dispersed in palm oil using laser ablation technique. J Nanomater
49. Sekeri SH, Ibrahim MNM, Umar K, Yaqoob AA, Azmi MN, Hussin MH, Othman MBH, Malik MFIA (2020) Preparation and characterization of nanosized lignin from oil palm (Elaeis guineensis) biomass as a novel emulsifying agent. Int J Biol Macromol 164:3114–3124
50. Sharma P, Pandey V, Sharma MMM, Patra A, Singh B, Mehta S, Husen A (2021) A review on biosensors and nanosensors application in agroecosystems. Nano Res Lett 16:136. https://doi.org/10.1186/s11671-021-03593-0
51. Siddiqi KS, Husen A (2017) Recent advances in plant-mediated engineered gold nanoparticles and their application in biological system. J Trace Elem Med Biol 40:10–23
52. Shen J, Zhu Y, Yang X, Li C (2012) Graphene quantum dots: emergent nanolights for bioimaging, sensors, catalysis and photovoltaic devices. Chem Commun 48:3686–3699
53. Sofiah AGN, Samykano M, Shahabuddin S, Kadirgama K, Pandey AK (2021) A comparative experimental study on the physical behavior of mono and hybrid RBD palm olein based nanofluids using CuO nanoparticles and PANI nanofibers. Int Commun Heat Mass Transf 120

54. Soni H, Pamidimukkala PS (2018) Green synthesis of N, S co-doped carbon quantum dots from triflic acid treated palm shell waste and their application in nitrophenol sensing. Mater Res Bull 108:250–254
55. Sudesh K, Bhubalan K, Chuah JA, Kek YK, Kamilah H, Sridewi N, Lee YF (2011) Synthesis of polyhydroxyalkanoate from palm oil and some new applications. Appl Microbiol Biotechnol 89:1373–1386
56. Sudhakar P, Soni H (2018) Catalytic reduction of Nitrophenols using silver nanoparticles-supported activated carbon derived from agro-waste. J Environ Chem Eng 6:28–36
57. Sun Y-P, Zhou B, Lin Y, Wang W, Fernando KAS, Pathak P, Meziani MJ, Harruff BA, Wang X, Wang H (2006) Quantum-sized carbon dots for bright and colorful photoluminescence. J Am Chem Soc 128:7756–7757
58. Tahir NAM, Bin AMF, Tamaldin N, Zin MRBM, Amiruddin H (2020) Optimisation of graphene grown from solid waste using CVD method. Int J Adv Manuf Technol 106:211–218
59. Umar MS, Jennings P, Urmee T (2014) Generating renewable energy from oil palm biomass in Malaysia: the feed-in tariff policy framework. Biomass Bioenerg 62:37–46
60. Usman AI, Aziz AA, Noqta OA (2019) Green sonochemical synthesis of gold nanoparticles using palm oil leaves extracts. Mater Today: Proc 7:803–807
61. Zafar S, Zafar A (2019) Biosynthesis and characterization of silver nanoparticles using Phoenix dactylifera fruits extract and their in vitro antimicrobial and cytotoxic effects. Open Biotechnol J 13:37–46
62. Zakaria NZJ, Rozali S, Mubarak NM, Ibrahim S (2022) A review of the recent trend in the synthesis of carbon nanomaterials derived from oil palm by-product materials. Biomass Conv Biorefinery 1–32
63. Zevallos Torres LA, Woiciechowski AL, de Andrade O, Tanobe V, Zandoná Filho A, Alves de Freitas R, Noseda MD, Saito Szameitat E, Faulds C, Coutinho P, Bertrand E, Soccol CR (2021) Lignin from oil palm empty fruit bunches: characterization, biological activities and application in green synthesis of silver nanoparticles. Int J Biol Macromol 167:1499–1507

Rice Straw Waste Utilization for Nanoparticles Synthesis and Their Various Applications

Daljeet Kaur, Amarjit Singh, Sunita Dalal, and Jitender Sharma

Abstract Massive open combustion of lignocellulosic agro-wastes has become a worldwide concern due to its direct connotation with the emission of noxious gases, pollution and severe ecosystem hazards. The ecological unbalancing necessitates the execution of sustainable management practices through innovation in technology to overcome the ecological footprint. The infield burning of rice straw is increasing substantially in Asian countries despite strict administrative legislations against this practice. Rice straw is one of the most economical and easily accessible raw materials and gaining acceptance for the formulation of nanoparticles. It is widely explored to manufacture the versatile nanostructures at low cost. Rice straw has been exploited to prepare nanomaterials which have wide range of applicability in pharmaceutical and biomedical sciences. The stubble is a significant organic source of silica which can be extracted from the crop and converted to biogenic nanosilica particles carrying the potential for heavy metals adsorption and pollution abatement. This waste can be utilized as a substrate and reduced to generate other nanoparticles like carbon-ZnO, rice straw/Ag and rice straw/Fe_2O_3, etc. The synthesis of economically viable nanoscale polymers for the next generation, for example, cellulose nanocrystals (CNC) from rice straw has significant potential in polymer, plastic and packaging industry. Rice straw is a suitable raw material for production of bionanocomposites consisting of biopolymer matrix armoured with nanoparticles of organic origin. These green rice straw-based bionanocomposites exhibit superior characteristics than biopolymers alone in terms of surface area, high aspect ratio, barrier and strength properties. This chapter elaborate the utilization of rice straw for nanoparticle synthesis, the techniques used to formulate the nanoparticles and various sectors describing the product utility at laboratory and commercial scale to attain a sustainable, green and ecofriendly future.

D. Kaur · S. Dalal · J. Sharma
Department of Biotechnology, Kurukshetra University, Kurukshetra, Haryana, India

A. Singh (✉)
Department of Botany, Guru Nanak Khalsa College, Yamuna Nagar, Haryana, India
e-mail: drajsgnkc@gmail.com

A. Husen (ed.), *Nanomaterials from Agricultural and Horticultural Products*,
Smart Nanomaterials Technology,
https://doi.org/10.1007/978-981-99-3435-5_7

Keywords Rice straw · Nanoparticles · Silica · Nanocellulose · Bionanocomposites

1 Introduction

Agriculture is one of the crucial industries that extracts its economy from different cultivated crops. Agriculture and farming activities release huge amounts of lignocellulosic-based waste biomass which acts as potential raw material for different industrial entities. Cereals (rice and wheat) are the main grain crops grown in India and contribute more than half of the total agricultural residues generated. Rice is a chief food utilized by about half the world's inhabitants, which makes it the second largest crop produced over the globe. Rice straw, rice husks and wheat straw are the major by-products of the agricultural industries and are considered as agrowastes. Farmers burn rice waste onsite due to mechanization, reduction in live-stock, time scarcity for next crop plantation, decomposition and inaccessibility of viable management strategies. This leads to significant environmental problems like pollution, deprivation of soil nutrients and health risks. It degrades the air quality by releasing harmful pollutants and gases such as oxides of carbon, nitrogen and sulphur, ammonia, methane, volatile organic matter, non-methylated hydrocarbons and particulate matter that exaggerate the respiratory and allergic diseases. It also affects visibility that may lead to road accidents. The vast cloud of smoke in the form of smog resulting from rice residues burning not only causes serious threats to human health, but also degrades soil and environmental quality.

Rice straw is a valuable agrowaste that mainly consists of polysaccharides, lignin and ash (mostly silica). This can be considered as a source for the production of new products which can upgrade its applicability in different sectors. The fibrous part of rice straw is essentially important for various value-added products like biopolymers, ethanol, oil production, biogas generation, etc. Rice straw contains ash concentrations of about $12.6 \pm 0.11\%$ of which 11.7% is silica. The content of silica collected in rice straw is greater than other agro residues, softwoods and hardwoods. Its ash contains more than 80% of silica that can be extracted from the ash for some potential use [47]. Rice straw is also widening its applicability in new research areas as nanotechnology [87], pollution abatement [74, 86, 88] and production of green chemicals [104] being biodegradable and organic in nature. Nanotechnology has arisen as a foremost field of science having remarkable utility in varied disciplines such as biotechnology, pharmaceutical, medical, environmental security and agriculture [31–35, 40, 53, 54, 64, 83]. The widespread applicability of the technology corresponds to the distinct physiochemical characteristics of nanoparticles specifically the large surface area to volume ratio which provides a gigantic reactive interface between nanoparticle and its environment. The rice straw has been explored for the production of graphene oxide, carbon nanotubes (CNTs) and carbon dots which have been described in the review. Silica extracted from rice straw if accessible in large quantities can be considered as a cost-effective substrate for synthesis of non-toxic and biocompatible nanoparticles which is the need of the hour for developing medical and pharmaceutical sciences. Other than silica nanoparticles, the literature supports the use of rice straw as substrate for synthesis of silver, zinc, iron and potassium nanoparticles. A vast number of technologies have been described in literature for synthesizing the nanocrystalline cellulose, but the most supported in literature is the acid hydrolysis [85, 108, 109]. The chapter includes a detailed study of raw straw production, its availability, characteristics for nanomaterial synthesis and potential uses of rice straw-based nanomaterials for commercial, economic and ecological benefits.

2 Rice Straw Scenario

Rice grain is another nutritional crop after corn which is used considerably to meet dietary needs, globally. The worldwide rice grain production in year 2021–22 was approximately 509.9 MT which was a bit lower than its consumption (510.3 MT) for the same session [18]. Asian countries contribute potentially to the total world's rice harvest. China is the highest rice producer in the world with a production of 211.86 million metric tonnes of rice in 2020 followed by India and Bangladesh [101]. In the United States, some provinces like Arkansas, California and Louisiana also stand in the list of top global rice producers. Kaur et al. [45] reported the generation of 731 million tons of rice straw during rice harvesting worldwide, in which only India released about 126.6 million tons. Literature found that for every 1 kg of rice grain production approximately 0.41–4.3 kg of straw is heaped [47, 58] (Van Hung et al. 2016). The grain/straw ratio investigated at International Rice Research Institute was found to be 7.5–8.0 tonne/ha. The climatic and topographic factors such as seed variability, soil integrity, rainfall and humidity, etc., greatly influence the quantity of rice straw generation. A small portion of rice straw is utilized for various applications while a huge percentage remains in-field named stubble. The uncut rice plant portion is usually burnt for fast disposal or integrated into the soil to maintain nutrient levels. This release of rice straw is problematic to handle and to manage for farmers, governments and environmental planners.

3 Characteristics Account

Rice straw exhibits chemical composition favourable for its transformation into sustainable, ecofriendly and biodegradable valuable products. Applications of rice straw are highly based on its morphological, chemical and thermal properties. A study by Chen et al. [14] through X-ray diffraction depicted the cellulosic-hemicellulosic manacles in cell walls of rice straw which were coupled with H-bonds. The fibre analysis of rice straw revealed the presence of fibres with length and width ranges from 0.5 to 3.0 mm and 7 to 14 mm, respectively [29, 76]. The particle dimension of rice straw is considered as a prime quality for its application because it strongly affects its flowability, thermal extension, diffusion, and reaction rate [50, 109]. Literature reported that the bulk density of rice straw ranges between 13 and 120 kg/m^{-3} [59, 67]. The thermal characteristics of rice straw include its heating value which ranges between 14.1 to 15.1 $MJKg^{-1}$. The high percentage of volatile matter in rice straw allows its rapid combustion and makes it suitable for energy production. The chemical analysis of rice straw showed that it contains polysaccharides interlinked with lignin, lipids, proteins, sugars, starch, water, hydrocarbons and ash. Kaur et al. [46] carried out the proximate analysis of rice straw and reported its composition as cellulose (33%), hemicellulose (27%), lignin (13%) and ash (12.6%) out of which major weight is of silica.

Cellulose is the most profuse natural component and contains linear chain of 4-O-β-D-glucopyranosyl-D-glucose and has a marked place in nanotechnology. Due to high crystallinity and polymerization of this component, the dissolution of rice straw into water becomes difficult [61, 62]. The second polymeric polysaccharide is hemicellulose and it exhibits low polymerization and poor crystallinity. It can be easily transformed into its monosaccharaides. Lignin is an important naturally occurring aromatic polyphenol (phenypropane) which constitutes sinapyl alcohol, coniferyl alcohol and p-coumaryl alcohol united with ester and carbon–carbon bonds. Such bonding makes a complicated web, which checks the enzymatic hydrolysis of polysaccharides [3, 70]. Silica in rice straw is vital for synthesis of carbohydrate, grain yield, synthesis of phenolic components and cell wall security. The silica can be extracted and used to formulate nanoparticles carrying potential applications.

4 Complications in Rice Straw Management

Rice straw has compact and tough morphology. The presence of silica and lignin is a prime limiting factor for extracting the benefits from rice straw. Due to high silica in cell walls of rice straw, animals are unable to digest it. Other than silica, lignification, silicification and high crystallinity of cellulose and other associated factors like lignocellulose complexes, etc. are the reasons for poor digestibility of straw [23]. Lignin is regarded as the most plentiful aromatic polymer that imparts strength and protection to rice straw from insects and mammals. Lignin inhibits the plant decay caused due to microbes. Silica prevents the colonization of microorganisms present in rumen, which affects the palatability and degradability of rice straw [2]. Therefore, the high level of lignification and silicification slow down the ruminal degradation of the carbohydrates. Reena and Patir [73] found the low nitrogen value in rice straw which decreases the nutritional value of ruminants feed. Rice straw lacks enough sugars, proteins and minerals which support microbial growth and that's why supplements are added to rice straw feed to enrich it with nutrient sources to enhance the performance of cattle. Due to these limitations and poor know-how about the rice straw potential, it remained unprocessed but, its composition can support its application in synthesizing the nanosized materials.

5 Prospects in Nanotechnology

Nanotechnology is a multidimensional arena, merging the basic sciences to their functional disciplines such as physics, biology, chemistry, mathematics, etc., with engineering, biotechnology, microbiology and environmental studies. The interdisciplinary nature of nanotechnology has emerged as a novel, sustainable and environmental friendly solution to global issues with collective efforts coming from experts of various disciplines. Nanotechnology has the efficiency to comprehend, influence and regulate matter at specific atoms and molecules level. The technique is used to construct nanostructures from bottom-up, with the help of scientific tools and techniques which impart high performance in different sectors. Due to wide applications, this branch of science in different disciplines such as medicines, drugs, as catalysts, energy generation, pollution abatement and resource management; it is designated as a revolutionary discipline. The nanomaterials exhibit advantages in terms of high surface area, solubility and dissolution rate, bioavailability, fast action mode and great results even at lower concentration of application. Despite of uncountable benefits of nanosynthetic materials, the extraction, purification and scaling up of the techniques and products used is still difficult.

Green nanotechnology is one of the important aspects of nanotechnology as it aims at reducing the environmental risks and promoting the sustainability of processes by manufacturing ecofriendly nanoproducts. Nanoparticles generated from green nanotechnical aspects are either based on plant extracts or on waste reuse and has advantage in terms of simplicity of the process, convenience, accessibility and process time. In the agricultural sector, the green nanotechnology has attained great importance in the synthesis of nano-fertilizers and nano-pesticides with efficient management of plant pathogens restrict use of agrochemicals and nutrient application [71]. The high surface area of nanomaterials makes them suitable candidates to solve problems, which are not possible by conventional physical, chemical and biological methods. The most advantageous revolution is the utilization of waste materials (agro wastes), for example, use of rice straw in the synthesis of carbon-based nanomaterials silica-nanoparticles, nanocellulose and other nanopolymers. Different studies have reported the application of nanotechnology in environmental conservation [24], nanoadsorbent [42], drafting supercapacitors [11], nanocellulose [43], nanosilica [89] and nanobiochar [38]. The agro-wastes must be treated effectively considering the principle of 3Rs (i.e. reduce, reuse and recycle), to reduce the emissions, consumption and achieve high efficiency of nanomaterials [80]. This review is presenting the effective contributions of nanotechnology for extracting the potential of rice straw in synthesis on nanoparticles for achieving a sustainable environment.

6 Graphene Sheets

Graphene is a two-dimensional copious carbon sheet in which the carbon atoms are crammed like a compact honeycomb web and provides a base for synthesis of numerous nanocarbons. The sheets are stacked over one another and structured to form three-dimensional units. These three-dimensional graphite structures are rolled along or enfolded to formulate carbon nanotubes and fullerenes, respectively [69]. The graphene sheets possess significant characteristics such as inertness, superhydrophobicity, high polarity and large carrier agility, etc., as shown in Fig. 1 [97]. These distinct properties make graphene an attractive choice in advanced research and technology [56]. Gadakh et al. [21] in their study reported that the production of graphene sheets is more economical than the formulation of other types of nanomaterials and occupy a key role in nano-science with varied applications. The functionalized (graphene oxide and its reduced form) and non-functionalized graphene (single-layer graphene and sheets) are the two major categories of graphene. The comparison between the properties of graphene, its oxide and reduced form has been depicted in Table 1.

Various studies reported the successful formulation of graphene using top–down and bottom–up techniques. The top-down techniques involve mechanical treatments, splitting of graphite racks, epitaxial growth on substrate and graphene oxide reduction

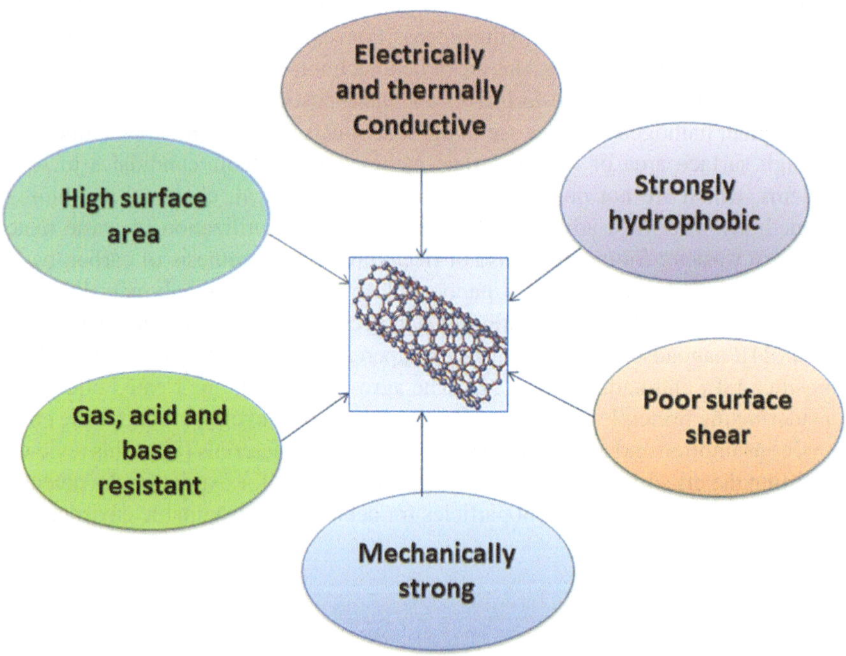

Fig. 1 Characteristics of graphene nanoparticles prepared from rice straw

Table 1 Comparison of graphene, graphene oxide and reduced graphene oxide

	Graphene	Graphene oxide	Reduced graphene oxide
Structure	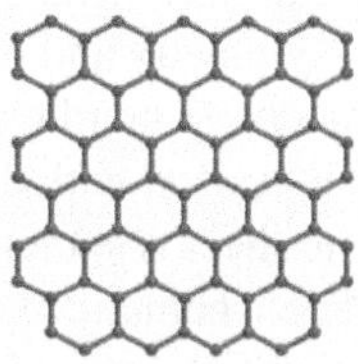	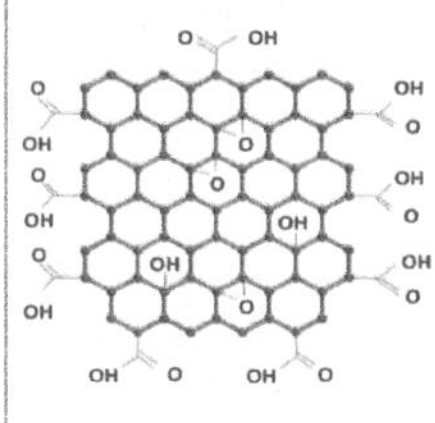	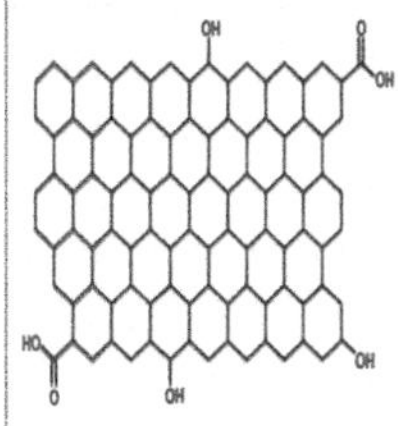
Formulation technique	CVD, thermal decomposition and graphite racks exfoliation	Graphite racks exfoliation/oxidation	Reducing the graphene oxide
Carbon: oxygen	Nil	Poor (2–4)	Wide range (8–250)
Young modulus of elasticity	1.0	0.25	0.25
Electron mobility	Highest (upto 50,000)	Insulator	Low (max. 200)
Capital cost	High	Low	Low

[47, 48]. But the most adapted method is bottom-up via chemical vapour deposition (CVD) due to its high scalability and commercial production [39].

Uda et al. [96] successfully created the graphene from rice straw ash using alkaline treatment in a muffle furnace at 700 °C followed by subsequent washing and drying. The study revealed the formulation of graphene sheets of size carrying up to 2000 nm. The graphene sheets were further utilized for creating multi-walled carbon nanotubes as described in the next section. The literature for the production of graphene from agricultural residues is scarce and needs to be explored thoroughly to achieve a sustainable progress in the field of green technology. Moreover, both the approaches for formulating graphene are time intensive and may cause an environmental hazard that's why this sector is area of research to be focussed on.

7 Carbon Nanotubes

Carbon nanotubes (CNTs) are the most prominent, one-dimensional carbon with distinctive structure which provides them super electronic, opto-electronic, physical and chemical characteristics. CNTs are the hexagonal arrangements of carbon atoms with a diameter 1 nm and length between 1 and 100 nm. The CNTs are gaining popularity due to their discrete properties like high electrical conductivity, stability, broad surface area, mechanical strength, optically sound, thermal resistance, electrochemical strength, biocompatibility, non-immunogenicity, non-toxicity and biodegradability [100]. These CNTs are designated to be semiconducting or metallic in nature. These are the biopolymers of pure carbon or can be extracted and altered by exploring and predicting the exceptionally ironic chemistry of carbon. These nanotubes may be of single layer or multilayer carrying a unique size, symmetry and properties [90]. The single-walled CNTs (SWCNTs) are comprised of a particular layer of graphene sheet whereas multi-walled CNTs (MWCNTs) are synthesized with different layers of graphene sheets over one other. The single graphene sheet is sometimes wrapped manifold in a way to generate multiple layered CNT. The schematic representation of sequential processes during CNT formulation is given in Fig. 2.

The formulation of CNT is known to process by techniques like chemical vapour deposition (CVD), laser ablation and arc discharge. The CNT formulation is greatly influenced by prime process factors like carbon feedstock, catalysts, substrate media and energy input. The gaseous and liquid hydrocarbons such as methane, ethylene, acetylene, alcohols and other solvents are comprehensively explored carbonaceous materials for CNTs synthesis. Besides organic materials, inorganic components like graphite, quartz, silicon carbide, silica, alumina, zeolite, calcium carbonate and magnesium oxide also have a great affinity for different catalysts in their mono/binary/tertiary state and are found to be supportive in carbon nanotubes synthesis. CNTs have a wide range of applicability as thermal conductors, insulating materials, electronic components, low-power electronic devices, fibres, fabrics, biomedical appliances, pollution abatement, ultrafiltration, hydrophobic papers and support material for different catalysts which has been explained in the last section of review (Fig. 2).

The demand for CNTs has enhanced greatly in various industrial, medical and environmental safety sectors which escalated the research thrust for establishing an advanced, sustainable and ecofriendly technology to scale up CNTs. The production of CNTs from waste materials is one of the most significant achievements in this sector which focussed on the use of renewable, eco-friendly biological materials for example camphor, petroleum wastes, de-oiled asphalt and tree oil (*Eucalyptus*

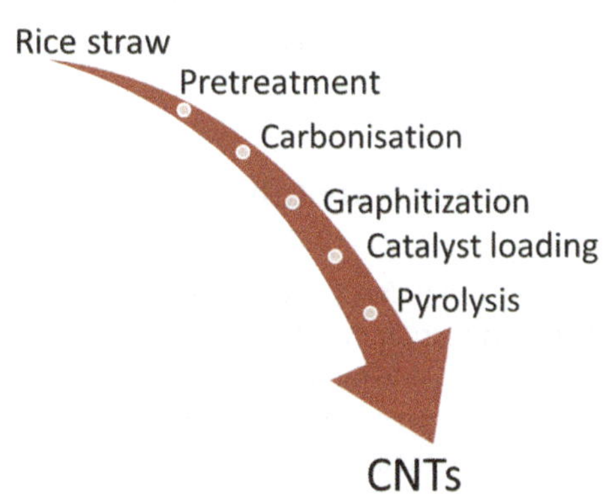

Fig. 2 Schematic presentation of CNTs production from rice straw

species), various vegetable and cooking oils as the starting materials for CNTs formulation. Only a few studies have reported the successful preparation of multi-walled nanotubes using rice straw via CVD process. There are numerous publications on CNTs from waste materials but only two studies by Fathy [19] and Lotfy et al. [60] were specifically oriented to produce MWCNTs from rice straw via CVD process. The general description of the formulation of CNTs from rice straw is represented in Fig. 3.

Fathy [19] used the hydrochar prepared from hydrothermally pretreated rice straw to support iron catalyst and a complex, RS-H/Fe was obtained. The RS-H/Fe was treated with ferrocene and nickel nitrate to produce a complex of rice straw hydrochar and catalyst. A CVD apparatus was used which was consisting of two separated chambers of furnaces and a stainless steel rod placed horizontally within the furnaces. The two alumina boats (one loading the camphor and other containing the RS-H/Fe–Ni) were used. The temperature profile was set from minimum room temperature to a maximum of 800 °C. The camphor was burned at 250 °C to get gaseous carbon substrate and passed to another furnace carrying catalyst-loaded rice straw under N_2 stream at a high temperature (800 °C) to formulate CNTs–Fe and CNTs–Fe–Ni. In another study, Fathy et al. [20] investigated the potential of two

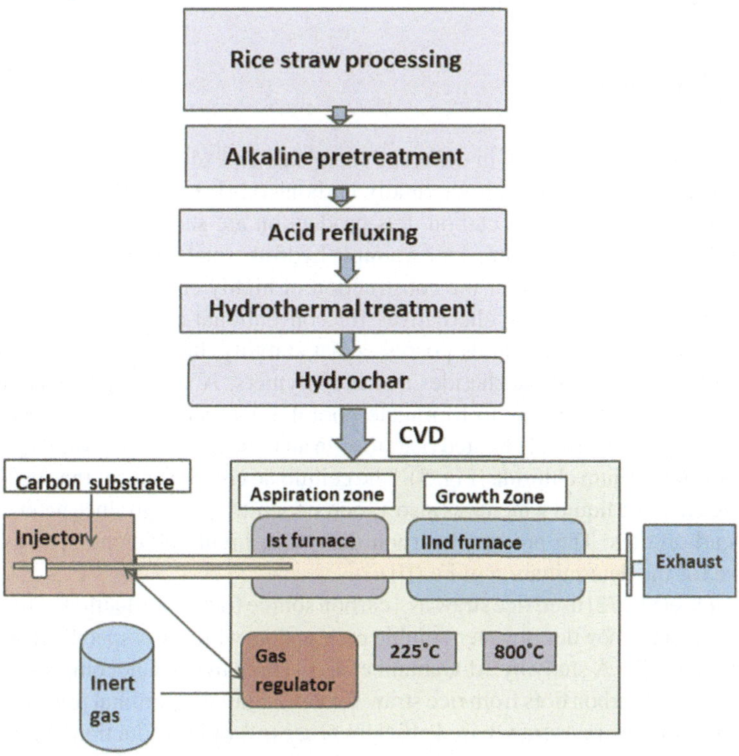

Fig. 3 Rice straw processing and CNTs formulation

different carbon supports (produced from rice straw and carbon-based xerogels) for formulating the MWCNTs via CVD process. The MWCNTs produced from rice straw yielded considerably with high characteristics like surface area and iodine number than carbon-based xerogels.

Another study by Lotfy et al. [60] depicted a cost-effective and environment-friendly method to formulate MWCNTs. The study reported that efficient formulation of CNTs varied in size and shape using a tube-shaped reactor and found the process more advantageous than CVD. The delignification of rice straw was carried out to produce three different types of pulp viz. alkaline, sulphite and neutral. The prepared pulps were used as substrates for CNTs production for which they were treated hydrothermally followed by pyrolysis in tubular steel reactor. The hydrothermal treatment involved the loading of various catalysts (nickel, iron and cobalt) on carbon sources by keeping the contents in stainless steel reactor which was subjected to heat in a muffle furnace at 180 °C. The produced hydrochar was washed thoroughly and dried overnight. The hydrochar was mixed with a catalyst, Fe-Ni oxides supported on Al_2O_3 in ratio 10:1 in the reactor under nitrogen stream. The reactor was kept in a furnace, which was heated gradually to the highest temperature of 830 °C and placed for 1 h at maximum temperature. The black colour product was washed with acids followed by hot water and dried subsequently.

8 Carbon Dots

The dissolution of rice straw in different chemicals and solvents is difficult which reduces the application of hydrothermally carbonized rice straw ash directly [68]. The sequential processes for carbon dots production are shown in Fig. 4. Jorn-am et al. [41] processed the rice straw via a simple hydrothermal technique to synthesize electrolyte and carbon dots for the construction of highly efficient supercapacitors. Ionic liquids are ecofriendly alternatives for conventional chemicals to solubilize the rice straw ash. These liquids possess great stability, high recycle rate and the potential to dissolve polysaccharides and biopolymers. A study by Sun et al. [91] reported an exceptional way to formulate photoluminescent carbon dots by using microwaved hydrothermally treated rice straw in an ionic liquid with formula, 1-allyl-3-methylimidazolium chloride (Fig. 5). The cellulosic components of the straw were solublized in ionic liquid which was also a source for nitrogen to produce heteroatom-doped carbon dots. The produced carbon dots were highly efficient, selective and sensitive for the determination of Fe (III).

Puvvada et al. [72] used rice straw as a carbon source for the production of luminescent, stable black powder of water-soluble carbon dots which was specific to analyse the bacterial cells. A study by Al-Qahtani et al. [4] prepared a nanocomposite called nitrogen-doped carbon dots from rice straw for anti-simulating ordinal printing. They formulated an easy rice straw-based efficient process for the production of luminous carbon dots using single vessel hydrothermal carbonization in NH_4OH medium. A similar study was also conducted by Alshareef et al. [6] for the successful preparation

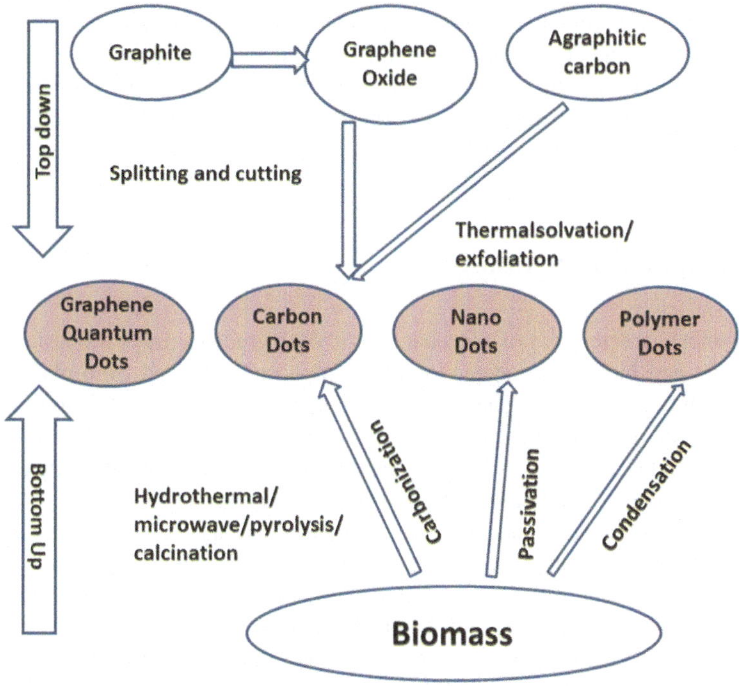

Fig. 4 General processes for the production of Carbon dots

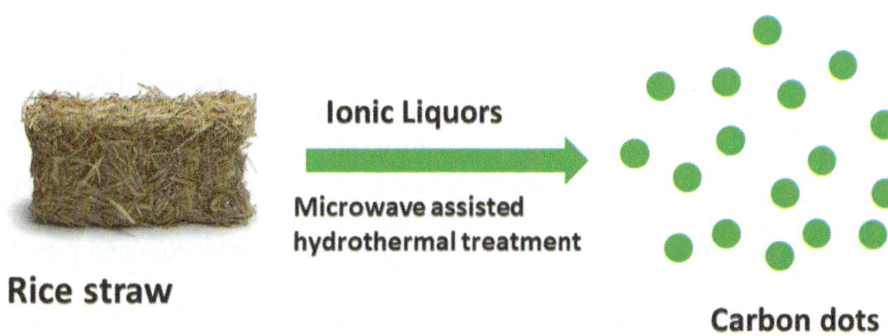

Fig. 5 Production of carbon dots from rice straw

of nitrogen-carbon dots from rice straw. Sun et al. [92] in their study reported utilization of chemically ($MgCl_2$) modified rice straw for the formulation of biomass-based carbon dots nanoenzyme showing peroxidase activity.

The study reported the use of carbon dots as markers for detecting bacteria and counting their number in any sort of environment. Brown coloured solid carbon dots with excellent photo-stability from rice straw were prepared by Thongsai et al. [93] using hydrothermal treatment. The synthesized carbon dots had significant sensitivity for alcoholic components at room temperature. The carbon dots were used as a sensing probe in the electronic nose to differentiate between alcoholic components and other volatile organic compounds. Kumari et al. [55] developed carbon dots for fabricating an ultra-sensitive incandescent marker for the detection of ellagic acid (EA).

9 Nanoparticles Synthesis

The distinct characteristics of silica nanoparticles upgrade its usage for designing the biosensors and biomarkers, detection of microparticles derived from platelets and cancer cells identification. The use of silica nanoparticles with superhydrophobic polymers will promote the plastic reduction. The silica nanoparticles exhibit the nucleation and pozzolanic properties which drop the setting time and extenuate the calcium percolation from cement-based products. The exclusive properties of silica nanoparticles upgrade its catalytic and photolytic activity for extracting the pesticides, heavy metals from effluent and colour reduction. The high porosity of silica nanomaterials allows imbedded particles into their nano matrix to retain the novel characteristics of primary molecules and add on the optical and magnetic properties to the material. Silica nanomaterials are non-biorefractory and highly compatible which enhances their development in field of drug and clinical sciences. Shen et al. [84] reported the chemical route of nanosilica using tetra-apoxysilanes for refining the vegetable oil, medicines, surfactants, epoxy resin, porcelains, and pesticides [84]. The utilization of agro-waste-derived silica nanoparticles for removing cationic dyes is a natural and promising route for treating the toxic effluents and an admirable alternative to cut down the chemicals for degradation of organic pollutants. The methods of extraction, purification and conversion of silica present in the earth are highly complex and complicated. The silica is usually yielded in the form of silicates which may be in different states like colloidal, fumed, fused, grounded, gel or precipitated silica. The extraction of silica from quartz sand and its conversion to amorphous silica is an energy-intensive process and involves high temperature and pressure conditions.

Agricultural waste is one of the cheapest and easily available materials for synthesis of nanoparticles and hence, plays a vital role in advancing the application of nanotechonology for sustainable future. Different methods have been reported in literature to prepare the nanoparticles from agro residues [5, 52]. The major disadvantage of rice straw is its high ash and silica content which limits its application for papermaking, biopolymer synthesis and energy production. The black liquor produced after cooking the fibres contains substantial lignin and silica which can be extracted by hydrolysis and separated for converting to high-quality products [63]. The extraction of silica and its conversion to nano-silica particles will be a win–win situation for 100% utilization of rice straw. A number of processes have been reported for the construction of silica nanoparticles as sol-gel, dissolution–precipitation, emulsification, pyrolysis, ball mulling and biological synthesis. A detailed

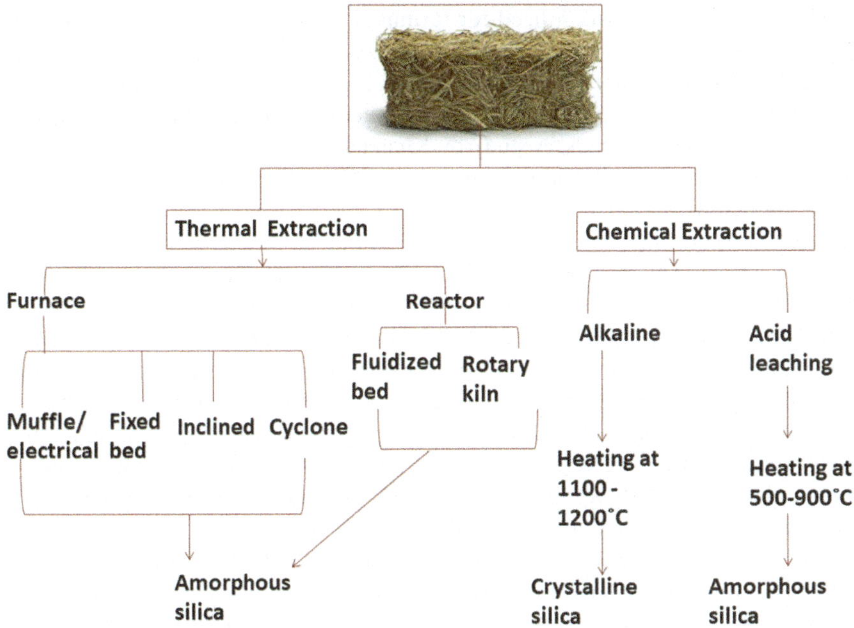

Fig. 6 Flowchart representing the extraction of silica from rice straw

description for the extraction of silica from rice straw is given in Fig. 6. Sol–gel process or the "Stöber method" is generally preferred due its cost-effectiveness [13].

Khorsand et al. [51] prepared silica nanoparticles from rice straw using dissolution–precipitation process and optimized the various process parameters for product success. Wibowo et al. [102] in their study investigated the sol-gel technique to synthesis the silica nanoparticles from rice straw for application as biofertilizers. The method involves the silica removal from rice straw using potassium hydroxide followed by acid treatment of extracted silica-forming nanoparticles. A clear silica gel from rice straw was obtained by El Sayed and El Gamal [79] for preparing the activated carbon for the effective removal of chromium from wastewater. Nandiyanto et al. [66] formulated the silica nanoparticles from rice straw using alkali dissolution method followed by acid precipitation at mild process conditions (atmospheric temperature, low reaction time and concentration). Robels-Jimarez et al. [74] synthesized a silica-based adsorbent from rice straw to reduce the nitrate concentration in water bodies. The authors also demonstrated the successful pilot and lab-scale trials from silica nanoparticles. Singh et al. [88] established a method to produce biogenic nanosilica particles from rice straw and utilized them for the degradation of dyes. Grimm et al. [22] used an approach to synthesize the biogenic silica from rice straw by its calcination at a temperature of 600 °C, which separated silica from its organic matrix.

The rice straw was also utilized for formulation of biocarbon and its activation using different nanoparticles. The rice straw is first carbonized followed by its activation with $ZnCl_2$ to prepare biocarbon nanoparticles [1]. The carbon–metal nanoparticles have high range of applicability in wastewater treatment, pesticides elimination and removal of heavy metals. A study by Khandanlou et al. [49] used rice straw for the first time to support the silver nanoparticles. For its preparation, the rice straw dipped in distilled water was mixed with urea followed by addition of silver with vigorous stirring. The contents were then heated at 70 °C and subsequent addition of alkali followed by continuous mixing for 1 h. The pH was maintained between 8.0 and 9.0 throughout the reactions. The product containing the rice straw/Ag was centrifuged and thoroughly washed with distilled water and solvent to remove silver ion residues and further dried at 60 °C. Li et al. [57] also utilized the rice straw biomass to synthesize rice straw/Ag nanoparticles from silver nitrate using light irradiation process at room temperature. The study focussed on the impact of light, time, rice straw amount and $AgNO_3$ concentration on the yield and quality of rice straw/Ag nanoparticles. The study resulted in a successful formulation of silver nanoparticles under different light conditions using rice straw biomass.

The rice straw ash was also explored to produce RS-KOH nanoparticles where both rice straw and KOH were mixed at 900 rpm and heated for 2 h at 338 K. A specially designed flame-assisted apparatus for spray-pyrolysis equipped with commercial LPG was used and to process the mixtures and synthesizing the RS-KOH nanoparticles [66]. Amirsoleimani and Ghorbani [7] manufactured the RS-ZnO nanoparticles using rice straw and zinc nitrate. During the process, zinc nitrate was added to rice straw extract at 80°C and yellow colour paste was formed. The paste was cooled at room temperature, dried and calcinated at 400 °C in muffle furnace for 2 h. Sangon et al. [78] developed mesoporous carbon-ZnO nanoparticles using rice straw and utilized the synthesized product as photo-catalytic adsorbents for the extraction of dyes (cationic and anionic) from wastewater. Areeshi [8] used the rice straw solution for formulating the iron oxide nanoparticles in an economical manner and utilized them for enhancing the thermal stability of endoglucanase.

The chemical processes for formulating the nanoparticles are neither economical nor environment-friendly, therefore, there is an urgent need for green technologies for a sustainable environment. The use of ionic liquid used for nanosilica production from agricultural waste is a green approach [14] but not well documented. A new area of research is emerging to investigate the potential of rice straw in nanotechnology by using ionic liquid and eutectic solvents for nanosilica production.

10 Nanoscale Biopolymers

Biopolymers are the most innovative and ecofriendly stuff that are potentially growing their market in the plastic and polymer industry. Bioplastics comprise a 1% share of the total plastic generation in a year. Bioplastics are becoming competitive in the market of plastic due to its biodegradability, low human health impact, good aroma barrier and molding strength. The major challenge for bioplastic production is the raw materials, which today are generally food crops that may compete with food security to fulfil the demand. The production of biopolymers from agricultural wastes (cereal straws, kenaf, bagasse and other non-woods) provides a good option as it does not have any pressure to the food resources. Such bioplastics are called the second-generation bioplastics. Cellulose is the most abundant and fascinating biopolymer derived from plants, animals and microorganisms due to its renewability, biocompatibility, biodegradability, high resistance, stiffness and recyclability. The derivatized monomers of cellulose are either applicable directly or polymerized for the production of novel polymers and biocomposites. The cellulose can be reinforced in the polymer matrix as unmodified fibres, cellulosic derivatives or the nanoscale fibres. The nanocellulose can be prepared or extracted from rice straw using different treatments such as pulping, bleaching, enzymatic treatment and acid hydrolysis. One of the easiest routes to convert rice straw into CNC is presented in Fig. 7.

Nanoscale cellulosic materials, cellulose nanocrystals, cellulose nanofibres and bacterial nanocellulose have significant potential for producing high-strength bionanocomposites for the next generation. The studies revealed that use of cellulose nanofiller in the renewable polymer matrix produce totally renewable bioplastics [10]. The polylectic acid (PLA) is an extensively explored biopolymer from the starch and cellulose-based raw materials. But, its biodegradability in the composting system is still under a question mark. The polyhydroxyalkanoates (PHA), a biodegradable

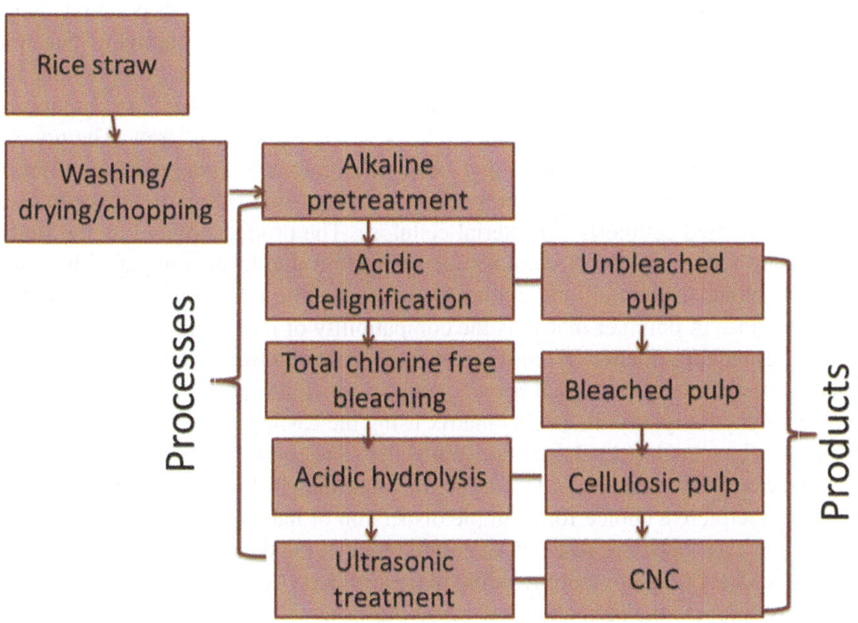

Fig. 7 Processes involved in the conversion of rice straw to CNC

polymer, has been put into the category of green product due to its production from the renewable resources with the involvement of micro-organisms. The major drawback of these two biopolymers, PLA and PHA, is their high production cost which limits its substitution to conventional plastics on a commercial scale.

Many studies have concentrated their efforts on reducing the material cost by using the waste as a potential source for biopolymer production [17, 98, 103]. Hu et al. [28] isolated the short fibres, fines and lignin from the aqueous suspension of rice straw and casted it into the nanocellulosic films to impart high thermal and water barrier properties. Rosa et al. [77] prepared pure cellulose whiskers of high crystallinity using a series of bleaching stages with environmentally friendly chemicals (hydrogen peroxide, ionic liquors and mixture of acids) and subsequently hydrolyzing with sulfuric acid to produce nanocellulose whiskers. The high-intensity ultrasonication method was also investigated to synthesize the morphologically uniform nanocellulosic fibres from rice straw [16]. A study by Yan et al. [107] described that pretreating the straw with steam explosion and its enzymatic treatment would be helpful in the production of nanofibrillated cellulose with high moisture absorption capacity.

The nanofibres isolated from rice straw were mixed with other biopolymers like cellulose acetate/chitosan/PHA/PLA to synthesize bionanocomposites with higher strength, enhanced thermal and water barrier properties [5, 25, 26, 106]. The prepared nanofibres, nanosilica, inorganic nanoparticles and bionanocomposites have a wide range of applications in food products, medicines, enzyme immobility, antimicrobial materials, plastic waste reduction and pollution abatement.

Bilo et al. [12] successfully prepared the biopolymer from rice straw with properties comparable to those of polystyrene, in its dry state and with polyvinylchloride in wet conditions. Rice straw can be efficiently used for isolation of bacteria that produce PHA and further as a carbon source for synthesizing biopolymer by isolated microorganism [94]. Huy and Khue [36] derived *Lactobacillus rhamnosus* from plants and utilized it to produce polylactic acid from rice straw. The review by Shaghaleh et al. [81] reported that fully renewable and biodegradable polymer composite can be prepared with biopolymers having low loading of nanocellulose, surface-modified cellulose or bacterial cellulose. The product formed by the study was found to be capable of commercial application due to biocompatability, efficient rheological, thermal and strength properties. The major limitation of using nanocellulose in polymer matrix is the compatibility of nanomaterial with polymer matrix, especially in case of reinforcement of nanocellulose in the microbial PHA matrix. A successful study had been conducted for reinforcement of nanowhiskers in the polyhydroxybutyrate (PHB) matrix using the casting or solvent method [15]. Martínez-Sanz et al. [61] also demonstrated the dispersion of nanowhiskers in PHA derived from waste materials. The drying of nanofillers followed by the polymerization is an attractive choice for a simple dispersion of nanomaterials in the polymer matrix [44]. An extensive research is required for making the nanocellulose produced from rice straw fully compatible in the polymer matrix so that these biocomposites can truly compete the petrochemical-based plastics.

11 Application of Rice Straw Nanomaterials

The nanomaterials prepared by using the rice straw occupy potential applicability in different sectors. The widespread applications of rice straw nanomaterials are provided in Fig. 8. The graphene, CNTs and silica nanoparticles can be utilized for constructing the highly permeable and selective next-generation filtration membranes. The application of biomass-based nanomaterials for the desalination of seawater is attractive due to high efficiency, cost-effectiveness and ecofriendlyness. Use of nanotechnology with a polymeric reverse osmosis membrane can prove to be highly significant in improving the quality of wastewater. The graphene oxide has a high capability of enhancing the separation efficiency of membranes [65]. The sheets upgrade the water permeation and mechanical strength of the membranes and protect them from degradation. The conventional techniques and adsorbents used for water purification have limited scope due to high cost, self-degradation, energy-intensive and poor efficiency for metal ions [9]. Tohamy et al. [95] reported the successful removal of Nickel (II) ion by using rice straw-based product and showed high efficiency at the initial stage which reduced after half an hour due to leaching.

The carbon-based nanomembranes are potentially known for removing organic pollutants from aqueous medium [82, 105]. Rocha et al. [75] explored the semiconducting type SWCNTs which were highly efficient for removing the toxic components from waste water such as 2,4-dichlorophenoxyacetic acid and pyrene butyric

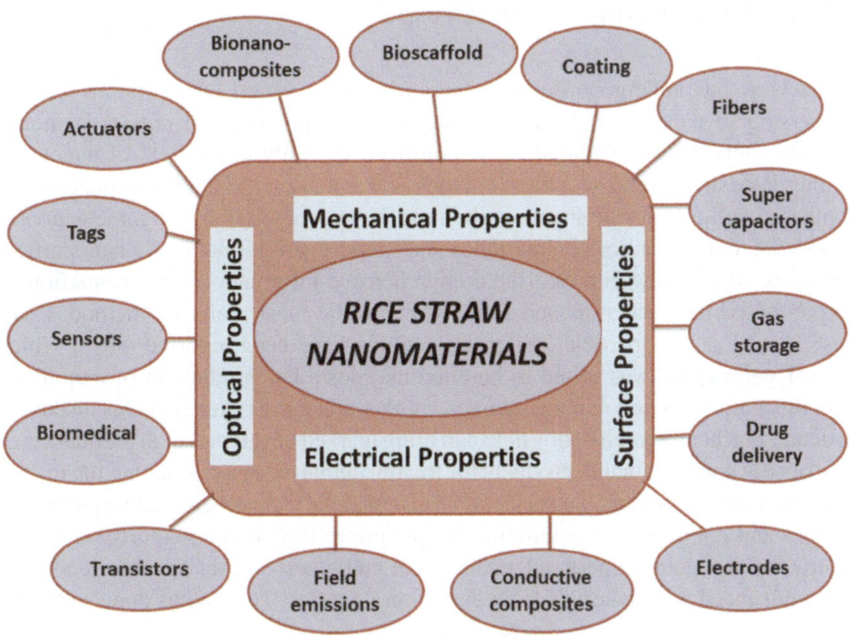

Fig. 8 Application of rice straw nanomaterials in different fields

acid. These nanomaterials possess a significant place in pharmaceuticals and medical research. Rice straw-based nanomaterials (CNTs) can be used in nanofilteration. Due to their large surface area, porosity, inertness, super-hydrophobicity, recyclability, selectivity and stability, these show high adsorption for oil [37]. The two-dimensional graphene oxides have a significant impact on active sites of bacterial cells which upgrade their role in formulating antimicrobial devices and medicines. A large number of studies showed that graphene oxides are effective in isolation of bacteria from environment and blocking their mode of nutrition which supress their growth. Silica nanoparticles can also act as a powerful weapon against high drug dose impacts. Other than these, the nanomaterials are selective and have high sensitivity for different solvents, proteins, organic and inorganic components which potentially make their space in ion detection devices, for treatment of cancer, heavy metals adsorption, electronic capacitors, agriculture etc. Wibowo et al. [102] produced the rice straw ash and prepared nanofertilizer for enhancing the soil quality.

Nanocellulose is also a fascinating biopolymer of biogenic origin due to its renewability, biocompatibility, biodegradability, high resistance, stiffness and recyclability. The nanobiopolymers and nanobiocomposites are either applicable directly or polymerized for application in various sectors like pulp, paper, textile, biocomposites, nanocellulose, bioplastics, biofilms, nanofiltration, ultrafiltration, reverse osmosis, gas permeation, drug control system and electronic devices.

12 Conclusions and Future Prospects

The review has undergone detailed literature survey based on rice straw and its conversion to nanomaterials. Rice straw has been produced in outsized amount over the globe but its potential is underutilized specially in the field of nanotechnology. Rice straw possesses characteristics, which makes it potential raw material for synthesizing the graphene sheets, carbon nanotubes, carbon dots inorganic nanoparticles especially the silica nanoparticles and support various metals for nanoparticle synthesis. The review depicted the comprehensive information for formulation of carbon-based nanomaterials and found CVD as the most preferred method. Likewise, for silica nanoparticles different methods have been reported out of which the sol–gel method was found to be cited the most. The production of nanofilers, nanobiocomposites and nanocellulose is another area of research that was discussed in detail. Further research in this field can bring marked development in production of ecofriendly and sustainable products for treating lethal diseases, reducing the microbial infections, upgrading pharmaceutical and medical technologies, energy-efficient devices and cut down the environmental pollution. Research is required to reduce efforts for producing superficial, economical and green courses for the creation of novel advanced nanomaterials from agricultural wastes. The studies done so far are limited to lab or pilot scales which must be scaled up at commercial level to ensure the success of rice straw-based nanomaterials. All efforts made by the researchers are still at sapling stage and need to be nurtured for the commercial exploitation of rice straw for nanoscale materials.

Acknowledgements The authors acknowledge the Principal, Guru Nanak Khalsa College and the Department of Biotechnology, Kurukshetra University, Kurukshetra for providing a congenial ambience, internet facility and infrastructure for the successful completion of this review paper. Further, authors acknowledge the support of Dr. Prabhjot Kaur, CSIR-RA, Department of Biotechnology, Kurukshetra University for structuring the review chapter.

References

1. Abioye AM, Ani FN (2015) Recent development in the production of activated carbon electrodes from agricultural waste biomass for supercapacitors: a review. Renew Sustain Energy Rev 1:1282–1293
2. Agbagla-Dohnani, A., Noziere, P., Gaillard-Martinie, B., Puard, M. and Doreau, M. (2003). Effect of silica content on rice straw ruminal degradation. The *Journal of Agricultural Science, 140*(2), 183–192.
3. Akhtar N, Gupta K, Goyal D, Goyal A (2016) Recent advances in pretreatment technologies for efficient hydrolysis of lignocellulosic biomass. Environ Prog Sustain Energy 35:489–511
4. Al-Qahtani SD, Hameed A, Snari RM, Shah R, Alfi AA, El-Metwaly NM (2022) Development of fluorescent carbon dots ink from rice straw waste toward security authentication. J Mol Liq 15:118927
5. Alcántara JC, González I, Pareta MM, Vilaseca F (2020) Biocomposites from rice straw nanofibers: morphology, thermal and mechanical properties. Materials 13:2138
6. Alshareef M, Snari RM, Alaysuy O, Aldawsari AM, Abumelha HM, Katouah H, El-Metwaly NM (2022) Optical detection of acetone using "Turn-Off" fluorescent rice straw based cellulose carbon dots imprinted onto paper dipstick for diabetes monitoring. ACS Omega
7. Amirsoleimani S, Ghorbani HR (2020) Preparation of antibacterial coating film using ZnO nanoparticles and epoxy resin. J Nanoanal 7:96–103
8. Areeshi MY (2022) Rice straw mediated green synthesis and characterization of iron oxide nanoparticles and its application to improve thermal stability of endoglucanase enzyme. Int J Food Microbiol 2:109722
9. Baby Shaikh R, Saifullah B, Rehman FU (2018) Greener method for the removal of toxic metal ions from the wastewater by application of agricultural waste as an adsorbent. Water 10:1316
10. Bhardwaj R, Mohanty AK (2007) Advances in the properties of polylactides based materials: a review. J Biobased Mater Bioenergy 1:191–209
11. Bhat VS, Jayeoye TJ, Rujiralai T, Sirimahachai U, Chong KF, Hegde G (2020) Influence of surface properties on electro-chemical supercapacitors utilizing Callerya atropurpurea pod derived porous nanocarbons: structure property relationship between porous structures to energy storage devices. Nano Select 1:226–243
12. Bilo F, Pandini S, Sartore L, Depero LE, Gargiulo G, Bonassi A, Federici S, Bontempi E (2018) A sustainable bioplastic obtained from rice straw. J Clean Prod 200:357–368
13. Bokov D, Turki Jalil A, Chupradit S, Suksatan W, Javed Ansari M, Shewael IH, Valiev GH, Kianfar E (2021) Nanomaterial by sol-gel method: synthesis and application. Adv Mater Sci Eng 24:2021
14. Chen X, Yu J, Zhang Z, Lu C (2011) Study on structure and thermal stability properties of cellulose fibers from rice straw. Carbohyd Polym 85:245–250

15. Dhar P, Bhardwaj U, Kumar A, Katiyar V (2014) Cellulose nanocrystals: a potential nanofiller for food packaging applications. In: Komolprasert V, Turowsk P (eds) Food additives and packaging. American Chemical Society, USA, pp 197–239
16. Dilamian M, Noroozi B (2019) A combined homogenization-high intensity ultrasonication process for individualizaion of cellulose micro-nano fibers from rice straw. Cellulose 26:5831–5849
17. Di Donato P, Taurisano V, Poli A, d'Ayala GG, Nicolaus B, Malinconinco M, Santagata G (2020) Vegetable wastes derived polysaccharides as natural eco-friendly plasticizers of sodium alginate. Carbohyd Polym 229:115427
18. Ers.usda.gov, as accessed in June, 2022
19. Fathy NA (2017) Carbon nanotubes synthesis using carbonization of pretreated rice straw through chemical vapor deposition of camphor. RSC Adv 7:28535–28541
20. Fathy NA, Lotfy VF, Basta AH (2017) Comparative study on the performance of carbon nanotubes prepared from agro-and xerogels as carbon supports. J Anal Appl Pyrol 128:114–120
21. Gadakh D, Dashora P, Wadhankar G (2019) A review paper on graphene coated fibres. Graphene. 8:53
22. Grimm AM, Dorsch LY, Kloess GH, Enke D, Roppertz A (2021) Transition metal promoted combustion of rice husk and rice straw towards an energy optimized synthesis of biogenic silica. Biomass Bioenerg 155:106282
23. Han YW (1978) Microbial utilization of straw (A review). Adv Appl Microbiol 23:119–153
24. Handojo L, Pramudita D, Mangindaan D, Indarto A (2020) Application of nanoparticles in environmental cleanup: production, potential risks and solutions. Emerg Eco-friendly Green Technol Wastewater Treat 2020:45–76
25. Hassan M, Abou Zeid RE, Abou-Elseoud WS, Hassan E, Berglund L, Oksman K (2019) Effect of unbleached rice straw cellulose nanofibers on the properties of polysulfone membranes. Polymers 11:938
26. Hassan ML, Mathew AP, Hassan EA, El-Wakil NA, Oksman K (2012) Nanofibers from bagasse and rice straw: process optimization and properties. Wood Sci Technol 46:193–205
27. He J, Che Y, Manthiram A (2018) MOF-derived cobalt sulfide grown on 3D graphene foam as an efficient sulfur host for long-life lithium-sulfur batteries. Iscience 4:36–43
28. Hu S, Gu J, Jiang F, Hsieh YL (2016) Holistic rice straw nanocellulose and hemicelluloses/lignin composite films. ACS Sustain Chem Eng 4:728–737
29. Hurter RW (2015) Nonwood fibres and moulded products. Paper Technol 14–17
30. Husen A, Siddiqi KS (2014) Phytosynthesis of nanoparticles: Concept, controversy and application. Nanosc Res Lett 9(229):1–24
31. Husen A (2022) Engineered nanomaterials for sustainable agricultural production, soil improvement and stress management. Elsevier Inc., Cambridge, USA. https://doi.org/10.1016/C2021-0-00054-7
32. Husen A Iqbal M (2019) N*anomaterials*and plant potential: an overview. In: Husen A, Iqbal M (eds) Nanomaterials and plant potential. Springer International Publishing AG, Cham pp 3–29. https://doi.org/10.1007/978-3-030-05569-1_1
33. Husen A, Jawaid M (2020) Nanomaterials for agriculture and forestry applications. Elsevier Inc., Cambridge, MA, USA. https://doi.org/10.1016/C2018-0-02349-X
34. Husen A, Siddiqi KS (2022) Advances in smart nanomaterials and their applications. Elsevier Inc., Cambridge, MA, USA. https://doi.org/10.1016/C2021-0-02202-1
35. Husen A (2023) Secondary metabolites based green synthesis of nanomaterials and their applications. Springer Nature Singapore Pte Ltd., Singapore. https://doi.org/10.1007/978-981-99-0927-8
36. Huy HN, Khue TN (2016) Lactic acid production from rice straw using plant-originated *Lactobacillus rhamnosus* PN04. J Chem Pharm Res 8:590–594
37. Jame SA, Zhou Z (2016) Electrochemical carbon nanotube filters for water and wastewater treatment. Nanotechnol Rev 5:41–50

38. Jiang C, Bo J, Xiao X, Zhang S, Wang Z, Yan G, Wu Y, Wong C, He H (2020) Converting waste lignin into nano-biochar as a renewable substitute of carbon black for reinforcing styrene-butadiene rubber. Waste Manage 102:732–742
39. Jiang Y, Carvalho-de-Souza JL, Wong R, Luo Z, Isheim D, Zuo X, Nicholls AW, Jung IW, Yue J, Liu DJ, Wang Y (2016) Heterogeneous silicon mesostructures for lipid-supported bioelectric interfaces. Nat Mater 15:1023–1030
40. Jin-Chul K, Madhusudhan A, Husen A (2021) Smart nanomaterials in biomedical applications. Springer Nature Switzerland AG, Cham, Switzerland. https://doi.org/10.1007/978-3-030-84262-8
41. Jorn-am T, Praneerad J, Attajak R, Sirisit N, Manyam J, Paoprasert P (2021) Quasi-solid, bio-renewable supercapacitor with high specific capacitance and energy density based on rice electrolytes and rice straw-derived carbon dots as novel electrolyte additives. Colloids Surf, A 628:127239
42. Kaliannan D, Palaninaicker S, Palanivel V, Mahadeo MA, Ravindra BN, Jae-Jin S (2019) A novel approach to preparation of nano-adsorbent from agricultural wastes (Saccharum officinarum leaves) and its environmental application. Environ Sci Pollut Res 26:5305–5314
43. Kamel R, El-Wakil NA, Dufresne A, Elkasabgy NA (2020) Nanocellulose: from an agricultural waste to a valuable pharmaceutical ingredient. Int J Biol Macromol 163:1579–1590
44. Kargarzadeh H, Mariano M, Huang J, Lin N, Ahmad I, Dufresne A, Thomas S (2017) Recent developments on nanocellulose reinforced polymer nanocomposites: a review. Polymer 132:368–393
45. Kaur D, Bhardwaj NK, Lohchab RK (2017) Prospects of rice straw as a raw material for paper making. Waste Manage 60:127–139
46. Kaur D, Bhardwaj NK, Lohchab RK (2018) A study on pulping of rice straw and impact of incorporation of chlorine dioxide during bleaching on pulp properties and effluents characteristics. J Clean Prod 170:174–182
47. Kaur D, Bhardwaj NK, Lohchab RK (2019) Impact of modifying conventional chlorine dioxide stage to hot chlorine dioxide during rice straw pulp bleaching on pulp, paper and effluent characteristics. Cellulose 26:7469–7482
48. Kenry (2018) Understanding the hemotoxicity of graphene nanomaterials through their interactions with blood proteins and cells. J Mater Res 33:44–57
49. Khandanlou R, Ngoh GC, Chong WT, Bayat S, Saki E (2016) Fabrication of silver nanoparticles supported on rice straw: in vitro antibacterial activity and its heterogeneous catalysis in the degradation of 4-nitrophenol. BioResources 11:3691–3708
50. Khir R, Pan Z (2019) Chapter 2: Rice in integrated processing technologies for food and agricultural by-products. In: Pan Z, Zhang R, Zicari S (eds) Integrated processing technologies for food and agricultural by-products. Academic Press, pp 51–56
51. Khorsand H, Kiayee N, Masoomparast AH (2013) Optimization of amorphous silica nanoparticles synthesis from rice straw ash using design of experiments technique. Part Sci Technol 31:366–371
52. Khoshnood Motlagh E, Asasian-Kolur N, Sharifian S (2020) A comparative study on rice husk and rice straw as bioresources for production of carbonaceous adsorbent and silica. Biomass Conv Biorefinery 16:1
53. Kumar A, Choudhary A, Kaur H, Guha S, Mehta S, Husen A (2022) Potential applications of engineered nanoparticles in plant disease management: a critical update. Chemosphere 295:133798. ISBN: 0045-6535, https://doi.org/10.1016/j.chemosphere.2022.133798
54. Kumar A, Choudhary A, Kaur H, Mehta S, Husen A (2021) Smart nanomaterial and nanocomposite with advanced agrochemical activities. Nanosc Res Lett 16 (156):1–26. ISBN: 1556-276X, https://doi.org/10.1186/s11671-021-03612-0
55. Kumari M, Chaudhary GR, Chaudhary S, Huang M, Guo Z (2022) Transformation of waste rice straw to carbon quantum dots and their potential chemical sensing application: green and sustainable approach to overcome stubble burning issues. Biomass Conv Biorefinery 6:1–2
56. Kumari N, Kumar K (2018) Mechanical behaviour of graphene and carbon fibre reinforced epoxy based hybrid nanocomposites for orthotic callipers. J Exp Nanosci 13:S14-23

57. Li J, Ma Q, Shao H, Zhou X, Xia H, Xie J (2017) Biosynthesis, characterization, and antibacterial activity of silver nanoparticles produced from rice straw biomass. BioResources 12:4897–4911
58. Lim JS, Manan ZA, Alwi SR, Hashim H (2012) A review on utilisation of biomass from rice industry as a source of renewable energy. Renew Sustain Energy Rev 16:3084–3094
59. Liu Z, Xu A, Long B (2011) Energy from combustion of rice straw: status and challenges to China. Energy Power Eng 3:325
60. Lotfy VF, Fathy NA, Basta AH (2018) Novel approach for synthesizing different shapes of carbon nanotubes from rice straw residue. J Environ Chem Eng 6:6263–6274
61. Martínez-Sanz M, Vicente AA, Gontard N, Lopez-Rubio A, Lagaron JM (2015) On the extraction of cellulose nanowhiskers from food by-products and their comparative reinforcing effect on a polyhydroxybutyrate-co-valerate polymer. Cellulose 22:535–551
62. Meneses IP, Novaes SD, Dezotti RS, Oliveira PV, Petri DF (2022) CTAB-modified carboxymethyl cellulose/bagasse cryogels for the efficient removal of bisphenol A, methylene blue and Cr (VI) ions: batch and column adsorption studies. J Hazard Mater 421:126804
63. Minu K, Jiby KK, Kishore VV (2012) Isolation and purification of lignin and silica from the black liquor generated during the production of bioethanol from rice straw. Biomass Bioenerg 39:210–217
64. Mondal R, Dam P, Chakraborty J, Paret ML, Katı A, Altuntas S, Sarkar R, Ghorai S, Mandal AK, Husen A (2022) Potential of nanobiosensor in sustainable agriculture: the state-of-art. Heliyon. https://doi.org/10.2139/ssrn.4134873
65. Mubarik S, Qureshi N, Sattar Z, Shaheen A, Kalsoom A, Imran M, Hanif F (2021) Synthetic approach to rice waste-derived carbon-based nanomaterials and their applications. Nanomanufacturing 1:109–159
66. Nandiyanto AB, Permatasari N, Sucahya TN, Abdullah AG, Hasanah L (2017) Synthesis of potassium silicate nanoparticles from rice straw ash using a flame-assisted spray-pyrolysis method. In: IOP conference series: materials science and engineering, vol 180. IOP Publishing, p 012133
67. Nguyen VH, Topno S, Balingbing C, Nguyen VC, Röder M, Quilty J, Jamieson C, Thornley P, Gummert M (2016) Generating a positive energy balance from using rice straw for anaerobic digestion. Energy Rep 2:117–122
68. Paula AJ, Ferreira OP, Souza Filho AG, Filho FN, Andrade CE, Faria AF (2022) Machine learning and natural language processing enable a data-oriented experimental design approach for producing biochar and hydrochar from biomass. Chem Mater 34:979–990
69. Penkov O (2020) Tribology of graphene: simulation methods, preparation methods, and their applications. Elsevier
70. Ponnusamy PG, Mani S (2022) Life cycle assessment of manufacturing cellulose nanofibril-reinforced chitosan composite films for packaging applications. Int J Life Cycle Assess 27:380–394
71. Pramanik P, Krishnan P, Maity A, Mridha N, Mukherjee A, Rai V (2020) Application of nanotechnology in agriculture. Environmental nanotechnology, vol 4. Springer, Cham, pp 317–348
72. Puvvada N, Kumar BP, Konar S, Kalita H, Mandal M, Pathak A (2012) Synthesis of biocompatible multicolor luminescent carbon dots for bioimaging applications. Sci Technol Adv Mater
73. Reena Y, Patir M (2021) Importance of rice straw and strategies to improve its nutritional value in ruminants feeding system. Vigyan Varta 2:26–30
74. Robles-Jimarez HR, Sanjuan-Navarro L, Jornet-Martínez N, Primaz CT, Teruel-Juanes R, Molins-Legua C, Ribes-Greus A, Campíns-Falcó P (2022) New silica based adsorbent material from rice straw and its in-flow application to nitrate reduction in waters: process sustainability and scale-up possibilities. Sci Total Environ 805:150317
75. Rocha JD, Rogers RE, Dichiara AB, Capasse RC (2017) Emerging investigators series: highly effective adsorption of organic aromatic molecules from aqueous environments by electronically sorted single-walled carbon nanotubes. Environ Sci: Water Res Technol 3:203–212

76. Rodríguez A, Sánchez R, Requejo A, Ferrer A (2010) Feasibility of rice straw as a raw material for the production of soda cellulose pulp. J Clean Prod 18:1084–1091
77. Rosa SM, Rehman N, de Miranda MI, Nachtigall SM, Bica CI (2012) Chlorine-free extraction of cellulose from rice husk and whisker isolation. Carbohyd Polym 87:1131–1138
78. Sangon S, Hunt AJ, Ngernyen Y, Youngme S, Supanchaiyamat N (2021) Rice straw-derived highly mesoporous carbon-zinc oxide nanocomposites as high performance photocatalytic adsorbents for toxic dyes. J Clean Prod 318:128583
79. El Sayed AM, El-Gamal S (2015) Synthesis and investigation of the electrical and dielectric properties of Co_3O_4/(CMC+ PVA) nanocomposite films. J Polym Res 22:1–2
80. Shaaban S, Nasr M (2020) Towards three R's agricultural waste in MENA: Reduce, reuse, and recycle. In: Waste management in MENA regions 2020. Springer, Cham, pp 337–353
81. Shaghaleh H, Xu X, Liu H, Wang S, Hamoud YA, Dong F, Luo J (2019) The effect of atmospheric pressure plasma pretreatment with various gases on the structural characteristics and chemical composition of wheat straw and applications to enzymatic hydrolysis. Energy 176:195–210
82. Shanmuganathan R, Sathishkumar G, Brindhadevi K, Pugazhendhi A (2020) Fabrication of naringenin functionalized-Ag/RGO nanocomposites for potential bactericidal effects. J Market Res 9:7013–7019
83. Sharma P, Pandey V, Sharma MMM, Patra A, Singh B, Mehta S, Husen A (2021) A review on biosensors and nanosensors application in agroecosystems. Nano Res Lett 16:136. https://doi.org/10.1186/s11671-021-03593-0
84. Shen Y, Zhao P, Shao Q (2014) Porous silica and carbon derived materials from rice husk pyrolysis char. Microporous Mesoporous Mater 188:46–76
85. Shi SC, Liu GT (2021) Cellulose nanocrystal extraction from rice straw using a chlorine-free bleaching process. Cellulose 28:6147–6158
86. Silveira MV, Zandoná G, Leitemberger A, Böhs L, Lopes TJ, Martins ML, Godoi M (2021) Water extract of rice straw ash: experimental design and evaluation of their activity in the hydrothiolation reaction. Waste Biomass Valorizat 12:5041–5050
87. Singh G, Dizaji HB, Puttuswamy H, Sharma S (2022) Biogenic nanosilica synthesis employing agro-waste rice straw and its application study in photocatalytic degradation of cationic dye. Sustainability 14:539
88. Singh A, Pal DB, Mohammad A, Alhazmi A, Haque S, Yoon T, Srivastava N, Gupta VK (2022) Biological remediation technologies for dyes and heavy metals in wastewater treatment: new insight. Biores Technol 343:126154
89. Singh SP, Endley N (2020) Fabrication of nano-silica from agricultural residue and their application. In: Husen A, Jawaid M (eds) Nanomaterials for agriculture and forestry applications, vol 1. Elsevier, pp 107–134
90. Sinha N, Yeow JW (2005) Carbon nanotubes for biomedical applications. IEEE Trans Nanobiosci 4:180–195
91. Sun X, Qian Y, Jiao Y, Liu J, Xi F, Dong X (2017) Ionic liquid-capped graphene quantum dots as label-free fluorescent probe for direct detection of ferricyanide. Talanta 165:429–435
92. Sun Q, Xu X, Yu J, Yin C, Wu M, Niu N, Chen L (2022). Rice straw-derived carbon based nanozyme sensor: application of identifying human urine xanthine content and study of active sites. Appl Surf Sci 29, 154372
93. Thongsai N, Tanawannapong N, Praneerad J, Kladsomboon S, Jaiyong P, Paoprasert P (2019) Real-time detection of alcohol vapours and volatile organic compounds via optical electronic nose using carbon dots prepared from rice husk and density functional theory calculation. Colloids Surf, A 560:278–287
94. Van Thuoc D, Chung NT, Hatti-Kaul R (2021) Polyhydroxyalkanoate production from rice straw hydrolysate obtained by alkaline pretreatment and enzymatic hydrolysis using Bacillus strains isolated from decomposing straw. Bioresourc Bioprocess 8:1–5
95. Tohamy HA, Anis B, Youssef MA, Abdallah AE, El-Sakhawy M, Kamel S (2020) Preparation of eco-friendly graphene oxide from agricultural wastes for water treatment. Desalin Water Treat 191:250–262

96. Uda MN, Gopinath SC, Hashim U, Halim NH, Parmin NA, Uda MN, Anbu P (2021) Production and characterization of graphene from carbonaceous rice straw by cost-effect extraction. 3 Biotech 11:1
97. Uda MN, Hashim U, Gopinath CS, Uda MA, Parmin NA, Adam H, Ibrahim NH (2020) Spectrophotometric analysis assay for the measurement of arsenic using nanocomposite of silica and graphene. In: IOP conference series: materials science and engineering, vol 743. IOP Publishing, p 012016
98. Valdés A, Fenollar O, Beltrán A, Balart R, Fortunati E, Kenny JM, Garrigós MC (2016) Characterization and enzymatic degradation study of poly (ε-caprolactone)-based biocomposites from almond agricultural by-products. Polym Degrad Stab 132:181–190
99. Van Hung N, Maguyon-Detras MC, Migo MV, Quilloy R, Balingbing C, Chivenge P, Gummert M (2020) Rice straw overview: availability, properties, and management practices. Sustain Rice Straw Manag 1
100. De Volder MF, Tawfick SH, Baughman RH, Hart AJ (2013) Carbon nanotubes: present and future commercial applications. Science 339:535–539
101. www.statista.com. Accessed July 2022
102. Wibowo FR, Saputra OA, Lestari WW, Koketsu M, Mukti RR, Martien R (2020) pH-triggered drug release controlled by poly (styrene sulfonate) growth hollow mesoporous silica nanoparticles. ACS Omega 5:4261–4269
103. Xie X, Gou G, Wei X, Zhou Z, Jiang M, Xu X, Wang Z, Hui D (2016) Influence of pretreatment of rice straw on hydration of straw fiber filled cement based composites. Constr Build Mater 113:449–455
104. Xu G, Chen J, Shi S, Tian A, Tang Q (2020) Shear behaviors of the intersurface between rice straw rope and dredger fill silt: experimental and mechanism studies. Adv Civil Eng 13:020
105. Xu L, He J, Yu Y, Chen JP (2017) Effect of CNT content on physicochemical properties and performance of CNT composite polysulfone membranes. Chem Eng Res Des 121:92–98
106. Xu K, Liu C, Kang K, Zheng Z, Wang S, Tang Z, Yang W (2018) Isolation of nanocrystalline cellulose from rice straw and preparation of its biocomposites with chitosan: physicochemical characterization and evaluation of interfacial compatibility. Compos Sci Technol 154:8–17
107. Yan J, Hu J, Yang R, Zhang Z, Zhao W (2018) Innovative nanofibrillated cellulose from rice straw as dietary fiber for enhanced health benefits prepared by a green and scale production method. ACS Sustain Chem Eng 6:3481–3492
108. Zhao G, Du J, Chen W, Pan M, Chen D (2019) Preparation and thermostability of cellulose nanocrystals and nanofibrils from two sources of biomass: rice straw and poplar wood. Cellulose 26:8625–8643
109. Zheng Y, Han X, Li Y, Liu S, Ji J, Tong Y (2020) Effects of mixed controlled release nitrogen fertilizer with rice straw biochar on rice yield and nitrogen balance in northeast China. Sci Rep 10:1

Wheat Straw Waste Utilization for Nanoparticles Synthesis and Their Various Applications

Aditi Sharma, Abhinav Sharma, Priyanka Kashyap, Payal Dhyani, and Manu Pant

Abstract Nanoparticles have established their advantages in multiple areas like medicine, diagnostics, pharmacology, agriculture, the food industry, and many more. With the advancements in nanotechnology, green synthesis of nanoparticles from plants has gained more attention as they offer the benefits of low-cost raw materials and a significant decline in the toxicity of the product. Green chemistry has, therefore, emerged as a cutting-edge eco-friendly technology. The utilization of plant extracts for producing metallic nanoparticles has led to the usage of plant waste material for the same. Wheat straw has drawn great interest lately as a potential filler for polymer matrices and as a plentiful and excellent biomass source for the construction of stable nanoparticles. The development of metal nanoparticles, such as silver (Ag), gold (Au), copper (Cu), nickel (Ni), and those based on silica, lignin, and cellulose, are some of the many potential uses for the wheat straw that have been researched. These wheat straw-based nanoparticles can be efficiently used in the field of medicine, water treatment, textile engineering, sensors, imaging, catalysis, food packaging and processing, recycling of agricultural waste, plant nutrition, plant propagation, etc. In this chapter, we aim to highlight the potential applications of wheat straw biomass, its treatment, and subsequent synthesis of nanoparticles.

Keywords Wheat straw · Green chemistry · Nanotechnology · Nanoparticles · Wheat straw utilization · Pretreatment

A. Sharma · P. Kashyap · P. Dhyani
Genetics and Tree Improvement Division, Forest Research Institute, Dehradun, Uttarakhand, India

A. Sharma
Department of PDP, Graphic Era Deemed to be University, Bell Road, Clement Town, Dehradun, Uttarakhand, India

M. Pant (✉)
Department of Biotechnology, Graphic Era Deemed to be University, Bell Road, Clement Town, Dehradun, Uttarakhand, India
e-mail: manupant@geu.ac.in

A. Husen (ed.), *Nanomaterials from Agricultural and Horticultural Products*, Smart Nanomaterials Technology,
https://doi.org/10.1007/978-981-99-3435-5_8

1 Introduction

Wheat is one of the important and popular cereal crops that is grown across the globe. Several hundred tonnes of wheat are produced annually worldwide. Improved agricultural techniques and breeding strategies have led to the development of superior wheat cultivars that have eventually led to a significant improvement in wheat production in different parts of the world. However, a serious concern remains to be the agricultural waste generated after the wheat grain harvest. The massive amount of post-harvest wheat straw remains that has to be disposed of and burning happens to be the easiest option. This waste is occasionally employed as a substantial amount of industrial raw material instead of being used as animal feed, domestic fuel, or raw materials for the paper sector [79]. Around 85% of the straw is burnt in situ to organize the fields for the upcoming sowing [67]. This leads to large emissions of carbon dioxide (CO_2), particulate matter (PM), and methane (CH_4) which contribute to degrading the air quality. In many regions, these pollutants have been reported to cause smog that has had detrimental effects on health and even affected normal daily routines [47]. It, therefore, becomes necessary to pretreat this waste and convert it into useful products before releasing it into the environment. One impressive option is the development of nanoparticles from wheat straw. Nanotechnology is an advanced field of contemporary material science that encompasses the development of nanoparticles exhibiting novel properties [6]. The constructed nanoparticles have found applications in different areas ranging from diagnostics, therapeutics, molecular imaging, food packaging, polymers, ceramics, agriculture, and many more [1, 37–41, 45, 48, 49, 56, 73, 78, 82, 105]. This chapter provides an insight into what makes wheat straw waste a sought-after material for nanoparticle synthesis, the methods of straw treatment and construction of nanoparticles, and how the technique holds strong potential for wheat straw waste management, reduction, reuse, and recycling.

2 Morphology and Composition of Wheat Straw

The wheat straw is made up of internodal segments (57%), nodes (10%), leaves (18%), chaffs (18%), and rachis (6%) on a mass basis [59]. Wheat straw possesses a lumen in the middle and concentric rings at the internodes. The epidermis is a thick, cellulose-rich layer found in the outermost ring of the body. On the surface of the epidermis, silica is concentrated. There is a loose layer below the epidermis that has vascular bundles and parenchyma in it [32]. Analogous to wood tracheid, the ultrastructure of wheat straw fibers is composed of the middle lamella, main wall, and secondary wall [107].

The chemical constituents differ between and within the wheat straw's anatomical components. Overall, the straw is made up of around 35–40% cellulose (linear and crystalline), non-crystalline heteropolysaccharides (hemicellulose ~20–35%), and lignin (~20%) [31, 85] (Table 1). Lignin is essentially a structural component that

Table 1 Lignocellulosic mass of wheat straw

Cellulose (%)	Lignin (%)	Hemicellulose (%)	Key references
38.2	19.1	36.4	[108]
33.7–40	11–22.9	21–26	[103]
32.0	16.2	26.9	[72]
41.7	7.9	28.05	[90]
39.5	10.53	29.36	[81]

gives cell walls more vigor and rigidity [5]. Lignin functions as a matrix alongside hemicelluloses for cellulose microfibrils, made up of organized polymer chains with closely packed, crystalline areas. A lengthy chain of glucose molecules, primarily joined by ß ($1 \rightarrow 4$) glycosidic bonds, is what makes up cellulose. Because of its simple structure, cellulose is biodegradable. Hemicellulose, a polysaccharide macromolecule made of several sugars, differs from cellulose in that it possesses low molecular weight and is not chemically homogenous. While cellulose has hydrolyzable oligomers, hemicellulose has branches with short lateral chains that contain several sugars [31].

Cellulose and hemicellulose are hydrophilic, while the other part is hydrophobic. However, because of hydrogen bonds and covalent bonds with lignin, they are essentially insoluble in water [97].

Besides the straw contains several different organic compounds, including proteins, extractives, wax that coats the straw's epidermis, and insoluble ash, including silica, sugars, and salts.

3 Utilization of Wheat Straw

Utilizing wheat straw wastes for worthwhile purposes is advantageous because an increase in their volume leads to health and environmental problems [33]. The in-situ burning of wheat straw entails significant emissions of pollutants that harm the atmosphere and worsen the air quality, leading to the prevalence of respiratory illnesses among the populace. Additionally, the energy in wheat straw is squandered when it is burned in the open air [27].

Since the straw contains cellulose and hemicelluloses, lignin, and other polysaccharides, it is well known natural composite material. It is also used as a heating fuel, animal feed, and bedding for domestic animals. Due to its low lignin content, it is also considered to be an excellent starting material for the generation of biofuel, bioethanol, and biogas [28, 35, 42, 92]. Reportedly, about 100 billion liters of bioethanol could be created annually from the 350 million tonnes of wheat straw formed globally [76]. However, the high production costs of the existing technology limit commercialization of the product. A wheat straw biorefinery is expected to provide a short-term solution for producing high-value products and bioethanol in a clean, effective, and financially feasible way [3].

However, these applications of wheat straw have limited users and a majority of waste continues to be burnt in the fields by the farmers. In this scenario, it becomes important to look for additional avenues for wheat straw waste management.

4 Wheat Straw Pretreatment

The lignocellulosic cell wall network exhibits resistance against enzymatic degradation. This is because cellulose in the biomass is well shielded by hemicellulose and lignin, preventing cellulase from reaching the reaction's active sites and yielding a lower-quality end. As a result, wheat straw has to be pretreated to loosen and break the bonding of lignocelluloses [62]. Pretreatment processes dissolve the hydrogen bonds between hemicellulose and lignin thereby increasing cellulose approachability for enzymatic hydrolysis (Fig. 1) [111]. The factors that are considered before selecting a pretreatment process include: (1) successfully breaking up the intricately linked fraction constituents, (2) improving cellulose availability and lignin elimination, (3) conserving the hemicellulose fraction, (4) reducing lignin solubility and enhancing the recovered purity of lignin, (5) minimizing side products, (6) cutting down on energy use, and (7) producing green, harmless, and sustainable target products.

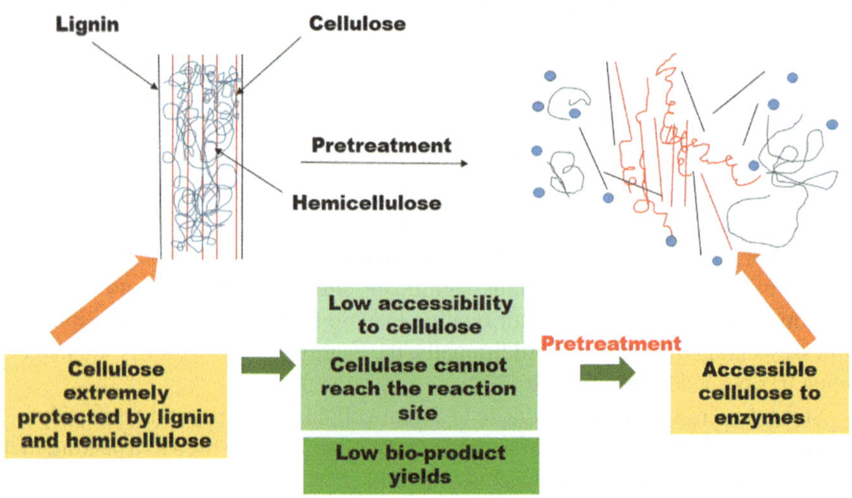

Fig. 1 Diagrammatic representation showing the pretreatment process

The pretreatment methods can be broadly categorized as (i) Physical Pretreatment (ii) Chemical Pretreatment (iii) Biological Pretreatment.

(i) **Physical Treatments**—These include high temperature or radiation treatment to the wheat straw biomass.

Pyrolysis—At temperatures exceeding 300°C, pyrolysis can break down cellulose in the wheat straw biomass into hydrogen, carbon monoxide, and leftover char. Biomass that has undergone pyrolysis treatment may have its cellulose structure disrupted, its calorific value & hydrophobicity improved, and its stability increased [34]. This being an endothermic process has less energy input, which has numerous benefits over other pretreatment approaches for straw biomass [4, 87].

Radiation treatment—This involves the use of gamma rays from radioisotopes like cobalt-60 or cesium-137 or an electron beam generated by an electron accelerator to destroy cellulose. The effectiveness of this method has been shown in studies where reducing the sugar yield of wheat straw could be increased to more than 80% by exposing enzyme and alkali-treated wheat straw to gamma radiations [88, 112].

(ii) **Chemical pretreatments**—In these methods, wheat straw biomass is exposed to different types of chemicals for further degradation.

Acid treatment—This method uses inorganic or organic acid to break down the bonds between cellulose, lignin, and hemicellulose, increasing the effectiveness of hemicellulose hydrolysis, lignin removal, and expediting the fermentation and saccharification processes [58]. Different concentrated and dilute acids like sulfuric acid, phosphoric acid, nitric acid, oxalic acid, and maleic acid have been utilized to treat the wheat straw biomass [4, 16, 50, 77, 110].

Alkali treatment—This is reflected to be an efficient and cost-effective process of dissolving the ester linkages and glycosidic bonds in the lignocellulosic cell wall [22]. The methods increase the surface area and porosity of processed biomass, while also changing the lignin structure thereby lowering the crystallinity and polymerization efficiency of cellulose, making it more obtainable [60]. The factors affecting alkali pretreatment are alkaline loading, reaction, and temperature duration [19, 104]. Sodium hydroxide is one of the most preferred alkaline components due to its strong delignification efficacy and is known to give better results when combined with grinding [71, 89]. Alkali pretreatment of wheat straw has the benefits of low operation costs, reduced energy consumption, lesser corrosion, and fewer inhibitors [12]. However, the drawbacks include the possible destruction of hemicellulose and cellulose and the persistence of alkaline constituents in the biomass after the treatment [58].

Hydrothermolysis—Also known as liquid hot water (LHW) treatment, aquasolv, solvolysis, or aqueous fractionation, this is an effective and green technology, it maintains the high synthesis of sugars, nearly neutral pH value, and lower corrosion while generating fewer inhibitors [26]. This treatment uses water in treating biomass at high pressures (up to 5 MPa) and temperatures (200 ± 20 °C) to maintain the liquid state of water and its role as an acid [50]. The drawback includes the requirement of a strict reactor setup [14], and high energy input [69]. Studies have also confirmed that a two-stage hydrothermal procedure results in high product yield with lower levels of byproducts [26, 64].

Ionic liquid (IL) treatment—This is a novel environmentally friendly solvent pretreatment strategy for fractionating, saccharifying, and fermenting lignocellulosic biomass [93]. The method might also be utilized to create additional byproducts, which would increase the overall economic advantages of pretreatment. The IL is made up of liquid organic cations and inorganic anions maintained at below 100 °C [15]. Certain distinctive physicochemical characteristics include: (i) low volatility melting point & vapor pressure; (ii) excellent recyclability; (iii) high solubility, stability & polarity; (iv) high ionic conductivity; (v) simple operation; (vi) less energy cost; (vii) low toxicity and hydrophobicity; (viii) inflammability and pollution resistance. Because of these distinctive qualities, it is a successful pretreatment technique for straw biomass [15, 50]. IL treatment of wheat straw biomass with cholinium taurate ([Ch] [Tau]) and enzymes, 1-butyl-3-methylimidazolium chloride solution are very effective in improving cellulose digestibility and lignin elimination from the biomass [70, 109]. Despite these advantages, high costs, high toxicity, more inhibitors, high cost of retrieval & recycling ability limit large-scale or marketable application of this technology [10, 50].

Deep eutectic solvents (DES) Treatment—These are unusual compounds that have hydrogen-bonding acceptor (HBA) and donor (HBD) components that can be combined into 3 or more mixes [63]. The intensive interaction between the HBD & the HBA not only significantly lowers the freezing or melting points of DESs than those of their mixtures, but also breaks up strong hydrogen bonds within straw biomass, increasing its solubility and conversion efficiency. Freeze-drying technology, heating and stirring, and evaporation are the three main methods used to prepare DESs. DESs have been known to rapidly replace ILs. They offer advantages of variety, design, affordability, greenness, high tunability, simplicity in synthesis, ease in recycling, high solubility, non-flammability, biocompatibility & biodegradability, and environmental safety [18, 63].

Organosolv (OS) treatment—This includes organic solvent treatment of wheat straw biomass, either alone or in combination with water and/ or with other organic solvents. It is a promising method for isolating cellulose, hemicellulose, and practically pure lignin from straw biomass.

Methanol, glycol, ethanol, acetic acid, butanol, formic acid, propionic acid, acetone, glycerin, formaldehyde dioxane, amines tetrahydrofuran, and phenol are among the mutual solvents used in the OS treatment. The frequently used OS systems include water/ethanol and H_2SO_4, water/methanol and alkaline, and OS paired with a steam explosion [55]. Atmospheric aqueous glycerol autocatalytic OS pretreatment (AAGAOP) has been revealed to increase the availability of cellulose, and break the ester & glycosidic bonds between lignin and hemicellulose, and facilitates the conversion and enzymatic digestion of straw biomass [84]. A recent study has shown that novel OS-dilute acid treatment can digest around 90% of the cellulose in the wheat straw substrate [43]. The advantages include low boiling point, flammability, high pressure, easy volatility of the solvents used, high lignin dissolution, fewer byproducts, along with maintaining the stability of β-O-4 linkages & preventing concentration for applications further down the line [25]. The main drawbacks include pricey investments, strong inhibitory products, and being eco-unfriendly [11].

Sulfite pretreatment to overcome the resistance of lignocellulose (SPORL)—The method has been reported to bring about significant bioconversion of straw biomass [96]. It is a two-stage process that includes pretreating biomass with calcium or magnesium sulfite to eliminate cellulose and break down hemicellulose under brief pretreatment conditions of 160–180 °C and pH 2–4 [50]. Other chemicals used are Na_2S, Na_2SO_3, Na_2CO_3, NaOH [101], and ammonium sulfite [66]. Advantages include cellulose decomposition and hemicellulose elimination, good hydrolysis efficacy, and reduced energy use. Sugar deterioration and high costs remain to be the limitations of the method.

Oxidative treatment—Oxygen, H_2O_2, peracetic acid, ozone, or air are utilized in the oxidative treatment of lignocellulosic biomass. The underlying mechanisms include oxidative cleavage, electrophilic substitution, and side-chain displacements of aromatic ring ether linkages. As a result of lignin oxidation and fragmentation, acids and inhibitory chemicals are produced, which may have an impact on the amount of fermentable sugar produced & the effectiveness of enzymatic digestion [50]. The wet oxidation method is executed under strict temperature, pressure, and time constraints. The process can break lignin, cause solubilization of hemicellulose, increase cellulose susceptibility, and reduce the production of byproducts. However, it necessitates a higher temperature (over 120 °C) and greater pressure (0.8–3.3 MPa) [16, 22, 100]. Moreover, the low efficacy of the treatment for producing fermentable sugars—which is attributable to the decomposition of a significant quantity of hemicellulose—is another significant disadvantage [54]. The effectiveness of straw biomass being transformed into fermentable sugars, the purity and stability of lignin are both increased by a novel oxidative treatment technique for fractionating straw biomass that uses both O_2 and H_2O_2 as co-oxidants under alkaline conditions [106].

Steam explosion (SE)—This is a frequent, efficient, and favorable approach used in industrial settings to handle straw biomass [21]. SE is often started at high temperatures (between 160 and 260 °C) and pressures (0.69–4.83 MPa) for a lesser period before the pressure is abruptly released. As soon as steam enters the lignocellulose, it swiftly decompresses, causing water to evaporate quickly and causing a blast inside the fibers. Because lignin is eliminated, hemicellulose is dissolved, and cellulose hydrolysis is strengthened during the entire process, the yield of fermentable sugars can be increased [95]. The drawbacks are the necessity for high temperature and pressure [83].

(iii) **Physico-Chemical Pretreatments**—These methods include a combination of physical and chemical means to treat wheat straw biomass.

Ultrasonic and mild alkaline oxidation treatments—This is a mild alkaline oxidation ultrasound-assisted method to achieve high lignin elimination in a quick turnaround time [102]. The process uses 1.54% NaOH concentration, 1160 W of ultrasonic power for 50 min, and 78–69 °C of beginning water temperature.

Ultrasonic and propionic acid treatment—In this method, a liquid ratio of 1:12, a propionic acid concentration of 900 g/L, a catalyst concentration of 3 g/L, and an ultrasonic power of 300 W during a 15 min reaction period are utilized for enzymatic hydrolysis of wheat straw biomass [2]. Propionic acid can prevent environmental contamination because it can be recycled using vacuum distillation or other techniques.

(iv) **Biological Pretreatments**—These methods involve the use of microorganisms (fungi, bacteria, and actinomycetes) or biobased products (enzymes to preferentially resolve lignin and hemicellulose, which speeds up the enzyme digestion process) [57]. These treatments offer advantages over chemical methods like high delignification ability, low energy input, eco-friendly behavior, reduced cost, and elimination of the need for harmful chemicals and associated byproducts. The process involves treating the biomass with microbial cultures which dissolve the lignin, even completely breaking it down to CO_2. White-rot fungi are known to be the potent organism that brings about the conversion [4, 13]. *Aspergillus niger, Aspergillus awamori, Escherichia coli, Saccharomyces cerevisiae, and Ceriporiopsis subvermispora* are some of the microorganisms that have been used to treat wheat straw biomass [20, 46].

5 Straw-Based Nanoparticle Development

The pretreated wheat straw biomass has the potential to be developed into nanoparticles that can have myriad applications. Nanoparticles have been observed to have multiple applications in the field of agriculture like breaking seed dormancy [80], improving soil properties [36, 44], improving uptake of nutrients in the soil [98], purifying groundwater, seawater, nanosensor in food packaging and food processing [99]. Apart from this nanotechnology offers an effective alternate strategy to recycle agricultural waste. Postharvest methods for the efficient production of nanocomposites, nano-celluloses, biochars, etc., from agricultural wastes are still being developed. The best examples of materials generated from a combination of cellulose from plant parts, lignin, and nanoparticles are called nano lignocellulosic materials. They are decomposable and readily address the issue of waste management. Around 110 million tonnes (Mt) of wheat waste after harvesting is the ideal raw material for renewable energy, nano silica, biochar, etc. The technical application of this waste mass opens the best route for its disposal and can be utilized to produce nanosilica employing nanotechnology methods. Apart from this pretreated wheat straw biomass can also be utilized in the food, pharmaceutical, and nutraceutical industry owing to its treated lignocellulosic component. Improvements in wheat straw-based CNCs have been reported to have the potential to be used as fillers for goods like food packaging films [61].

6 Nanoparticles Synthesis Using Various Wastes

Agricultural wastes have been effectively used in nanoparticle development and offer a potential alternative to waste treatment. The different categories of nanoparticles that can be formulated with wheat straw biomass are described as under.

Nanocellulose (NC)—Agricultural wastes are used to create nanoscale cellulose materials for the durable materials sector. Nano-fibrillated cellulose (NFC) and cellular nanocrystals are the two primary types of nano-celluloses (NC) (CNC). For oil-based products, they are frequently denoted as second-generation renewable resources. These materials have gained surplus attention because of their biogas characteristics, strong mechanical properties, and renewability [68]. Future utilization of NC in biological sciences is anticipated to increase with enhanced extraction of agricultural wastes like wheat straw [74]. NC has gained much attention due to its renewable accessibility and several beneficial qualities [30]. However, the effectivity of NC varies with the factors like (i) the structure of lingo-cellulosic biomass, (ii) the impact of biomass on NC characteristics, (iii) the effect of pretreatments on biomass, (iv) nano-cellulose extraction procedures [65].

Nanolignin—Nanolignin (NL) is primarily lignin-based nanoparticles. These can be prepared in a wheat straw biorefinery by using precipitation techniques. Solvent shifting and pH shifting are the popular precipitation methods where lignin solubility is decreased by reducing solvent concentration and raising pH [9]. However, direct precipitation of lignin nanoparticles from OS pretreatment extracts (OSE) has also been performed on a static mixture which potentially lowered the overall solvent consumption of the procedure and the size of the product [7]. The method is appropriate for large-scale commercial production of NL by overcoming the challenges of the influence of lignin extraction procedures [29] and large particle size [8]. The size of NL can be further adjusted by manipulating the level of lignin super saturation [52], hydrodynamic conditions of the process, anti-solvent to OSE ratios flow rates in the static mixer, and the pH of the fluid [52].

Nanosilica (NS)—Nano-structured silica with a high surface area can be successfully developed from agricultural residue like wheat straw [17]. Phytoliths, amorphous hydrated silica, can be extracted from the wheat straw and used in nanoparticle construction. Porous nanostructured silica can be developed by incineration and acid leaching. The extracted phytoliths vary in size, surface area, pore diameter, and volume depending on the treatment, temperature, and duration. The silica with a nanostructure can agglomerate extensively as the temperature rises.

Silver nanoparticles (AgNP)—Stable AgNPs can be produced from wheat straw by employing a green chemical biosynthesis technique which is affordable, safe, and very effective in nanoparticle synthesis, recycling, and reducing the buildup of agricultural waste. Ma et al. [53] used a light irradiation process to construct wheat straw-based AgNPs with an average diameter of 17.2 nm and zeta potential of −21.6 mV. The constructed AgNPs proved to be efficient in significantly inhibiting the growth of *B. subtilis* and *E. coli* bacteria.

Gold nanoparticles (GNP)—Due to their distinctive optical characteristics, gold nanoparticles (GNPs) stand out among other metal nanoparticles and are frequently used in biomedical, environmental, and catalytic studies [75]. Wheat straw-based GNPs can be constructed and used in systems where GNPs have shown their efficiency. For example, a study by Venzhik et al. [94] reported that GNP function as adaptogens that increase plant freezing tolerance by altering plant growth intensity, the accumulation of soluble sugars in seedlings, the activity of the photosynthetic apparatus, and oxidative processes.

Nanohydrogel—Cellulose-based hydrogels have significant advantages as green materials [23]. Hydrogels have been shown to bring about the catalytic reduction of different organic contaminants. It has recently been revealed that wheat straw can be used to create a smart cellulose-based semi-interpenetrating polymer network (semi-IPN hydrogel known as WSC-g-PAA/PVA), which has great capabilities in separation, water retention, and metal ions adsorption. Such wheat-straw-based hydrogel exhibits strong gel mechanical strength and swelling capacity [61]. Nanotechnology can be utilized to create nanogels for more efficient use of hydrogels. Ding et al. [24] showed that metal nanoparticles (Copper or Nickle) can be fixed using hydrogel as a carrier. When using wheat straw-based nanogels, the technique can be potentially used for recycling metal ions from aqueous solutions and making good use of waste materials.

7 Conclusion

Green synthesis of nanoparticles using natural resources is driven by the expanding trends in green chemistry and nanotechnology. Nanomaterials made of wheat straw have shown promise while also minimizing the considerable negative effects of disposing of wheat straw, including significant health risks. These nanoparticles have many benefits over traditional techniques since they are economical, non-toxic, and environment-friendly. It is crucial to pretreat the biomass with an appropriate technique, like physical, chemical, or biological treatments, before developing wheat-straw-based nanoparticles. With this processed biomass, various types of nanoparticles, such as silver nanoparticles, cellulose nanoparticles, lignin nanoparticles, etc., can be created and used for several purposes. Given the significance of wheat straw, which is otherwise discarded, a larger population must have access to the knowledge database on straw biomass composition, characterization, and prospective uses. For this, comprehensive testing, protocol standardization, industry and research cooperation, and active public–private partnerships are required. The use of biobased goods will not only help to address the significant issues of waste production and environmental pollution but will also support the development of nanotechnological solutions to many pertinent problems.

References

1. Abate L, Bachheti A, Bachheti RK, Husen A, Pandey DP, Megra MB (2022) Coir fiber-based cellulose, nanocellulose, and their cutting-edge applications. In: Jawaid M (ed) Coir fiber and its composites. Elsevier Inc., Cambridge, MA, USA, pp 309–325
2. Ahamed A, Vermette P (2010) Effect of mechanical agitation on the production of cellulases by Trichoderma reesei RUT-C30 in a draft-tube airlift bioreactor. Biochem Eng J 49:379–387
3. Ain BU, Huma FS, Ahmed A, Khan MA, Niaz B, Tufail T (2019) Improving the physicochemical properties of partially enhanced soluble dietary fiber through innovative techniques: a coherent review. J Food Process Preserv 43(4):13917
4. Akhtar N, Gupta K, Goyal D, Goyal A (2016) Recent advances in pretreatment technologies for efficient hydrolysis of lignocellulosic biomass. Environ Prog Sust Energ 35:489–511
5. Baurhoo B, Ruiz-Feria CA, Zhao X (2008) Purified lignin: nutritional and health impacts on farm animals—a review. Animal Feed Sci Technol 144(3–4):175–184
6. Bayada S, Muhammad A, Tiziano T, Marco C, Flavio R (2020) The history of nanoscience and nanotechnology: from chemical-physical applications to nanomedicine. Molecules 25:112
7. Beisl S, Adamcyk J, Friedl A (2020) Direct precipitation of lignin nanoparticles from wheat straw organosolv liquors using a static mixer. Molecules 25(6):1388
8. Beisl S, Loidolt P, Miltner A, Harasek M, Friedl A (2018) Production of micro- and nanoscale lignin from wheat straw using different precipitation setups. Molecules 23:633
9. Beisl S, Miltner A, Friedl A (2017) Lignin from micro- to nanosize: production methods. Int J Mol Sci 18:1244
10. Bhatia R, Lad JB, Bosch M, Bryant DN, Leak D, Hallett JP, Franco TT, Gallagher JA (2021) Production of oligosaccharides and biofuels from miscanthus using combinatorial steam explosion and ionic liquid pretreatment. Bioresour Tech 323:124625
11. Borand MN, Karaosmanoglu F (2018) Effects of organosolv pretreatment conditions for lignocellulosic biomass in biorefinery applications: a review. J Renew Sust Energ 10:033104
12. Cai D, Li P, Luo Z, Qin P, Chen C, Wang Y, Wang Z, Tan T (2016) Effect of dilute alkaline pretreatment on the conversion of different parts of corn stalk to fermentable sugars and its application in acetone-butanol-ethanol fermentation. Bioresour Tech 211:117–124
13. Canilha L, Chandel AK, Suzane dos Santos Milessi T, Antunes FAF, Luiz da Costa Freitas W, das Graças Almeida Felipe M, Silvério da Silva S (2012) Bioconversion of sugarcane biomass into ethanol: an overview about composition, pretreatment methods, detoxification of hydrolysates, enzymatic saccharification, and ethanol fermentation. J Biomed Biotechnol 1–15
14. Capolupo L, Faraco V (2016) Green methods of lignocellulose pretreatment for biorefinery development. Appl Microbiol Biotechnol 100:9451–9467

15. Chang KL, Chen XM, Han YJ, Wang XQ, Potprommanee L, Ning XA, Liu JY, Sun J, Peng YP, Sun SY, Lin YC (2016) Synergistic effects of surfactant-assisted ionic liquid pretreatment rice straw. Bioresour Tech 214:371–375
16. Chen H, Liu J, Chang X, Chen D, Xue Y, Liu P, Hualin L, Sheng H (2017) A review on the pretreatment of lignocellulose for high-value chemicals. Fuel Process Tech 160:196–206
17. Chen H, Wang F, Zhang C, Shi Y, Jin G, Yuan S (2010) Preparation of nano-silica materials: the concept from wheat straw. J Solids 356:2781–2785
18. Chen Y, Mu T (2019) Application of deep eutectic solvents in biomass pretreatment and conversion. Green Energ Environ 4:95–115
19. Cheng YS, Zheng Y, Yu CW, Dooley TM, Jenkins BM, VanderGheynst JS (2010) Evaluation of high solids alkaline pretreatment of rice straw. Appl Biochem Biotechnol 162:1768–1784
20. Cianchetta S, Di Maggio B, Burzi PL, Galletti S (2014) Evaluation of selected white-rot fungal isolates for improving the sugar yield from wheat straw. Appl Biochem Biotechnol 173:609–623
21. Dai NH, Huynh KTT, Nguyen TAD, Do VVT, Van Tran M (2021) Hydrothermal and steam explosion pretreatment of bambusa stenostachya bamboo. Waste Biomass Valor. 12(7):4103–4112
22. Das N, Jena PK, Padhi D, Kumar Mohanty M, Sahoo G (2021) A Comprehensive review of characterization, pretreatment and its applications on different lignocellulosic biomass for bioethanol production. Biomass Conv Bioref 1–5
23. de Oliveira JP, Pinheiro Bruni G, Oliveira Lima K, Lisie Mello El Halal S, Silveira da Rosa G, Renato Guerra Dias A, da Rosa Zavareze E (2017) Food Chem 221:153–160
24. Ding J, Qian Li, Zhao L, Li X, Yue Q Gao B (2017) A wheat straw cellulose based semi-IPN hydrogel reactor for metal nanoparticles preparation and catalytic reduction of 4-nitrophenol. RSC Adv 7:17599
25. Dong C, Meng X, Yeung CS, Tse HY, Ragauskas AJ, Leu SY (2019) Diol pretreatment to fractionate a reactive lignin in lignocellulosic biomass biorefineries. Green Chem 21:2788–2800
26. Douyong M, Rui-sen X, Zhe H, Jia-qing L, Caoxing H, Yongcan J, Qiang Y (2015) Minimizing inhibitors during pretreatment while maximizing sugar production in enzymatic hydrolysis through a two-stage hydrothermal pretreatment. Cellulose 22:1253–1261
27. Fahad S, Hussain S, Matloob A, Khan F, Khaliq A, Saud S, Hassan S, Shan D, Khan F, Ullah N, Faiq M, Khan M, Tareen A, Khan A, Ullah A, Ullah N, Huang J (2015) Phytohormones and plant responses to salinity stress: a review. Plant Growth Regul 75:391–404
28. Ferreira ME, Prochasson P, Berndt KD, Workman JL, Wright AP (2009) Activator-binding domains of the SWI/SNF chromatin remodeling complex characterized in vitro are required for its recruitment to promoters in vivo. The FEBS J 276(9):2557–2565
29. Gao W, Fatehi P (2019) Lignin for polymer and nanoparticle production: current status and challenges. Can J Chem Eng 97:2827–2842
30. García A, Gandini A, Labidi J, Belgacem N, Bras J (2016) Industrial and crop wastes: a new source for nanocellulose biorefinery. Ind Crops Prod 93:26–38
31. Ghaffar SH (2016) Straw fibre-based construction materials. In: Fan M, Fu F (eds) Advanced high strength natural fibre composites in construction. Woodhead, Duxford, UK, pp 257–283
32. Ghaffar SH, Fan M (2015) Revealing the morphology and chemical distribution of nodes in wheat straw. Biomass Bioenerg 77:123–134
33. Giuntini E, Bazzicalupo M, Castaldini M, Fabiani A, Miclaus N, Piccolo R, Ranalli G, Santomassimo F, Zanobini S, Mengoni A (2006) Genetic diversity of dinitrogen-fixing bacterial communities in soil amended with olive husks. Ann Microbiol 56(2):83–88
34. Gogate PR (2013) Intensification of enzymatic hydrolysis of lignocellulose using ultrasound for efficient bioethanol production: a review. Ind Eng Chem Res 52:11816–11828
35. Himmel ME, Bayer EA (2009) Lignocellulose conversion to biofuels: current challenges, global perspectives. Curr Opin Biotechnol 20(3):316–317
36. Huang Y, Wang L (2016) Experimental studies on nanomaterials for soil improvement: a review. Environ Earth Sci 75:497

37. Husen A (2022) Engineered nanomaterials for sustainable agricultural production, soil improvement and stress management. Elsevier Inc., Cambridge, MA, USA. https://doi.org/10.1016/C2021-0-00054-7
38. Husen A, Iqbal M (2019) Nanomaterials and plant potential: an overview. In: Husen A, Iqbal M (eds) N*anomaterials* and plant potential. Springer International Publishing AG, Cham, pp 3–29. https://doi.org/10.1007/978-3-030-05569-1_1
39. Husen A, Jawaid M (2020) Nanomaterials for agriculture and forestry applications. Elsevier Inc., Cambridge, MA, USA. ISBN 9780128178522, https://doi.org/10.1016/C2018-0-02349-X
40. Husen A, Siddiqi KS (2014) Phytosynthesis of nanoparticles: concept, controversy and application. Nanoscale Res Lett 9(229):1–24
41. Husen A, Siddiqi KS (2022) Advances in smart nanomaterials and their applications. Elsevier Inc., Cambridge, MA, USA. https://doi.org/10.1016/C2021-0-02202-1
42. Irfan M, Asghar U, Nadeem M, Nelofer R, Syed Q, Shakir HA, Qazi JI (2016) Statistical optimization of saccharification of alkali pretreated wheat straw for bioethanol production. Waste Biomass Valori 7(6):1389–1396
43. Islam MK, Thaemngoen A, Lau CY, Guan J, Yeung CS, Chaiprapat S, Leu SY (2021) Staged organosolv pretreatment to increase net energy and reactive lignin yield in whole oil palm tree biorefinery. Bioresour Tech 326:124766
44. Jamal A, Mohd T, Ali F, Gp P (2016) Potential of using nanocarbons to stabilize weak soils. Appl Environ Soil Sci 1–9
45. Jin-Chul K, Madhusudhan A, Husen A (2021) Smart nanomaterials in biomedical applications. Springer Nature Switzerland AG, Cham, Switzerland. https://doi.org/10.1007/978-3-030-84262-8
46. Jönsson LJ, Martín C (2016) Pretreatment of lignocellulose: formation of inhibitory by-products and strategies for minimizing their effects. Bioresour Tech 199:103–112
47. Kapil S (2019) Public health emergency declared in Delhi due to air pollution. Down to earth.
48. Kumar A, Choudhary A, Kaur H, Guha S, Mehta S, Husen A (2022) Potential applications of engineered nanoparticles in plant disease management: a critical update. Chemosphere 295:133798. ISBN: 0045-6535, https://doi.org/10.1016/j.chemosphere.2022.133798
49. Kumar A, Choudhary A, Kaur H, Mehta S, Husen A (2021) Smart nanomaterial and nanocomposite with advanced agrochemical activities. Nanosc Res Lett 16(156):1–26. ISBN: 1556-276X, https://doi.org/10.1186/s11671-021-03612-0
50. Kumar AK, Sharma S (2017) Recent updates on different methods of pretreatment of lignocellulosic feedstocks: a review. Bioresour Bioproc 4–7
51. Lee HV, Hamid SBA, Zain SK (2014) Conversion of lignocellulosic biomass to nanocellulose: structure and chemical process. Sci World 1–20
52. Lewis A, Seckler MM, Kramer H, van Rosmalen G (2015) Agglomeration. In: Industrial crystallization. Cambridge University Press, Cambridge, UK, pp 130–150
53. Ma Q, Li J, Xu L, Sun Q, Yong B, Wang Y (2016) Biosynthesis of silver nanoparticles using wheat straw biomass under light radiation and their antibacterial activity. BioResources 11
54. Machineni L (2020) Lignocellulosic biofuel production: review of alternatives. Biomass Conv Bioref 10:779–791
55. Matsakas L, Raghavendran V, Yakimenko O, Persson G, Olsson E, Rova U, Olsson L, Christakopoulos P (2019) Lignin-first biomass fractionation using a hybrid organosolv—steam explosion pretreatment technology improves the saccharification and fermentability of spruce biomass. Bioresour Tech 273:521–528
56. Mondal R, Dam P, Chakraborty J, Paret ML, Katı A, Altuntas S, Sarkar R, Ghorai S, Mandal AK, Husen A (2022) Potential of nanobiosensor in sustainable agriculture: the state-of-art. Heliyon. https://doi.org/10.2139/ssrn.4134873
57. Moodley P, Sewsynker-Sukai Y, Gueguim Kana EB (2020) Progress in the development of alkali and metal salt catalysed lignocellulosic pretreatment regimes: potential for bioethanol production. Bioresour Tech 310:123372

58. Mosier N, Wyman C, Dale B, Elander R, Lee YY, Holtzapple M (2005) Features of promising technologies for pretreatment of lignocellulosic biomass. Bioresour Tech 96:673–686
59. Motte JC, Escudié R, Beaufils N, Steyer JP, Bernet N, Delgenès DC (2014) Morphological structures of wheat straw strongly impacts its anaerobic digestion. Ind Crops Prod 52:695–701
60. Nargotra P, Sharma V, Gupta M, Kour S, Bajaj BK (2018) Application of ionic liquid and alkali pretreatment for enhancing saccharification of sunflower stalk biomass for potential biofuel-ethanol production. Bioresour Tech 267:560–568
61. Oun AA, Rhim JW (2016) Isolation of cellulose nanocrystals from grain straws and their use for the preparation of carboxymethyl cellulose-based nanocomposite films. Carbohydr Polym 5(150):187–200
62. Paes G, Habrant A, Ossemond J, Chabbert B (2017) Exploring accessibility of pretreated poplar cell walls by measuring dynamics of fluorescent probes. Biotechnol Biofuels 10:15
63. Paiva A, Craveiro R, Aroso I, Martins M, Reis RL, Duarte ARC (2014) Natural deep eutectic solvents—solvents for the 21st century. ACS Sust Chem Eng 2:1063–1071
64. Pérez JA, Ballesteros I, Ballesteros M, Sáez F, Negro MJ, Manzanares P (2008) Optimizing liquid hot water pretreatment conditions to enhance sugar recovery from wheat straw for fuel-ethanol production. Fuel 87:3640–3647
65. Pires JRA, Souza VGL, Fernando AL (2019) Ecofriendly strategies for the production of nanocellulose from agro-industrial wastes. In: European biomass conference and exhibition proceedings, pp 1781–1784
66. Qi G, Xiong L, Tian L, Luo M, Chen X, Huang C, Li H, Chen X (2018) Ammonium sulfite pretreatment of wheat straw for efficient enzymatic saccharification. Sustain Energ Tech Assess 29:12–18
67. Quintero M, Moncada A (2008) Contaminación y control de las quemas agrícolas en Imperial, California, y Mexicali, Baja California. Región y Sociedad 20:3–24
68. Rajinipriya M, Nagalakshmaiah M, Robert M, Elkoun S (2018) Importance of agricultural and industrial waste in the field of nanocellulose and recent industrial developments of wood based nanocellulose: a review. ACS Sustain Chem Eng 6(3):2807–2828
69. Rajput AA, Zeshan Visvanathan C (2018) Effect of thermal pretreatment on chemical composition, physical structure and biogas production kinetics of wheat straw. J Environ Manag 221:45–52
70. Ren H, Zong MH, Wu H, Li N (2016) Efficient pretreatment of wheat straw using novel renewable cholinium ionic liquids to improve enzymatic saccharification. Ind Eng Chem Res 55:1788–1795
71. Rezania S, Oryani B, Cho J, Talaiekhozani A, Sabbagh F, Hashemi B, Fatemeh RP, Akbar MA (2020) Different pretreatment technologies of lignocellulosic biomass for bioethanol production: an overview. Energy 199:117457
72. Rio JCD, Rencoret J, Prinsen P, Martinez AT, Ralph J, Gutierrez A (2012) Structural characterization of wheat straw lignin as revealed by analytical pyrolysis, 2D-NMR & reductive cleavage method. J Agric Food Chem 60:5535–5922
73. Rothemund PWK (2006) Folding DNA to create nanoscale shapes and patterns. Nature 440:297–302
74. Sangeetha J, Thangadurai D, Hospet R, Purushotham P, Manowade KR, Mujeeb MA, Mundaragi A, Jogaiah S, David M, Thimmappa SC, Prasad R, Harish ER (2017) Production of bionanomaterials from agricultural wastes. In: Nanotechnology: an agricultural paradigm. Nanomaterials Nanostructes, pp 33–58
75. Sarfraz N, Khan I (2021) Plasmonic gold nanoparticles (AuNPs): properties, synthesis and their advanced energy, environmental and biomedical applications. Chem Asian J 16:720–742
76. Sarkar N, Ghosh SK, Bannerjee S, Aikat K (2012) Bioethanol production from agricultural wastes: an overview. Renew Energy 37(1):19–27
77. Satari Baboukani B, Vossoughi M, Alemzadeh I (2012) Optimisation of dilute-acid pretreatment conditions for enhancement sugar recovery and enzymatic hydrolysis of wheat straw. Biosyst Eng 111:166–174

78. Sharma P, Pandey V, Sharma MMM, Patra A, Singh B, Mehta S, Husen A (2021) A review on biosensors and nanosensors application in agroecosystems. Nano Res Lett 16:136. https://doi.org/10.1186/s11671-021-03593-0
79. She D, Sun RC, Jones GL (2010) Chemical modification of straw as novel materials for industries. Cereal straw as a resource for sustainable biomaterials and biofuels. Elsevier, pp 209–217
80. Shinde S, Paralikar P, Ingle A, Rai M (2018) Promotion of seed germination and seedling growth of Zea mays by magnesium hydroxide nanoparticles synthesized by the filtrate from Aspergillus niger. Arab J Chem 13
81. Shrivastava B, Jain KK, Kalra A, Kuhad RC (2014) Bioprocessing of wheat straw into nutritionally rich & digested cattle feed. Sci Rep 4:6360
82. Siddiqi KS, Husen A (2017) Recent advances in plant-mediated engineered gold nanoparticles and their application in biological system. J Trace Elem Med Biol 40:10–23
83. Smichi N, Messaoudi Y, Allaf K, Gargouri M (2020) Steam explosion (SE) and instant controlled pressure drop (DIC) as thermo-hydro-mechanical pretreatment methods for bioethanol production. Bioproc Biosyst Eng 43:945–957
84. Sun F, Chen H (2008) Enhanced enzymatic hydrolysis of wheat straw by aqueous glycerol pretreatment. Bioresour Technol 99:6156–6161
85. Sun R, Lawther JM, Banks WB (1996) Fractional and structural characterization of wheat straw hemicelluloses. Carbohydr Polym 29(4):325–331
86. Tan J, Li Y, Tan, X, Wu H, Li H, Yang S (2021) Advances in pretreatment of straw biomass for sugar production. Front Chem
87. Tan L, Jin YL, Sun ZY, Tang YQ, Kida K (2020) Production of bioethanol from unwashed-pretreated rapeseed straw at high solid loading. Bioresour Technol 303:122949
88. Tang H, Ha Y, Wang F (2011) Effect of y-rays radiation pretreatment on enzymatic hydrolysis of corn straw for producing sugar. J Radiat Res Radiat Process 29:300–313
89. Tang X, Zuo M, Li Z, Liu H, Xiong C, Zeng X, Sun Y, Hu L, Liu S, Lei T, Lin L (2017) Green processing of lignocellulosic biomass and its derivatives in deep eutectic solvents. Chem Sus Chem 10:2696–2706
90. Thakur S, Shrivastava B, Ingale S, Kuhad RC, Gupte A (2013) Degradation & selective ligninolysis of wheat straw & banana stem for an efficient bioethanol production using fungal & chemical pretreatment. Biotechnology 3:365–372
91. Tufail T, Saeed F, Afzaal M, Ain HBU, Gilani SA, Hussain M, Anjum FM (2021) Wheat straw: a natural remedy against different maladies. Food Sci Nutr 9(4):2335–2344
92. Tufail T, Saeed F, Arshad MU, Afzaal M, Rasheed R, Bader Ul Ain H, Imran M, Abrar M, Farooq MA, Shahid MZ (2020) Exploring the effect of cereal bran cell wall on rheological properties of wheat flour. J Food Process Preserv 44(3):14345
93. Usmani Z, Sharma M, Gupta P, Karpichev Y, Gathergood N, Bhat R, Gupta VK (2020) Ionic liquid based pretreatment of lignocellulosic biomass for enhanced bioconversion. Bioresour Technol 304:123003
94. Venzhik Y, Deryabin A, Popov V, Dykman L, Moshkov I (2022) Gold nanoparticles as adaptogens increazing the freezing tolerance of wheat seedlings. Environ Sci Pollut Res Int 29(36):55235–55249
95. Vochozka M, Marouˇsková A, Váchal J, Straková J (2016) Economic and environmental aspects of steam-explosion pretreatment. Waste Biomass Valor 7:1549–1554
96. Wang GS, Zhu JY, Gleisner RD (2009) Sulfite pretreatment to overcome recalcitrance of lignocellulose (Sporl) for robust enzymatic saccharification of hardwoods. Biotechnol Prog 25:1086–1093
97. Xiao B, Sun XF, Sun R (2001) Chemical, structural, and thermal characterizations of alkalisoluble lignins and hemicelluloses, and cellulose from maize stems, rye straw, and rice straw. Polym Degrad Stab 74(2):307–319
98. Xiaolin C, Bernard R, Zhang SY, Wan H (2007) Formation and properties of nanocomposites made up solid aspen wood, melamine-urea-formaldehyde, and clay. Holzforschung 61(2):148–154

99. Xiqi Z, Xiaoyong Z, Bin Y, Junfeng H, Meiying L, Zhenguo C, Siwei L, Jiarui X, Yen W (2013) Facile preparation and cell imaging applications of fluorescent organic nanoparticles that combine AIE dye and ring-opening polymerization. Polym Chem 5:318–322
100. Xu H, Peng J, Kong Y, Liu Y, Su Z, Li B, Song X, Liu S, Tian W (2020) Key process parameters for deep eutectic solvents pretreatment of lignocellulosic biomass materials: a review. Biorescour Technol 310:123416
101. Yang L, Cao J, Mao J, Jin Y (2013) Sodium carbonate-sodium sulfite pretreatment for improving the enzymatic hydrolysis of rice straw. Ind Crops Prod 43:711–717
102. Yang Y, Yang H, Li J, Pan CW, Wei SQ (2000) Alkalized pretreatment of ethanolized maize stalk accelerated by ultrasonic wave. J Southwest Univ (Nat Sci Edn) 29:149–152
103. Yasin M, Bhutto AW, Bazmi AA, Karim S (2010) Efficient utilization of rice-wheat straw to produce value added composite products. Int J Energ Environ Engin 1(2):136–143
104. Yu H, Du W, Zhang J, Ma F, Zhang X, Zhong W (2010) Fungal treatment of cornstalks enhances the delignification and xylan loss during mild alkaline pretreatment and enzymatic digestibility of glucan. Bioresour Technol 101:6728–6734
105. Yuan Y, Gu Z, Yao C, Luo D Yang D (2019) Nucleic acid-based functional nanomaterials as advanced cancer therapeutics 15:1900172
106. Yuan Z, Klinger GE, Nikafshar S, Cui YB, Fang Z, Alherech M, Goes S, Anson C, Singh SK, Bals B, Hodge DB, Nejad M, Stahl SS, Hegg EL (2021) Effective biomass fractionation through oxygen-enhanced alkaline oxidative pretreatment. ACS Sustain Chen Eng 9:1
107. Zhai H, Lee Z (1989) Ultrastructure and topochemistry of delignification in alkaline pulping of wheat straw. J Wood Chem Technol 9(3):387–406
108. Zhang J, Deng H, Lin L, Sun Y, Chunsheng P, Liu S (2010) Isolation & characterization of wheat straw lignin with a formic acid process. Bioresour Technol 101:2311–2316
109. Zhang ZY, O'Hara IM, Doherty WOS (2013) Effects of pH on pretreatment of sugarcane bagasse using aqueous imidazolium ionic liquids. Green Chem 15:431–438
110. Zhao J, Tao XX, Li JF, Jia YS, Shao T (2021) Enhancement of biomass conservation and enzymatic hydrolysis of rice straw by dilute acid-assisted ensiling pretreatment. Bioresour Technol 320:12431
111. Zhao Z, Chen X, Ali MF, Abdeltawab AA, Yakout SM, Yu G (2018) Pretreatment of wheat straw using basic ethanolamine-based deep eutectic solvents for improving enzymatic hydrolysis. Bioresour Technol 263:325–333
112. Zou A, Shen C, Zhao L Dai G (2011) Effect of microwave pretreatment on component fractionation and saccharification of corn stalk. Trans Chin Soc Agric Eng 20:204–269

Maize Waste Utilization for Nanoparticles Synthesis and Their Various Application

Harshita Shand, Rittick Mondal, Suvankar Ghorai, and Amit Kumar Mandal

Abstract Human consciousness for agricultural production increment is on the rise, arising because of the increasing human population. Agricultural activities generate waste, in large amounts leading to environmental pollution. Maize is a globally grown and consumed cereal. The maize waste is rich in flavonoids, steroids, terpenoids, holocellulose (α-, β- and γ-Cellulose), pentosans, and klason lignin. The maize waste constituents can be used in the production as well as the capping and reduction of nanoparticles (NPs). Metallic oxide NPs like CuO-NPs, NiO-NPs, Fe_2O_3-NPs, etc., and metallic-NPs (MNPs) can be produced from maize waste and the characterization can be done via scanning electron microscopy (SEM), high-resolution transmission electron microscope (HRTEM), X-ray diffraction (XRD), and Fourier transform infrared-attenuated total reflection (FTIR-ATR). NPs acquired from maize waste possess anti-microbial activity, and have pseudo-capacitive energy storage potential and supercabattery energy storage ability. Above all, the production of NPs from waste material help in managing waste and gives an eco-friendly way of synthesizing NPs.

Keywords Maize · Nanoparticles · TEM · Agro-waste · XRD · Application

H. Shand and R. Mondal: Contributed equally.

H. Shand · S. Ghorai
Department of Microbiology, Raiganj University, North Dinajpur 733134, West Bengal, India

R. Mondal · A. K. Mandal (✉)
Chemical Biology Laboratory, Department of Sericulture, Raiganj University, North Dinajpur 733134, West Bengal, India
e-mail: amitmandal08@gmail.com

A. Husen (ed.), *Nanomaterials from Agricultural and Horticultural Products*, Smart Nanomaterials Technology,
https://doi.org/10.1007/978-981-99-3435-5_9

1 Introduction

Nanotechnology is a developing stream of science with huge implementation in biomedicine, pharmaceuticals, sensors, catalysis, cosmetics, agriculture, textile products, etc. [1, 9–14, 19]. Firstly, American physicist Richard Feynman mentioned the concept of nanotechnology [2, 31]. Nanotechnology refers to the understanding and manipulation of matter at atomic and molecular levels. The ultra-small size of nanoparticles (NPs) (1–100 nm) and their high surface-to-volume ratio brings researchers to the utility of NPs for the development of humans. NPs synthesis techniques are time-consuming, overpriced, and toxic to the ecosystem due to the use of harmful compounds [18, 30]. Because of these researchers are trying to develop an environment-friendly way to synthesize NPs which includes using plant parts as well as agricultural waste materials to make the process simple, economical, and sustainable [32]. The world's population is expected to rise to 9.8 billion and 11.2 billion in the year 2050 and 2100 respectively (https://www.un.org/en/desa/world-population-projected-reach-98-billion-2050-and-112-billion-2100). The rapidly growing demand of the increasing population generates tons of waste to meet the food supply needs. The last five decades have seen an increasing global population with an increase in the generation of agricultural waste to meet the global food demand [7]. Maize (*Zea mays*) is a major crop that is cultivated all over the world. The USA is the highest producer and exporter of maize followed by China, Brazil, Argentina, Ukraine, and India. In 2020, maize production in the USA was 360,252 thousand tons which accounts for globally 34.28% of the maize cultivation. The world's total maize production was estimated at 1.05 million thousand tons in 2020 (https://knoema.com/atlas/topics/Agriculture/Crops-Production-Quantity-tonnes/Maize-production). Maize belongs to the Poaceae grass family. The male flowers terminate from the central line of the stem whereas female flowers mature and transform into edible ear cobs [29]. The cobs are enclosed in modified leaves (husks) with a tread-like material called silk. This silk is rich in various phytochemicals like proteins, tannins, flavonoids, steroids, carbohydrates, and terpenoids [6, 8]. Studies have revealed that the silk of maize can minimize oxidative stress and have anti-bacterial and anticancer properties. Hence it is used in traditional medicine as preventive and curative measures against various diseases like cardiovascular disease (CVDs), cancer, diabetes, atherosclerosis, and neurodegenerative disease [28]. The maize husk plant is also rich in anthocyanins and used as a noise-minimizing material due to its acoustic absorbing prospects [16]. This agricultural waste is a part of solid waste and possess a significant threat to humans. After the removal of the edible part, the remains are dumped and left to wither. This waste can have a huge impact on our environment clogging the drainage system and causing water stagnation, which in turn becomes a breeding ground for mosquitos. Agricultural waste is burnt or dumped in landfill causing air, soil, and water pollution [3]. In developing countries, agro-waste is burnt which produces greenhouse gases like N_2O, CO_2, CH_4, etc., volatile organic compounds, air pollutants like CO, NH_3, SO_2, and particulate matter causing air pollution [17]. This waste can also lead to flooding which may turn catastrophic. Almost 998 million tons of agro-waste are produced [26]. The highest amount of crop waste is produced from rice (straw and husk), maize (leaves, cobs, and maize stalk), barley or wheat (straw), sugarcane (bagasse), and cotton (stalk) [5, 27]. Alteration of this maize stalk (MS) into something useful will reduce the burden of this waste that it has on the environment. Despite all the advantages of silk and husk maize, it is treated as a waste product. Researchers are employing maize waste in synthesizing metallic nanoparticles. Researchers are employing agro-waste to produce nanomaterials like nano-cellulose, nanocomposite, nano-adsorbent, nano silica, and nano-cementitious. The silk and husk that forms the outer covering of maize and water for steaming the maize which is treated as discarded materials can be employed in functional transition metal oxide NP production. The phytochemicals in MS waste products vary depending upon the product made from it and have an important role in metal salt reduction as a capping agent, and aid in the purpose of the shape, size, and crystalline structure determination of the metal oxide NPs synthesized. Although the structure of the phytochemicals and antioxidant of diverse portions of the maize crop varies as revealed by studies but studies have indicated that each metal oxide synthesized from the various parts of the maize plant are similar [20]. The merging of green chemistry and sustainable development can help reduce the burden of agricultural waste. Table 1 and Fig. 1 depict various NPs synthesized from maize waste and their application.

Table 1 NPs synthesized from maize waste and their application

Maize waste	Type of nanoparticles	Size	Application	Key reference
Maize husk	CuO	10–90 nm	Potent anti-bacterial activity against *E. coli*, *S. aureus*, Cu_2O is a better inhibitor of *B. licheniformis* and *P. aeruginosa*	Nwanya et al. [24]
Maize husk	CuO	20–39 nm	Psuedo-capacitive energy storage potential	Nwanya et al. [23]
Maize silk	NiO	10–20 nm	Supercabattery energy storage	Nwanya et al. [22]
Maize cob	MNPs (AgNPs, AuNPs)	2–28, 5–50 nm	Anti-bacterial activity against *B. cereus*, *S.aureus*, and *S. typhimurium*	Doan et al. [4]
Maize cob biochar	Biochar based nanocomposite	–	Helpful in sustainable agricultural activity	Lateef et al. [15]
Corn leaf	AgNPs	45.26 nm	Effective anticandidal, antioxidant and antibacterial agent	Patra and Baek [25]

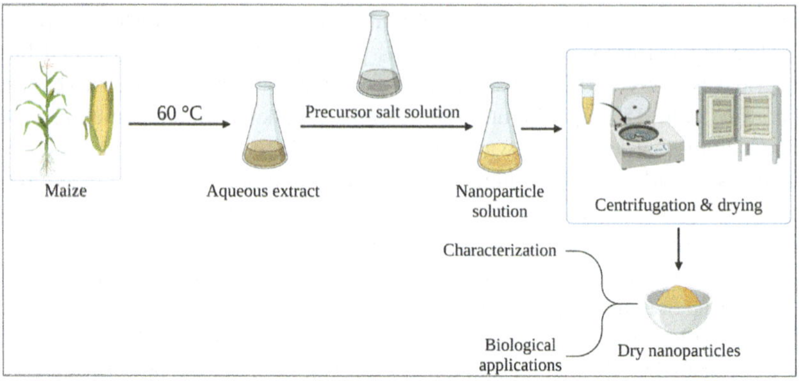

Fig. 1 Maize waste-mediated synthesis of nanoparticles

2 Metallic Oxide Nanoparticle Synthesis

2.1 Copper Oxide Nanoparticle

Maize husk extract has been utilized for the bio-production of copper oxide NPs. At 600 °C Cu_2O-NPs can be thermally oxidized to pure monoclinic CuO-NP, XRD studies confirmed the phases of copper oxide synthesis. HR-TEM analysis showed the NP sizes as 36–73, 10–26 nm for the unannealed Cu_2O-NPs, and 30–90 nm for 300 and 600 °C annealed CuO-NP. The band gap energy values obtained through diffuse reflectance of NPs are 2.0, 1.30 for unannealed, 300, and 600 °C CuO-NPs respectively and 1.42 eV for the annealed CuO-NPs. Under visible light irradiation, the CuO-NPs showed 90 and 91% of degradation of textile effluents and methylene blue dyes. This copper-synthesized CuO-NPs has shown anti-bacterial properties against *Escherichia coli*, *Staphylococcus aureus*, Cu_2O is a better inhibitor of *Bacillus licheniformis* and *Pseudomonas aeruginosa* [24].

2.2 Nickel Oxide Nanoparticle

Nickel oxide nanoparticles (NiO-NPs) have been synthesized using an environment-friendly green synthesis route from maize silk extract (dry). XRD and HRTEM studies revealed that the NiO-NPs have a cubic structure (space group: Fm3m) with spherical and quasi-spherical NPs having an average diameter of 10–20 nm. Galvanostatic charge–discharge cycles, electrochemical impedance spectroscopy, and cyclic voltammetry were used to determine the supercabattery properties of the NiO-NPs. A capacity of 54C/g was gained with a scan rate of 5 mV/s and capacity retention of 60% after 2000 GCD cycles. The synthesized NiO-NPs obtained from maize silk aqueous extract are competent in electrochemical energy storage devices [22].

2.3 *Zinc Oxide Nanoparticle*

Zinc oxide nanoparticles (ZnO-NPs) are synthesized from fresh maize silk and husk extracts and dry husk extracts. The TEM study revealed the shape (spherical) and size (diameter to be 20–70 nm) of ZnO-NPs synthesized from fresh maize silk extract. The ZnO-NPs produced by dry husk extracts were spherical in shape with the diameter ranging from 12 to 67 nm and the diameter of highest density was NP of 30 nm. Concentric rings of bright spots were obtained on ZnO-NPs in selected area diffraction (SAED) that indicated the poly nanocrystalline nature of synthesized NPs [21].

2.4 *Chromium Oxide Nanoparticles*

Chromium oxide nanoparticles (Cr_2O_3-NPs) have been synthesized from the dry silk extract of maize. The Cr_2O_3-NPs absorb strongly in the UV region (331 nm) of the solar spectrum. TEM studies revealed quasi-spherical NPs of varying sizes with clustering of the NPs. The particle size distribution revealed that the maximum frequency of NPs has diameters ranging from 20 to 25 nm [21].

2.5 *Cadmium Oxide Nanoparticles*

Dry maize silk extract is used to synthesize Cadmium oxide nanoparticles (CdO-NPs). CdO-NPs showed well-defined peaks. CdO-NP shows absorbance in the spectral range (360–700 nm). CdO-NPs produced via maize silk extract show high absorption in the UV region, and transparency in NIR, and visible region making it a suitable material for transparent conduction oxide (TCO). TEM studies have revealed the size (quasi-spherical) of the synthesized NPs and varying diameters predominantly diameter ranging from 40 to 60 nm, with some clusters of various sizes of up to 140 nm [21].

2.6 *Iron Oxide Nanoparticles*

Iron Oxide Nanoparticles (Fe_2O_3-NP) have been synthesized using dry maize silk extracts. TEM studies showed that Fe_2O_3-NPs were clustered and the shape (spherical and quasi-spherical) with varying sizes but the highest density of the NPs ranged from 10 to 15 nm [21].

3 Metallic Nanoparticles (MNPs)

3.1 Silver and Gold Nanoparticles

Researchers have fabricated silver and gold nanoparticles from aqueous extract of waste corn cobs. UV–Vis method was employed to optimize the NPs generation. FTIR spectroscopy demonstrated the chemical groups that are present in MNPs. XRD and SAED confirmed the crystalline structure of MNPs. AgNPs showed high antimicrobial action on gram (−) ve and gram (−) ve bacteria [4].

4 Application of Nanoparticles Produced from Agricultural Waste

The rapidly growing population requires increased agricultural expansion to fulfill the food demand. The demand-to-supply ratio of food is increasing agro-waste production as well. Maize is an essential crop consumed worldwide and produces huge agro-waste (husk, cob, stalk, silk). Agricultural waste management is extremely important as it can make the environment clean and green. The nanoparticles produced from agricultural waste have huge applications. Worldwide research is trying to develop a new way to manage agro-waste and producing NPs from waste is one such way. CuO-NPs produced from maize husk waste are potent inhibitors of gram (+) ve and gram (−) ve bacteria [24]. Researchers have also found pseudo-capacitive energy storage ability of CuO-NP [23]. NiO-NPs produced from maize silk have superca-battery energy storage potential [22]. Biochar-based nano-composite obtained from maize cob biochar can be used as nano-fertilizer for sustainable agricultural activity [15]. MNPs produced from waste corn cob aqueous solution have activity against *B. cereus*, *S. aureus*, and *S. typhimurium* [4].

5 Conclusion

The fabrication of NPs from non-toxic plant food processing waste emerging as a significant focus of nanotechnology. The utilization of natural resources for the manufacturing of NPs is sustainable, eco-friendly, cheap, and devoid of chemical pollutants, making it a viable option for usage in a variety of disciplines such as food packaging, food preservation, pharmaceutical, drug delivery, cosmetic and health care. Furthermore, as compared to physicochemical approaches, the NPs synthesized using the green way are more stable and effective.

References

1. Abate L, Bachheti A, Bachheti RK, Husen A, Pandey DP, Megra MB (2022) Coir fiber-based cellulose, nanocellulose, and their cutting-edge applications. In: Jawaid M (ed) Coir fiber and its composites. Elsevier Inc., Cambridge, USA, pp 309–325
2. Bhattacharyya D, Singh S, Satnalika N, Khandelwal A, Jeon SH (2009) Nanotechnology, big things from a tiny world: a review. Int J u-e- Serv Sci Technol 2:29–38
3. de Souza AG, Barbosa RF, Rosa DS (2020) Nanocellulose from industrial and agricultural waste for further use in PLA composites. J Polym Environ 28:1851–1868
4. Doan VD, Luc VS, Nguyen TL, Nguyen TD, Nguyen TD (2020) Utilizing waste corn-cob in biosynthesis of noble metallic nanoparticles for antibacterial effect and catalytic degradation of contaminants. Environ Sci Pollut Res Int 27:6148–6162
5. Elbasiouny H, Elbanna BA, Al-Najoli E, Alsherief A, Negm S, Abou El-Nour E, Nofal A, Sharabash S (2020) Agricultural waste management for climate change mitigation: some implications to Egypt. In: Negm AM, Shareef N (eds) Waste management in MENA regions. Springer, pp 149–169
6. El-Ghorab A, El-Massry KF, Shibamoto T (2007) Chemical composition of the volatile extract and antioxidant activities of the volatile and nonvolatile extracts of Egyptian corn silk (*Zea mays* L.). J Agric Food Chem 55:9124–9127
7. Food and Agriculture Organization of the United Nations (FAO) (2017) Strategic work of FAO for sustainable food and agriculture. http://www.fao.org/3/a-i6488e.pdf. Accessed 13 Aug 2020
8. Hasanudin K, Hashim P, Mustafa S (2012) Corn silk (*Stigma maydis*) in healthcare: a phytochemical and pharmacological review. Molecules 17:9697–9715
9. Husen A (2022) Engineered nanomaterials for sustainable agricultural production, soil improvement and stress management. Elsevier Inc., Cambridge, MA, USA. https://doi.org/10.1016/C2021-0-00054-7
10. Husen A, Iqbal M (2019) Nanomaterials and plant potential: an overview. In: Husen A, Iqbal M (eds) N*anomaterials* and plant potential. Springer International Publishing AG, Cham, pp 3–29. https://doi.org/10.1007/978-3-030-05569-1_1
11. Husen A, Jawaid M (2020) Nanomaterialsfor agriculture and forestry applications. Elsevier Inc., Cambridge, MA, USA. https://doi.org/10.1016/C2018-0-02349-X
12. Husen A, Siddiqi KS (2014) Phytosynthesis of nanoparticles: concept, controversy and application. Nanoscale Res Lett 9(229):1–24
13. Husen A, Siddiqi KS (2022) Advances in smart nanomaterials and their applications. Elsevier Inc., Cambridge, MA, USA. https://doi.org/10.1016/C2021-0-02202-1
14. Jin-Chul K, Madhusudhan A, Husen A (2021) Smart nanomaterials in biomedical applications. Springer Nature Switzerland AG, Cham, Switzerland. https://doi.org/10.1007/978-3-030-842 62-8
15. Lateef A, Nazir R, Jamila N, Alam S, Shah R, Khan MN, Saleem M, Rehman S (2019) Synthesis and characterization of environmental friendly corncob biochar based nano-composite—a potential slow release nano-fertilizer for sustainable agriculture. Environ Nanotechnol Monit Manag 11:100212
16. Li CY, Kim HW, Won SR, Min HK, Park KJ, Park JY, Ahn MS, Rhee HI (2008) Corn husk as a potential source of anthocyanins. J Agric Food Chem 56:11413–11416
17. Li S, Chen G (2020) Agricultural waste-derived superabsorbent hydrogels: preparation, performance and socioeconomic impacts. J Clean Prod 251:119669
18. Mondal R, Yilmaz MD, Mandal AK (2021) Green synthesis of carbon nanoparticles: characterization and their biocidal properties In: Handbook of greener synthesis of nanomaterials and compounds. Elsevier, pp 277–306
19. Mondal R, Dam P, Chakraborty J, Paret ML, Katı A, Altuntas S, Sarkar R, Ghorai S, Mandal AK, Husen A (2022) Potential of nanobiosensor in sustainable agriculture: the state-of-art. Heliyon. https://doi.org/10.2139/ssrn.4134873

20. Nawaz H, Muzaffar S, Aslam M, Ahmad S (2018) Phytochemical composition: antioxidant potential and biological activities of corn. In: Corn-production and human health in changing climate. Intechopen, 10, pp 49–68
21. Nwanya AC, Botha S, Ezema FI, Maaza M (2021) Functional metal oxides synthesized using natural extracts from waste maize materials. Curr Res Green Sustain Chem 4:100054
22. Nwanya AC, Ndipingwi MM, Ikpo CO, Obodo RM, Nwanya SC, Botha S, Ezema FI, Iwuoha EI, Maaza M (2020) Zea mays lea silk extract mediated synthesis of nickel oxide nanoparticles as positive electrode material for asymmetric supercabattery. J Alloys Compd 822:153581
23. Nwanya AC, Ndipingwi MM, Mayedwa N, Razanamahandry LC, Ikpo CO, Waryo T, Ntwampe SK, Malenga E, Fosso-Kankeu E, Ezema FI, Iwuoha EI (2019) Maize (Zea mays L.) fresh husk mediated biosynthesis of copper oxides: potentials for pseudo capacitive energy storage. Electrochim Acta 301:436–448
24. Nwanya AC, Razanamahandry LC, Bashir AK, Ikpo CO, Nwanya SC, Botha S, Ntwampe SK, Ezema FI, Iwuoha EI, Maaza M (2019) Industrial textile effluent treatment and antibacterial effectiveness of *Zea mays* L. Dry husk mediated bio-synthesized copper oxide nanoparticles. J Hazard Mater 375:281–289
25. Patra JK, Baek KH (2017) Antibacterial activity and synergistic antibacterial potential of biosynthesized silver nanoparticles against foodborne pathogenic bacteria along with its anticandidal and antioxidant effects. Front Microbiol 8:167
26. Peerzada JG, Chidambaram R (2021) A statistical approach for biogenic synthesis of nano-silica from different agro-wastes. Silicon 13:2089–2101
27. Prasad M, Ranjan R, Ali A, Goyal D, Yadav A, Singh TB, Shrivastav P, Dantu PK (2020) Efficient transformation of agricultural waste in India. In: Contaminants in agriculture. Springer, pp 271–287
28. Rahman NA, Rosli WIW (2014) Nutritional compositions and antioxidative capacity of the silk obtained from immature and mature corn. J King Saud Univ Sci 26:119–127
29. Solihah MA, Wan Rosli WI, Nurhanan AR (2012) Phytochemicals screening and total phenolic content of Malaysian Zea mays hair extracts. Int Food Res J 19:1533–1538
30. Some S, Das S, Mondal R, Gangopadhyay M, Basak GK (2021) Medicinal plant extract mediated green synthesis of metallic nanoparticles: a review. Int J Plant Env 7:119–132
31. Some S, Mondal R, Dam P, Mandal AK (2022) Synthesis of biogenic silver nanoparticles using medicinal plant extract: a new age in nanomedicine to combat multidrug-resistant pathogens. In: Abd-Elsalam KA (eds) Green synthesis of silver nanomaterials. Elsevier, pp 359–387
32. Some S, Sarkar B, Biswas K, Jana TK, Bhattacharjya D, Dam P, Mondal R, Kumar A, Deb AK, Sadat A, Saha S, Kati A, Ocsoy I, Franco OL, Mandal A, Mandal S, Mandal AK, Ince IA (2020) Bio-molecule functionalized rapid one-pot green synthesis of silver nanoparticles and their efficacy toward the multidrug resistant (MDR) gut bacteria of silkworms (*Bombyx mori*). RSC Adv 10:22742–22757

Various Metabolites and or Bioactive Compounds from Vegetables, and Their Use Nanoparticles Synthesis, and Applications

Noureddine Chaachouay, Abdelhamid Azeroual, Bouchaib Bencherki, Allal Douira, and Lahcen Zidane

Abstract Green chemistry refers to producing chemical products and processes that eliminate or minimize the production of harmful chemicals. In recent decades, chemical constituents, mainly plant secondary compounds, have been studied for their ability to develop green nanoparticles. The demand for green nanoparticle synthesis is increasing, and plant-mediated synthesis is gaining enormous popularity. The universal bioactive compounds of plants, such as alkaloids, terpenoids, flavonoids, and phenolics acids, play a crucial role in synthesizing nanoparticles. Green nanoparticles are found in items that come into touch with the human body and are employed in various scientific and pharmacological domains. Thus, the nanoparticles plant-derived secondary metabolites are environmentally friendly, inexpensive, more energy efficient, biocompatible method, safer, and less poisonous than their chemically manufactured counterparts. As a result, this sector is continuously evolving, and new techniques for enhancing the properties of green nanoparticles are continually being developed. The present chapter summarizes the latest findings in synthesizing green nanoparticles using plant-derived secondary metabolites. The procedures required to characterize produced green nanoparticles are also highlighted. Furthermore, this chapter discusses the benefits of green nanoparticles in various applications, including pharmaceutical and biomedical science, the food industry, cosmetics, agriculture, renewable energy, and wastewater treatment.

Keywords Application · Biosynthesis · Metabolites compounds · Nanoparticles · Vegetable

---

N. Chaachouay (✉) · A. Azeroual · B. Bencherki
Agri-Food and Health Laboratory (AFHL), Hassan First University, Address: Po Box 382, 26000 Settat, Morocco
e-mail: noureddine.chaachouay@uhp.ac.ma

A. Douira · L. Zidane
Plant, Animal Productions and Agro-Industry Laboratory, Department of Biology, Faculty of Sciences, Ibn Tofail University, B.P. 133 14000, Kenitra, Morocco

A. Husen (ed.), *Nanomaterials from Agricultural and Horticultural Products*, Smart Nanomaterials Technology,
https://doi.org/10.1007/978-981-99-3435-5_10

1 Introduction

In past few years, to promote the expansion of the commercial and industrial sectors, nanotechnology has generated a variety of goods with added value for use in daily life. Nanotechnology is a field that integrates the life sciences, and it has expanded to include nanomaterial creation and energy production [76, 81]. It entails the fabrication, manipulation, characterization, production, and application of devices, systems, and structures by manipulating their size and shape on a nanoscale (1–100 nm) [116]. Nanotechnology investigation has accelerated, notably for specific applications in medicine, energy production, molecular computing, the environment, food packaging, water treatment, agriculture and forestry sectors [53, 54, 58–60, 74, 87, 124, 125]. Nanoparticles have considerably different chemical and physical characteristics than macroscale or bulk objects made of the same substance. The approach utilized to synthesize nanomaterials may be divided into bottom-up and top-down. Bulk particles are progressively reduced to smaller particles using physical techniques [63]. Bottom-up approaches include the assembly of atoms and molecules of a material to make nanoparticles of varying sizes and shapes. This method often creates nanomaterials with homogeneous size, shape, and distribution. Based on the nature of the synthesis, the bottom-up method may be classed as a chemical, physical, and biological process [120]. Physical techniques use thermal energy, electrical energy, mechanical pressure, and high-energy radiations to form nanoparticles by inducing material abrasion, melting, evaporation, or condensation. Chemical techniques are ideal for synthesizing metal and metal oxide nanoparticles, in which metal ions are reduced by chemical reducing agents and capping agents to stabilize the nanoparticles [123]. Biological techniques used mostly bioactive compounds from vegetables (alkaloids, terpenoids, tannins, polyphenols, etc.), micro-organisms (algae, bacteria, fungus, etc.), and biomolecules as models as decreasing and stabilizing mechanisms [50, 65, 123].

The biosynthesis of nanoparticles employing bioactive compounds from vegetables is a well-designed, simple, one-step process that produces nanoparticles of the appropriate size and form. Yet, no reductants or stabilizing agents are used in the bio-reduction phases of the biosynthetic approach [19]. Recently, multiple biological substances such as plants, mushrooms, bacteria, and fungi have made more effort to synthesize nanoparticles using green chemistry approaches. Still, biologically mediated nanoparticles may manage a variety of acute and chronic disorders [12, 76]. The green synthesis technique is practical, cost-effective, eco-friendly, and productive [16, 23]. Numerous investigations have suggested the green production of nanoparticles, and their usage is widespread in the biomedical, pharmacological, environmental, cosmetic, and food industries [23]. In this chapter, we investigated the various vegetable metabolites compounds and their use in nanoparticle synthesis and applications.

2 The Importance of Plant-Based Bioactive Compounds

Metabolites are the intermediate metabolism molecules catalyzed by various enzymes that occur naturally inside cells. Typically, only tiny compounds are referred to as metabolites [8]. In addition to fuel, structure, enzymatic activity, signaling, defense, and interactions with other species, metabolites provide a variety of additional activities. Generally, plant metabolites may be divided into two categories: Primary metabolites and secondary metabolites [21, 83, 150] (Table 1). Primary Metabolites are the chemical substances created throughout the growth and development process. Organic acids, nucleotides, alcohol, antioxidants, amino acids, polyols and vitamins comprise the primary metabolites essential for the plant's general growth and development [150]. They also participate in the basic metabolic activities of respiration, translocation, and photosynthesis. The primary metabolites are generated during the growth period. They are considered central metabolites because they run the body's physiological processes. They are intermediary products of anabolic metabolism that cells employ to synthesize important macromolecules [115].

Secondary metabolites are chemical molecules generated by plants that are not important for reproduction, proliferation, or development [14, 27]. Some of them have been identified as defensive chemicals against parasites, pathogens, predators, oxidants, and UV radiation, to aid reproduction and interspecies competition [83, 152]. Furthermore, secondary metabolites are used in signaling pathways. Thus, metabolites are helpful fuels, medications, cosmetics, dietary supplements, and chemicals with various uses in agriculture, ecology, and medicine [8]. Around 2,140,000 known secondary metabolites are often categorized according to their wide structural, functional, and biosynthetic variety. There are many major categories of secondary metabolites, including alkaloids, steroids and terpenoids, enzyme cofactors, fatty acid-derived substances and polyketides, and nonribosomal polypeptides [61, 85].

2.1 Alkaloids

Alkaloids are a diverse class of moderate compounds containing nitrogenous bases that are highly bioactive and bitter-tasting. Chemically, alkaloids are often composed of one or more carbon rings containing a nitrogen atom, and the vast majority are generated from amino acids [99]. Moreover, its derivation from diverse plant families or various classes of alkaloids causes the position of the nitrogen atom in the carbon ring to vary, resulting in the alkaloid family's distinctive structures and activities [108]. More than 12,000 naturally occurring alkaloids from over 318 plant species have been found. Additionally, alkaloids are extracted from animals, fungi, and microorganisms [84, 152].

Table 1 Different categories and examples of primary and secondary metabolites

Metabolite types	Categories	Examples	Role
Primary metabolites	Amino acids	Methionine, valine, histidine, isoleucine, tryptophan, leucine, phenylalanine, lysine, threonine, glutamic acid, and aspartic acid	Physiological processes: Respiration, photosynthesis, development, growth, translocation, and reproduction of the organism
	Organic acids	Acetic acid, formic acid, citric acid, lactic acid, malic acid, oxalic acid, tartaric acid, uric acid	
	Polyols	Erythritol, isomalt, lactitol, glycerol, sorbitol, mannitol, maltitol, maltitol syrup, xylitol	
	Alcohols	Ethanol, methanol, 1-pentanol, 1-propanol, 1-butanol, 1-hexanol, cyclohexanol, 1-heptanol	
	Nucleotides	Adenilyc acid, guanylic acid, uridylic acid	
	Vitamins	Vitamins A, C, D, E, and K, vitamin B6, vitamin B12, choline, niacin, biotin, thiamin, pantothenic acid, riboflavin, and folic acid	
	Antioxidants	Isoascorbic acid, selenium, beta-carotene, lycopene, lutein, and zeaxanthin	
Secondary metabolites	Alkaloids	Atropine, berberine, caffeine, colchicine, codeine, coniine, ephedrine, morphine, nicotine, strychnine, quinine, scopolamine, hyoscamine, and sangunarine	Ecological function: Defense mechanisms, antibiotics, and production of pigments
	Terpenoids and steroids	Camphor, cetrol, citral, limonene, menthol, nerol, pinene, nerolidol, farnesol, cortisone, methylprednisolone, and prednisone	
	Nonribosomal polypeptides	Actinomycin, ACV-tripeptide, bacitracin, bleomycin, ciclosporin, enterobactin, epothilone, microcystins, cyanophycin	

(continued)

Table 1 (continued)

Metabolite types	Categories	Examples	Role
	Fatty acid-derived substances and polyketides	Fatty alkanes, fatty alcohols, fatty acid ethyl/methyl esters, hydroxy fatty acids, macrolides, pikromycin, erythromycin A clarithromycin, and azithromycin	
	Enzyme cofactors	Flavin, folic acid, thiamine, iron–sulfur clusters, and metal ions Mg^{2+}, Cu + , Mn^{2+}	

2.2 *Terpenoids and Steroids*

Terpenoids and steroids constitute one of the biggest natural substance groups. They have a carbon skeleton composed of C5 isoprene units as a unifying characteristic. These are composed of isopentenyl diphosphate, which is generated from either mevalonic acid or methylerythritol phosphate [138, 140]. There are currently approximately 35,000 identified terpenoid and steroid molecules. Terpenoids contain a range of unconnected structures, while steroids have a tetracyclic carbon skeleton derived from triterpene lanosterol [18, 69].

2.3 *Fatty Acid-Derived Substances and Polyketides*

Fatty acids are essential ingredients of lipids (the fat-soluble constituents of biological systems) in animals, plants, and microorganisms. Fatty acids have an even number of carbon atoms in a straight chain, hydrogen atoms throughout the length of the chain and at one end, and a carboxyl group (−COOH) at the other end [62, 144, 148]. Polyketide chains are composed of malonyl CoA units and a starting unit, often acetyl CoA, but may also be various CoA esters. Some of the keto groups are often reduced during the biosynthesis of derivatives; however, this happens after the creation of polyketide chains. Due to the multiple secondary changes, including aromatic ring breakage, polyketide-derived compounds show a broad scope of structural variety [141]. Approximately 10,000 molecules have been found and biosynthesized from simple acyl precursors such as methylmalonyl CoA, acetyl CoA, and propionyl CoA [142].

2.4 Enzyme Cofactors

Enzyme cofactors are non-protein chemical compounds necessary for the biological action of proteins. Numerous enzymes need cofactors for optimal functioning. Cofactors may be seen as "assistant molecules" that aid enzymes in their function [85]. Cofactors may consist of ions or organic compounds (called coenzymes). Frequently, organic cofactors consist of or are derived from vitamins. Small amounts of these vitamins are required to function our enzymes properly [39, 107].

2.5 Nonribosomal Polypeptides

Nonribosomal peptides are secondary metabolites generated by nonribosomal peptide synthetases in the absence of ribosomal machinery and messenger RNA [36]. Nonribosomal peptides are generated spontaneously by microorganisms like bacteria and fungus and by the symbionts of higher eukaryotes [38, 136]. These peptides contain about 20 commercially available medications, including anticancer substances (bleomycin), antibacterial (penicillin, vancomycin), and immunosuppressants (cyclosporine) [119, 134, 136].

3 Synthesis of Nanoparticles from Bioactive Compounds Derived from Plants

Generally, nanoparticles may be created by two distinct techniques. The top-down procedure consists of physical techniques such as sonication, laser ablation, radiation, and thermal degradation [57]. Using the top-down synthesis procedure, nanoparticles are generated by lowering the size of a macroscale material to form agglomerates containing particles of the proper size [47]. The primary weaknesses of this approach are the polydispersity of the final product and the creation of defects. Furthermore, this manufacturing procedure needs a large quantity of energy and specialized laboratory equipment, which is costly [89, 100].

The second method, bottom-up, employs a nanometric structure composed of atomic and molecular constituents [47]. Nanoparticles are formed by biological and chemical synthesis. Chemical synthesis employs electrochemistry, the sol–gel method, vapor flux condensation, and chemical reduction. The latter is arguably the most extensively used approach for producing nanoparticles and employs chemical lowering agents such as sodium citrate [146]. However, chemical processes often involve the use of many chemical species or molecules. These substances may enhance particle toxicity and reactivity and negatively impact public health and the environment due to the disintegration of chemical groups, the production of byproducts, and high energy consumption [40, 133].

Biological synthesis, also known as "green synthesis", employs plant extracts, fungi, viruses, and microorganisms to produce nanoparticles [96]. Recently, green approaches to nanoparticle synthesis that use plant derivates have been explored as an option for conventional physical and chemical processes. Several metal nanoparticles such as Gold (Au), Silver (Ag), Cobalt (Co), Palladium (Pd), Magnesium (Mg), Lead (Pb), Copper (Cu), Zinc (Zn), Manganese (Mn), and Iron (Fe) are employed in green synthesis [26]. Extracts from diverse plant parts, such as flower, leaves, stem, rhizome, bark, fruit, and seed, produce nanoparticles of varying shapes and sizes [76]. Especially, plants may create an abundance of secondary metabolites, including alkaloids, phenolic chemicals, and terpenoids. Employing natural plants to make nanoparticles is environmentally friendly, more cost-effective, user-friendly, and convenient than conventional physical or chemical techniques. Plant extracts with this composition have been utilized to make nanoparticles because they contain reducing and stabilizing properties that might be useful in nanoparticle manufacturing [50]. Furthermore, plants are biological substrates rich in important phytochemicals, which decreases the need for chemicals to reduce, capping, and stabilize agents when producing nanoparticles from their precursor solutions [127].

The following methods are substantially involved in the typical plant-derivates nanoparticle production. Initially, plant materials are gathered and thoroughly washed twice to three times using double-distilled water [74, 153]. Typically, distilled water or ethanol is employed as a solvent for manufacturing plant preparations from their different sections to produce green nanoparticles. Generally, the green routes for cobalt nanoparticle production utilizing natural plant extracts are segmented according to the sequential processing stages. The first step in preparing plant extracts is washing, drying, and chopping or grinding chosen plant parts into powder form [127]. Generally, certain plant components were rinsed with distilled water and tap to dismiss dirt, epiphytes, and dust particles from their surfaces. Next step, the powdered plant components are boiled at a certain temperature with a precise dose of needed solvents (methanol, deionized water, ethanol, etc.) [47]. To separate phytochemicals (alkaloids, phenolic acids, flavonoids, terpenoids and steroids, sugars and their derivatives molecules with carboxylic, hydroxyl, amino, allyl, alkoxy and sulfhydryl groups) in appropriate solvents, filtering is necessary after a particular period of boiling. Consequently, resulting plant extracts were combined with the precursor of metal salts in a precise concentration and volume, respectively. In this procedure, phytochemicals included in the extracts may serve as stabilizing, reducing, and capping agents in the formation of nanoparticles from the precursor solution. Usually, the resulting mixture of the metal precursor and plant extract is calcined in the muffle furnace at a predetermined temperature in the presence of air or an inert condition. After calcination, the product must be washed again to eliminate contaminants from the nanoparticles' surface (Fig. 1). It has been found that additional active chemical components are present in the plant extract at varying quantities and function as bioreductors, capping agents, and stabilizers of the produced nanoparticles [1, 104, 110]. Due to the difficulties in detecting the precise quantity during nanoparticle manufacturing, antioxidant agents and phytochemicals in plant extracts cannot be precisely

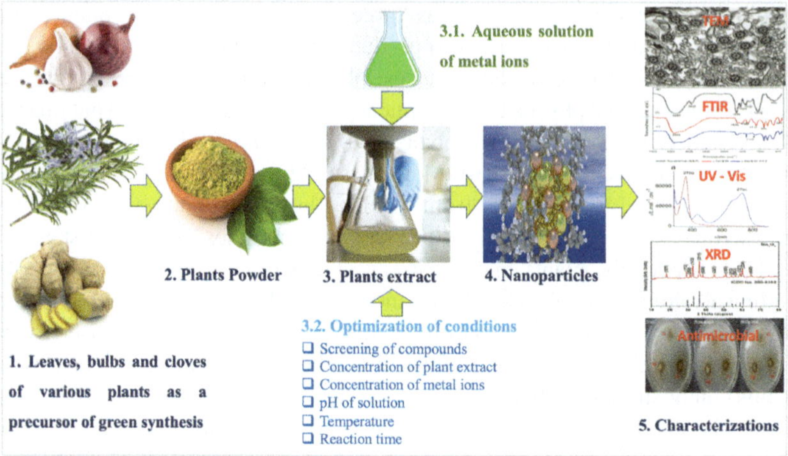

Fig. 1 A schematic illustration of nanoparticle creation utilizing natural plant compounds

quantified. Besides, the size of the generated nanoparticles changes depending on the kind and dose of phytochemicals contained in the chosen plant parts [118, 127].

4 Factors Influencing the Synthesis of Green Nanoparticles

Nanoparticle synthesis, characterization, and use are all influenced by several variables. Numerous studies have demonstrated that the characteristics of synthesized nanoparticles vary depending on the activity of the catalysts and the type of adsorbate utilized in the synthesis procedure [97, 130]. Most of them have documented the dynamic character of the nanoparticles with varying forms of signs and repercussions due to environmental and time-dependent changes, etc. [3, 97]. Other key elements that influence nanoparticle production include the temperature, pH of the solution, concentration of the extracts utilized, size, the concentration of the raw materials used, and, most importantly, the protocols employed throughout the synthesis method [10].

4.1 Temperature

Temperature is a crucial factor influencing nanoparticle creation in all three approaches. Physical procedures demand the maximum temperature (>350 °C), whereas chemical techniques require temperatures below 350 °C. In most instances, temperatures below 100 °C or room temperature are needed to produce nanoparticles utilizing environmentally friendly processes. The reaction medium's temperature specifies the type of nanoparticles produced [82, 103].

4.2 Pressure

Pressure is essential for the production of green nanoparticles. The pressure in the reaction media influences the size and shape of the nanoparticles. The speed of metal ion decrease employing natural mechanisms has been discovered to be substantially quicker under ambient pressure settings [122, 139].

4.3 pH

pH is a significant component that affects the production of nanoparticles using green technology. Studies have demonstrated that the pH of the solution medium affects the texture and size of produced nanoparticles [95, 97, 137]. Consequently, nanoparticle size may be influenced by adjusting the pH of the solution medium. Prakash and Soni proved that the pH influences the form and size of produced silver nanoparticles [131].

4.4 Time

The incubation duration of the response medium substantially affects the quality and kind of nanoparticles manufactured utilizing green technology [28]. Also, the features of the generated nanoparticles change with period and are significantly impacted by the light exposure, storage conditions, product method, etc. [73]. Divergences in time may appear in various forms, including the accumulation of particles during the long-term hold, the shrinking or enlargement of particles during long-term storage, the existence of shelf life, etc., all of which impact the potential [88].

4.5 Environment

The environmental condition has a substantial effect on the features of nanoparticles generated. In numerous situations, a single nanoparticle rapidly transforms into core–shell nanoparticles by interacting or absorbing with various environmental components through oxidation or corrosion [117]. Generated nanoparticles create a covering in a physical system that makes them wider and more significant in measurement [80]. Further, the environment influences the chemistry structure and physical of produced nanoparticles. Few instances demonstrate how the environment affects the nature of produced nanoparticles. Similarly, the chemical composition of cerium nitrate nanoparticles is affected by the peroxide concentration in the fluid in which they are suspended [73, 88, 97].

4.6 Particle Shape and Size

Particle size is a critical factor in defining the characteristics of nanoparticles. For example, it has been found that the melting point of nanoparticles decreases as their size approaches the nanoscale scale [6]. Various nanoparticle structures have comparable energy, facilitating their form modification [147]. The energy often employed during the study of nanoparticles induces a transformation in the nanoparticles' form. The dynamic nature and structure of produced nanoparticles significantly influence their chemical characteristics [9].

4.7 Other Factors

Diverse biological systems, including plants, are abundant in secondary metabolites that serve as stabilizing and decreasing agents for the creation of nanoparticles. Furthermore, the content of these metabolites changes by plant species, plant portion, and extraction method [94, 97]. The cost-effectiveness of the synthesis procedure is an important element influencing nanoparticle products. Although the chemical synthesis process produces an increased output in a brief moment, it is not a cost-effective process. Consequently, synthesis by physical and chemical processes may be restricted, while biological production of nanoparticles is scalable and less expensive [97]. The porosity of produced nanoparticles significantly impacts the nanoparticles' quality and applications. It has proven possible to immobilize biomolecules onto nanoparticles to expand their usage in medication delivery and biological research [111]. Besides, various microbes generate varying amounts of diverse extracellular and intracellular enzymes that influence nanoparticle production [46]. Moreover, the procedure utilized to purify the produced nanoparticles might affect the quantity and quality of the nanoparticles. Centrifugation is used in certain circumstances to separate nanoparticles established on gravitational power [10].

5 Application of Nanoparticles

Because of their unique features, nanoparticles are utilized in everyday life. they have been employed in developing optical goods, textiles, fabric cleaners, cosmetics, nanoantibiotics, optics, energy and composites agricultural products, diagnostics and biosensors, electronics, disinfectants, food packaging materials, medicine, and so on [1, 2, 23, 52, 55, 56, 74, 101, 121, 127]. Furthermore, biological and pharmacological investigations are the most appropriate fields for nanoparticles [43, 104, 127]. In addition to their uses in medicinal products, nanoparticles derived from noble metals such as silver, gold, platinum, and palladium are frequently utilized in consumer items such as lotions, soaps, shoes, and toothpaste shampoos (Fig. 2).

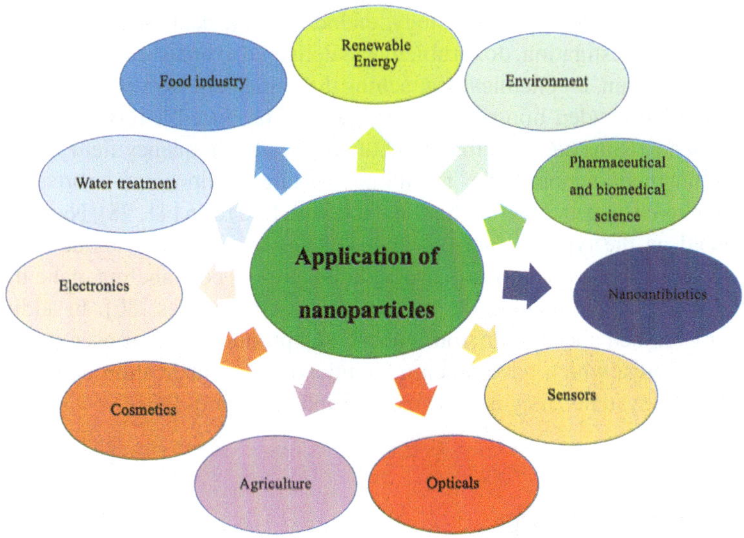

Fig. 2 Bio-prospective applications of plant-derived nanoparticles

5.1 In Pharmaceutical and Biomedical Science

The pharmaceutical use of nanoparticles is acquiring prominence, with a rising number of nanoparticle-based medicines under clinical outcome. Nanotechnology is now being implemented in diagnostics, medication delivery methods, bone replacement implants, and biosensing and bioimaging technologies [7, 22, 25, 105, 149]. As drug delivery vehicles, nanoparticles may boost therapeutic efficacy, reduce adverse effects associated with current medications, and allow a new class of therapies [143]. In addition to their general uses, nanoparticles have various applications, such as targeted medication delivery, imaging, optical imaging, photoablation treatment, and hyperthermia [86].

Recent achievements of nanoparticle therapies have piqued the attention of academic and industrial researchers in nanomedicine. In the last ten years, the growth rate in the discovery rate has led to the creation of increasingly sophisticated nanoparticle systems. These comprise an expanding number of nanoscale vehicles with different biological, chemical, and physical features for various therapeutic reasons [37, 45, 151]. Metal nanoparticles have been widely employed in pharmaceutical applications, including illness diagnosis, medicine, genetic engineering, medication delivery processes, tissue engineering, etc. Metal nanoparticles are frequently utilized as catalysts and in several biomedical activities in combination with other nanomaterials such as core–shell, bimetallic nanostructures, and alloy. Also, metal nanoparticles have numerous applications, such as catalysis, sensing, electrocatalysis, and plasmonic wave directing [24]. These nanoparticles offer therapeutic promise in almost every medical specialty, including cardiology, neurology, cancer, pulmonary medicine, immunology, ophthalmology, endocrinology, dentistry, and orthopedics [37]. In mouse investigation, doxorubicin was delivered to breast cancer cells through a nanoparticle chain. By chemically attaching three magnetic iron-oxide nanospheres to a doxorubicin-loaded liposome, the researchers produced a 100 nm-long chain. After the nano chains had penetrated the tumor, a radio-frequency field was used to cause the magnetic nanoparticles to vibrate, causing the liposome to break and the medication to be dispersed throughout the tumor in free form [41, 98]. Nanoparticles also depend on the type of nanoparticle used; for instance, iron oxide nanoparticles have important applications in drug administration, cell labeling, gene transfer, hyperthermia, and bioimaging due to their various properties [86]. In addition to the uses mentioned above, numerous nano-based products such as user-synthesized hybrid nanowires, which have potential applications in medication administration and diagnostic systems, may now be used in optical devices [49].

5.2 As Nanoantibiotics

Resistance to antibiotics has been progressively common over the last several decades, resulting in many human fatalities. Globally, the emergence and re-emergence of infections have evolved into a severe public health issue, as has the fast rise of antibiotic-resistant Gram-negative and Gram-positive pathogenic bacteria [113, 123, 127]. The extensive list of drug-resistant microorganisms includes macrolide-resistant *Streptococcus pyogenes*, sulfonamide, multi-drug resistant *Mycobacterium tuberculosis*, penicillin, methicillin-resistant *Staphylococcus aureus*, *K. pneumoniae*, *E. coli*, *Shigella flexneri*, vancomycin-resistant *Enterococcus*, penicillin-resistant *Streptococcus pneumoniae*, *P. aeruginosa*, penicillin-resistant *Neisseria gonorrhoeae*, *Vibrio cholerae*, *Salmonella enterica*, *E. cloacae*, *Acinetobacter baumannii*, and beta-lactamase- expressing *Haemophilus influenzae* [127, 135].

5.3 In the Food Industry

Nano-Ag is used in diverse mechanical systems because silver metal is a very effective heat conductor. It is primarily used in heat-sensitive instruments, such as the PCR lid and UV-spectrophotometer. Coated nanosilver is used to create instrumentation components. It is stable at high temperatures and does not affect samples [77, 90, 132, 145]. Due to their multiple open-scale activities, such as production, processing, and transportation of raw materials, food products in the food industry are highly contaminated with microorganisms. Consequently, it is necessary to design a cost-effective biosensor for evaluating the quality of items. As biosensors, metallic nanoparticles have been created; they successfully detect pathogens and cheaply monitor the various phases of contamination [15, 109].

5.4 In Cosmetics

In the cosmetic and food fields, nanoparticles are employed as a preservative. Nanoparticles of metal are utilized in different commercial applications, including food preservatives, cosmetics, and pharmaceutical coatings [67, 68]. Nanoscale metal nanoparticles, such as silver, gold, and platinum, are widely used in various commercial items, including detergent, shampoo, shoes, and soap. Most of the chemical components are manufactured, and they adversely impact humans [77]. Thus, green metallic nanoparticles are a substitute for preservation compounds in the healthcare and food sectors.

5.5 In Agriculture

Due to their comprehensive physiochemical features in shape, size, surface energy, surface area, agglomeration, charge, aggregation, crystallinity, and chemical composition, nanoparticle-based antimicrobials are effective. In previous studies, the microbicidal properties of many inorganic nanoparticles, including Titanium dioxide (TiO_2), Silver (Ag), Copper oxide (CuO), Magnesium oxide (MgO), and Zinc oxide (ZnO), were analyzed independently or in combination with biopolymer [78, 101]. Consequently, it is critical to develop novel green synthesis-based nanoparticles qualified for regulating fungal phytopathogens using biofunctionalized antimicrobial nanoparticles to cover plants in an ecologically friendly, long-term way, and cost-effective [13].

5.6 Wastewater Treatment

Nowadays days, nanoproducts have several uses in everyday day. There are also several very effective eco-friendly nanoproducts on the store, including homemade products, water purifiers, teeth and bone cement, and beauty lotion [35, 72]. Silica, platinum, and silver nanoparticles, for example, have different uses in cosmetics and personal care and are utilized as components in many goods, including toothpaste, hair care products, sunscreens, anti-ageing creams, and mouthwashes, and scents [75, 128]. Silica nanoparticles are utilized as components in a variety of commercial goods. Furthermore, adjusted silica nanoparticles are employed as effective pesticide management in several non-agricultural applications.

5.7 Other Applications

Probably one of the best advantages of nanoparticles in the environmental field is the bioremediation and treatment of water using a variety of procedures, principally by the adsorption of dangerous substances, the elimination of pathogenic organisms, heavy metals, and other contaminants, and the transformation of toxic into a non-toxic or less harmful form [42]. By manufacturing distinct nanomaterials, scientists and researchers realize their objective of creating a healthy world free of pollution. Using silver nano-catalysts inhibits or reduces the generation of by-products during the synthesis of propylene oxide, a common chemical used in detergents, paints, brake fluid, plastics, etc. It has been discovered that most iron and iron-containing nanomaterials serve as catalysts for the photodegradation-based elimination of toxic elements from organic pollutants to purify groundwater [29].

Silver nanoparticles derived from plants are also utilized to address various environmental issues, including air disinfection, wastewater treatment, groundwater treatment, and surface disinfection [151]. Duran and his colleagues discovered in 2007 that *Fusarium oxysporum* created silver nanoparticles that may be employed in textile textiles to reduce bacterial infections such as *Staphylococcus aureus* [34].

Due to their unique optical and electrical effects, nanoparticles may be employed to construct biosensors. Yeast cells have been reported to synthesize Au–Ag nanoparticles to produce vanillin sensors responsive to chemical and electrical stimulation [112]. Using a vanillin sensor based on an Au–Ag alloy nanoparticle-modified carbon electrode could boost the electrochemical response of vanillin by at least a factor of five, according to a thorough analysis [64].

Nanoparticles have played a crucial role in developing more efficient systems for collecting wind energy, solar energy, and fuel cells and harvesting and storing power in clean or renewable energy. Nanoparticles may be used to develop materials that are heat resistant, flexible, and high-performance electrodes for lithium-ion batteries. Nanoporous materials like zeolites, which might be employed as heat storage in residential and industrial areas, could potentially improve thermal energy storage [4, 51, 71]. In electronics, nanotechnology has helped eliminate obstacles and circumvent limits. Nanoelectronics is the application of nanotechnology to electrical devices, most notably transistors. Nanoelectronics may enhance production nets on electronic devices, modernize several electronic goods and processes, and lower their weight, power consumption, and integrated circuit transistor size [70, 71, 102]. It has also been noted that plant-derived nanoparticles have potential benefits in mosquito control. Salunkhe et al. [114] and Banu and Balasubramanian [11] were able to test silver and gold nanoparticles against malaria (*Anopheles stephensi*) and the mosquito that spreads dengue (*Aedes aegypti*) without any problems.

6 Nanoparticles and Their Harmful Effects on Human Health and the Environment

The possible damage of nanoparticles to organs and systems in the body has steadily been identified, which may affect nanoparticle biomedical applications [129]. The extensive usage of nanoparticles in agriculture products, industrial, and consumer will drastically increase human exposure to particles in this size range [20]. The developing area of nanotoxicology is informed by research on the negative health consequences of exposure to tiny particles like welding fumes, asbestos, coal and silica dust, and air pollution. Still, much more investigation is required to comprehend the health dangers posed by nanoparticles utilized in hundreds of items throughout the globe [5, 31]. The toxic effects of nanoparticles are often attributed to their exceedingly tiny size. Nanoparticles possess a more extensive responsive contact area than bigger particles, are more chemically reactive, and generate a more significant number of reactive oxygen species, including free radicals [92]. Reactive oxygen species have been discovered in nanoparticles, including metal oxides, carbon fullerenes, and carbon nanotubes. This is one of the critical causes of nanoparticle toxicity; it may cause oxidative stress, inflammation, damage to membranes, DNA, and proteins [91, 106]. The human body far more easily absorbs nanoparticles than larger-sized particles due to their incredibly tiny size. Nanoparticles can traverse cellular membranes and access cells, tissues, and organs, but bigger particles cannot [48]. Following inhalation or consumption, nanoparticles may penetrate the bloodstream. At least some nanoparticles can penetrate the skin, mainly when the skin is flexed [30, 66, 92]. Nanoparticles may be carried throughout the system and absorbed by organs and tissues such as the heart, kidneys, brain, bone marrow, liver, spleen, and nervous system. Nanoparticles are toxic to cell cultures and human tissue, causing an increase in oxidative stress, the generation of inflammatory cytokines, and cell death [66, 92]. To lighten the toxicity effects of nanoparticles, it is necessary to examine the individual experiments individually, paying close attention to interactions between nanocarriers and biological systems. This method of evaluating the existing data on the toxicity of nanoparticles is essential for establishing the dependability of delivery methods. To properly transition nanoformulations from the laboratory level to their clinical use, it is necessary to establish their safety profiles, including immunotoxicity investigations. Size is undoubtedly an essential aspect in evaluating a particle's potential toxicity. Furthermore, it is not the only element to consider. Other nanoparticle features that affect toxicity include surface structure, aggregation, chemical composition, surface charge, shape, and solubility, as well as the existence of active classes from other compounds. Particular focus should be placed on the relationship between the physicochemical and functional properties of nanoparticles and the degree to which these characteristics assist in achieving the intended available potential and mitigating any health hazards [126].

The variety of impact data and nanoparticles makes determining the environmental dangers of individual nanomaterials challenging. Most of the existing material is in the aquatic environment, and there is very little data regarding the risks of nanoparticles in soils and sediments [17]. The possible detrimental consequences of transformation products created following the introduction of a nanomaterial into the environment are receiving more attention [32]. Models that characterize nanoparticle release, dispersion in the environment, and exposure of live species are currently limited, as are data to verify the models. Progress in applying computational instruments and methodologies for determining and measuring nano-characteristics in different conditions is required to acquire insights into the existence and exposure to nanoparticles [33, 44, 79].

7 Conclusion and Future Prospective

Recently, a series of investigations have been documented about nanoparticle synthesis using a green procedure based on plant-derived secondary metabolites. Green synthesis is gaining popularity as a cost-effective nanoparticle synthesis method due to its many benefits in rapid synthesis, biocompatibility, non-toxic reagents, ease of processing, low energy requirement, long-term stability and easy scale compared to wet physical processing or chemical ways of synthesis. The plant metabolites promote the creation of environmentally benign green nanoparticles. Synthesis of eco-friendly nanoparticles via manipulating plant crude extracts and purified metabolites offers a potential substrate for industrial-scale production. The plant-mediated nanoparticles have the potential to be exploited in diverse disciplines, including pharmacology, medicine, agriculture, cosmetics, electronics, renewable and sustainable energy, and commercial products. But further significant investigations are required to explore and evaluate the long-term effects of nanoparticle plant-derivatives; if they are safe, they could be employed as alternatives to conventional products in different applications. It is also necessary to precisely investigate the interaction between plant-derived nanoparticles and other elements that may increase toxic nanoparticle effects, such as application method, plant age and species, ambient variables, exposure time, nanoparticle concentration and shape, and others.

References

1. Abinaya S, Kavitha HP, Prakash M, Muthukrishnaraj A (2021) Green synthesis of magnesium oxide nanoparticles and its applications: a review. Sustain Chem Pharm 19:100368
2. Abate L, Bachheti A, Bachheti RK, Husen A, Pandey DP, Megra MB (2022) Coir fiber-based cellulose, nanocellulose, and their cutting-edge applications. In: Coir Fiber and its Composites, Jawaid M (ed) Elsevier Inc., 50 Hampshire Street, 5th Floor, Cambridge, MA 02139, United States, pp 309–325
3. Adewale OB, Davids H, Cairncross L, Roux S (2019) Toxicological behavior of gold nanoparticles on various models: influence of physicochemical properties and other factors. Int J Toxicol 38(5):357–384

4. Ahmadi MH, Ghazvini M, Alhuyi Nazari M, Ahmadi MA, Pourfayaz F, Lorenzini G, Ming T (2019) Renewable energy harvesting with the application of nanotechnology: a review. Int J Energy Res 43(4):1387–1410
5. Ai J, Biazar E, Jafarpour M, Montazeri M, Majdi A, Aminifard S, Zafari M, Akbari HR, Rad HG (2011) Nanotoxicology and nanoparticle safety in biomedical designs. Int J Nanomed 6:1117
6. Akbari B, Tavandashti MP, Zandrahimi M (2011) Particle size characterization of nanoparticles—a practicalapproach. Iran J Mater Sci Eng 8(2):48–56
7. AlMasoud N, Alomar TS, Awad MA, El-Tohamy MF, Soliman DA (2020) Multifunctional green silver nanoparticles in pharmaceutical and biomedical applications. Green Chem Lett Rev 13(4):316–327
8. Altaf-Ul-Amin MD, Kanaya S, Mohamed-Hussein Z-A (2019) Investigating metabolic pathways and networks
9. Baer DR, Engelhard MH, Johnson GE, Laskin J, Lai J, Mueller K, Munusamy P, Thevuthasan S, Wang H, Washton N (2013) Surface characterization of nanomaterials and nanoparticles: important needs and challenging opportunities. J Vac Sci Technol, A: Vac, Surf Films 31(5):050820
10. Baker S, Rakshith D, Kavitha KS, Santosh P, Kavitha HU, Rao Y, Satish S (2013) Plants: emerging as nanofactories towards facile route in synthesis of nanoparticles. BioImpacts: BI 3(3):111
11. Banu AN, Balasubramanian C (2014) Myco-synthesis of silver nanoparticles using Beauveria bassiana against dengue vector, Aedes aegypti (Diptera: Culicidae). Parasitol Res 113(8):2869–2877
12. Becker RO (1999) Silver ions in the treatment of local infections. Met-Based Drugs 6(4–5):311–314
13. Begum SJ, Pratibha S, Rawat JM, Venugopal D, Sahu P, Gowda A, Qureshi KA, Jaremko M (2022) Recent advances in green synthesis, characterization, and applications of bioactive metallic nanoparticles. Pharmaceuticals 15(4):455
14. Bennett RN, Wallsgrove RM (1994) Secondary metabolites in plant defence mechanisms. New Phytol 127(4):617–633
15. Berekaa MM (2015) Nanotechnology in food industry; advances in food processing, packaging and food safety. Int J Curr Microbiol App Sci 4(5):345–357
16. Bhattarai B, Zaker Y, Bigioni TP (2018) Green synthesis of gold and silver nanoparticles: challenges and opportunities. Curr Opin Green Sustain Chem 12:91–100
17. Biswas P, Wu C-Y (2005) Nanoparticles and the environment. J Air Waste Manag Assoc 55(6):708–746
18. Brassell SC, Eglinton G, Maxwell JR (1983) The geochemistry of terpenoids and steroids. Portland Press Ltd.
19. Carlson C, Hussain SM, Schrand AM, Braydich-Stolle L., Hess, K. L., Jones, R. L., and Schlager, J. J. K (2008) Unique cellular interaction of silver nanoparticles: size-dependent generation of reactive oxygen species. J Phys Chem B 112(43):13608–13619
20. Chaachouay N, Abdelhamid A, Benkhnigue O, Zidane L, (2023) Alkaloids: a suitable precursor for nanomaterials synthesis, and their various applications. In: Husen A (eds) Secondary metabolites based green synthesis of nanomaterials and their applications. Smart Nanomaterials Technology. Springer, Singapore: 23-48. https://doi.org/10.1007/978-981-99-0927-8_2
21. Chaachouay N, Douira A, Hassikou R, Brhadda N, Dahmani J, Belahbib N, Ziri R, Zidane L (2020) Etude floristique et ethnomédicinale des plantes aromatiques et médicinales dans le Rif (Nord du Maroc)". PhD thesis, Département de Biologie—Université Ibn Tofail—Kénitra. https://tel.archives-ouvertes.fr/tel-03376377
22. Chaachouay N, Douira A, Zidane L (2021) Herbal medicine used in the treatment of human diseases in the Rif, Northern Morocco. Arab J Sci Eng: 1–23
23. Chaachouay N, Azeroual A, Ansari MKA, Zidane L (2023) Use of plants as medicines and aromatics by indigenous communities of Morocco: pharmacognosy, ecology and conservation. Plants as medicine and aromatics:33-44

24. Chaachouay N, Douira A, Zidane L (2021) ACOVID-19, prevention and treatment with herbal medicine in the herbal markets of Salé Prefecture, North-Western Morocco. Eur J Integr Med 42:101285
25. Choi J, Wang NS (2011) Nanoparticles in biomedical applications and their safety concerns. Biomed Eng Theory Appl 29:486
26. Chowdhury A, Ara J, Islam MS (2021) Green Synthesis of Phytochemical Nanoparticles and their Antimicrobial Activity, A Review Study. Biomed J Sci & Tech Res 34(4):26929–26935
27. Costa T, Vieira RF, Bizzo HR, Silveira D, Gimenes MA (2012) Secondary metabolites
28. Darroudi M, Ahmad MB, Zamiri R, Zak AK, Abdullah AH, Ibrahim NA (2011) Time-dependent effect in green synthesis of silver nanoparticles. Int J Nanomed 6:677
29. Das SK, Dickinson C, Lafir F, Brougham DF, Marsili E (2012) Synthesis, characterization and catalytic activity of gold nanoparticles biosynthesized with Rhizopus oryzae protein extract. Green Chem 14(5):1322–1334
30. Dick Bell D, Senjen R, Kimbrell G (2006) A report prepared for friends of the Earth Australia and friends of the Earth United States May 2006
31. Donaldson K, Poland CA (2013) Nanotoxicity: challenging the myth of nano-specific toxicity. Curr Opin Biotechnol 24(4):724–734
32. Dreher KL (2004) Health and environmental impact of nanotechnology: toxicological assessment of manufactured nanoparticles. Toxicol Sci 77(1):3–5
33. Dunphy Guzmán KA, Taylor MR, Banfield JF (2006) Environmental risks of nanotechnology: national nanotechnology initiative funding, 2000– 2004. ACS Publications
34. Durán N, Marcato PD, De Souza GI, Alves OL, Esposito E (2007) Antibacterial effect of silver nanoparticles produced by fungal process on textile fabrics and their effluent treatment. J Biomed Nanotechnol 3(2):203–208
35. Esakkimuthu T, Sivakumar D, Akila S (2014) Application of nanoparticles in wastewater treatment. Pollut Res 33(03):567–571
36. Evans BS, Chen Y, Metcalf WW, Zhao H, Kelleher NL (2011) Directed evolution of the nonribosomal peptide synthetase AdmK generates new andrimid derivatives in vivo. Chem Biol 18(5):601–607
37. Farokhzad OC, Langer R (2006) Nanomedicine: Developing smarter therapeutic and diagnostic modalities. Adv Drug Deliv Rev 58(14):1456–1459
38. Finking R, Marahiel MA (2004) Biosynthesis of nonribosomal peptides. Annu Rev Microbiol 58:453
39. Fischer JD, Holliday GL, Thornton JM (2010) The CoFactor database: organic cofactors in enzyme catalysis. Bioinformatics 26(19):2496–2497
40. Fu PP, Xia Q, Hwang H-M, Ray PC, Yu H (2014) Mechanisms of nanotoxicity: generation of reactive oxygen species. J Food Drug Anal 22(1):64–75
41. Garde D (2012) Chemo bomb'nanotechnology effective in halting tumors. Fierce Drug Delivery
42. Gaur N, Flora G, Yadav M, Tiwari A (2014) A review with recent advancements on bioremediation-based abolition of heavy metals. Environ Sci Process Impacts 16(2):180–193
43. Ghiassi S, Sedaghat S, Mokhtary M, Kefayati H (2018) Plant-mediated bio-synthesis of silver–montmorillonite nanocomposite and antibacterial effects on gram-positive and-negative bacteria. J Nanostructure Chem 8(3):353–357
44. Grieger KD, Linkov I, Hansen SF, Baun A (2012) Environmental risk analysis for nanomaterials: review and evaluation of frameworks. Nanotoxicology 6(2):196–212
45. Gu FX, Karnik R, Wang AZ, Alexis F, Levy-Nissenbaum E, Hong S, Langer RS, Farokhzad OC (2007) Targeted nanoparticles for cancer therapy. Nano Today 2(3):14–21
46. Haverkamp RG, Marshall AT, van Agterveld D (2007) Pick your carats: nanoparticles of gold–silver–copper alloy produced in vivo. J Nanopart Res 9(4):697–700
47. Hernández-Díaz JA, Garza-García JJ, Zamudio-Ojeda A, León-Morales JM, López-Velázquez JC, García-Morales S (2021) Plant-mediated synthesis of nanoparticles and their antimicrobial activity against phytopathogens. J Sci Food Agric 101(4):1270–1287

48. Holsapple MP, Farland WH, Landry TD, Monteiro-Riviere NA, Carter JM, Walker NJ, Thomas KV (2005) Research strategies for safety evaluation of nanomaterials, part II: toxicological and safety evaluation of nanomaterials, current challenges and data needs. Toxicol Sci 88(1):12–17
49. Hu X, Fagone P, Dong C, Su R, Xu Q, Dinu CZ (2018) Biological self-assembly and recognition used to synthesize and surface guide next generation of hybrid materials. ACS Appl Mater Interfaces 10(34):28372–28381. https://doi.org/10.1021/acsami.8b09421
50. Huang J, Lin L, Sun D, Chen H, Yang D, Li Q (2015) Bio-inspired synthesis of metal nanomaterials and applications. Chem Soc Rev 44(17):6330–6374
51. Hussein AK (2015) Applications of nanotechnology in renewable energies—A comprehensive overview and understanding. Renew Sustain Energy Rev 42:460–476
52. Husen A (2022) Engineered nanomaterials for sustainable agricultural production, soil improvement and stress management. Elsevier Inc., 50 Hampshire St., 5th Floor, Cambridge, MA 02139, USA. https://doi.org/10.1016/C2021-0-00054-7
53. Husen A, Iqbal M (2019) Nanomaterials and plant potential: an overview. In: Husen A, Iqbal M (eds) N*anomaterials* and plant potential. Springer International Publishing AG, Gewerbestrasse 11, 6330 Cham, pp 3–29 https://doi.org/10.1007/978-3-030-05569-1_1
54. Husen A, Jawaid M (2020) Nanomaterialsfor agriculture and forestry applications. Elsevier Inc., 50 Hampshire St., 5th Floor, Cambridge, MA 02139, USA ISBN 9780128178522 https://doi.org/10.1016/C2018-0-02349-X
55. Husen A, Siddiqi KS (2014) Phytosynthesis of nanoparticles: Concept, controversy and application. Nanoscale Res Lett 9(229):1–24
56. Husen A, Siddiqi KS (2022) Advances in smart nanomaterials and their applications. Elsevier Inc., 50 Hampshire St., 5th Floor, Cambridge, MA 02139, USA. https://doi.org/10.1016/C2021-0-02202-1
57. Iqbal P, Preece JA, Mendes PM (2012) Nanotechnology: The "top-down" and "bottom-up" approaches. Supramol Chem: Mol Nanomater
58. Jin-Chul K, Madhusudhan A, Husen A (2021) Smart nanomaterials in biomedical applications. Springer Nature Switzerland AG, Gewerbestrasse 11, 6330 Cham, Switzerland https://doi.org/10.1007/978-3-030-84262-8
59. Kumar A, Choudhary A, Kaur H, Guha S, Mehta S, Husen A (2022) Potential applications of engineered nanoparticles in plant disease management: A critical update. Chemosphere 295:133798. https://doi.org/10.1016/j.chemosphere.2022.133798
60. Kumar A, Choudhary A, Kaur H, Mehta S, Husen A (2021) Smart nanomaterial and nanocomposite with advanced agrochemical activities. Nanoscale Res Lett 16(156):1–26. https://doi.org/10.1186/s11671-021-03612-0
61. Kabera JN, Semana E, Mussa AR, He X (2014) Plant secondary metabolites: Biosynthesis, classification, function and pharmacological properties. J Pharm Pharmacol 2(7):377–392
62. Kachroo A, Kachroo P (2009) Fatty acid-derived signals in plant defense. Annu Rev Phytopathol 47(1):153–176
63. Khalaj M, Kamali M, Costa MEV, Capela I (2020) Green synthesis of nanomaterials—A scientometric assessment. J Clean Prod 267:122036. https://doi.org/10.1016/j.jclepro.2020.122036
64. Khandel P, Shahi SK (2018) Mycogenic nanoparticles and their bio-prospective applications: Current status and future challenges. J Nanostructure Chem 8(4):369–391
65. Kharissova OV, Dias HR, Kharisov BI, Pérez BO, Pérez VMJ (2013) The greener synthesis of nanoparticles. Trends Biotechnol 31(4):240–248
66. Kharlamov A, Skripnichenko A, Gubareny N, Bondarenko M, Kirillova N, Kharlamova G, Fomenko V (2011) Toxicology of nano-objects: Nanoparticles, nanostructures and nanophases. In Biodefence. pp 23–32, Springer
67. Kim BS, Song JY (2009) 28 Biological synthesis of metal nanoparticles. Biocatal Agric Biotechnol 399
68. Kokura S, Handa O, Takagi T, Ishikawa T, Naito Y, Yoshikawa T (2010) Silver nanoparticles as a safe preservative for use in cosmetics. Nanomedicine: Nanotechnol, Biol Med 6(4):570–574

69. Kolak U, Topcu G, Birteksöz S, Ötük G, Ulubelen A (2005) Terpenoids and steroids from the roots of Salvia blepharochlaena. Turk J Chem 29(2):177–186
70. Komatsu H, Ogasawara A (2005) Applying nanotechnology to electronics-recent progress in Si-LSIs to extend nano-scale
71. Korkin A, Krstić PS, Wells JC (2010) Nanotechnology for electronics, photonics, and renewable energy. vol. 78, Springer
72. Kouvaris P, Delimitis A, Zaspalis V, Papadopoulos D, Tsipas SA, Michailidis N (2012) Green synthesis and characterization of silver nanoparticles produced using Arbutus Unedo leaf extract. Mater Lett 76:18–20
73. Kuchibhatla SV, Karakoti AS, Baer DR, Samudrala S, Engelhard MH, Amonette JE, Thevuthasan S, Seal S (2012) Influence of aging and environment on nanoparticle chemistry: Implication to confinement effects in nanoceria. J Phys Chem C 116(26):14108–14114
74. Kumar S, Basumatary IB, Sudhani HP, Bajpai VK, Chen L, Shukla S, Mukherjee A (2021) Plant extract mediated silver nanoparticles and their applications as antimicrobials and in sustainable food packaging: A state-of-the-art review. Trends Food Sci Technol 112:651–666
75. Kumar V, Yadav SK (2009) Plant-mediated synthesis of silver and gold nanoparticles and their applications. J Chem Technol & Biotechnol: Int Res Process, Environ & Clean Technol 84(2):151–157
76. Kuppusamy P, Ichwan SJ, Parine NR, Yusoff MM, Maniam GP, Govindan N (2015) Intracellular biosynthesis of Au and Ag nanoparticles using ethanolic extract of Brassica oleracea L. and studies on their physicochemical and biological properties. J Environ Sci 29:151–157
77. Kuppusamy P, Yusoff MM, Maniam GP, Govindan N (2016) Biosynthesis of metallic nanoparticles using plant derivatives and their new avenues in pharmacological applications–An updated report. Saudi Pharm J 24(4):473–484
78. Küünal S, Rauwel P, Rauwel E (2018) Plant extract mediated synthesis of nanoparticles. In Emerging applications of nanoparticles and architecture nanostructures. pp 411–446, Elsevier
79. Linkov I, Steevens J, Adlakha-Hutcheon G, Bennett E, Chappell M, Colvin V, Davis JM, Davis T, Elder A, Foss Hansen S (2009) Emerging methods and tools for environmental risk assessment, decision-making, and policy for nanomaterials: Summary of NATO advanced research workshop. J Nanopart Res 11(3):513–527
80. Lynch I, Cedervall T, Lundqvist M, Cabaleiro-Lago C, Linse S, Dawson KA (2007) The nanoparticle–protein complex as a biological entity; a complex fluids and surface science challenge for the 21st century. Adv Coll Interface Sci 134:167–174
81. Madkour LH (2019) Introduction to nanotechnology (NT) and nanomaterials (NMs). In Nanoelectronic Materials. pp 1–47, Springer
82. Makarov VV, Love AJ, Sinitsyna OV, Makarova SS, Yaminsky IV, Taliansky ME, Kalinina NO (2014) "Green" nanotechnologies: Synthesis of metal nanoparticles using plants. Acta Naturae (Англоязычная Версия) 6(1 (20)): 35–44
83. Makkar HP, Siddhuraju P, Becker K (2007) Plant secondary metabolites. vol. 393, Springer
84. Manske RHF, Holmes HL (2014) The alkaloids: Chemistry and physiology. Elsevier
85. McMurry JE (2014) Organic chemistry with biological applications. Cengage Learning
86. McNamara K, Tofail SA (2017) Nanoparticles in biomedical applications. Adv Phys: X 2(1):54–88
87. Mondal R, Dam P, Chakraborty J, Paret ML, Katı A, Altuntas S, Sarkar R, Ghorai S, Mandal AK, Husen A (2022) Potential of nanobiosensor in sustainable agriculture: The state-of-art. Heliyon. https://doi.org/10.2139/ssrn.4134873
88. Mudunkotuwa IA, Pettibone JM, Grassian VH (2012) Environmental implications of nanoparticle aging in the processing and fate of copper-based nanomaterials. Environ Sci Technol 46(13):7001–7010
89. Murphy CJ (2002) Nanocubes and nanoboxes. Science 298(5601):2139–2141
90. Naseer B, Srivastava G, Qadri OS, Faridi SA, Islam RU, Younis K (2018) Importance and health hazards of nanoparticles used in the food industry. Nanotechnol Rev 7(6):623–641
91. Nel A, Xia T, Madler L, Li N (2006) Toxic potential of materials at the nanolevel. Science 311(5761):622–627

92. Oberdörster G, Maynard A, Donaldson K, Castranova V, Fitzpatrick J, Ausman K, Carter J, Karn B, Kreyling W, Lai D, Olin S, Monteiro-Riviere N, Warheit D, Yang H, A report from the ILSI Research Foundation/Risk Science Institute Nanomaterial Toxicity Screening Working Group (2005) Principles for characterizing the potential human health effects from exposure to nanomaterials: Elements of a screening strategy. Part Fibre Toxicol 2(1): 8https://doi.org/10.1186/1743-8977-2-8
93. Orch H, Chaachouay N, Douiri EM, Faiz N, Zidane L, Douira, A (2021) Use of medicinal plants in dermato-cosmetology: An ethnobotanical study among the population of Izarène. Jordan Journal of Pharmaceutical Sciences 14(3):323-340
94. Park Y, Hong YN, Weyers A, Kim YS, Linhardt RJ (2011) Polysaccharides and phytochemicals: A natural reservoir for the green synthesis of gold and silver nanoparticles. IET Nanobiotechnol 5(3):69–78
95. Parsons JG, Peralta-Videa JR, Gardea-Torresdey JL (2007) Use of plants in biotechnology: Synthesis of metal nanoparticles by inactivated plant tissues, plant extracts, and living plants. Dev Environ Sci 5:463–485
96. Parveen K, Banse V, Ledwani L (2016) Green synthesis of nanoparticles: Their advantages and disadvantages. AIP Conf Proc 1724(1):020048
97. Patra JK, Baek K-H (2014) Green nanobiotechnology: Factors affecting synthesis and characterization techniques. J Nanomater
98. Peiris PM, Bauer L, Toy R, Tran E, Pansky J, Doolittle E, Schmidt E, Hayden E, Mayer A, Keri RA (2012) Enhanced delivery of chemotherapy to tumors using a multicomponent nanochain with radio-frequency-tunable drug release. ACS Nano 6(5):4157–4168
99. Pelletier SW (1983) Alkaloids: Chemical and biological perspectives
100. Pirtarighat S, Ghannadnia M, Baghshahi S (2019) Green synthesis of silver nanoparticles using the plant extract of Salvia spinosa grown in vitro and their antibacterial activity assessment. J Nanostructure Chem 9(1):1–9
101. Qamar SUR, Ahmad JN (2021) Nanoparticles: Mechanism of biosynthesis using plant extracts, bacteria, fungi, and their applications. J Mol Liq 334:116040
102. Rae A (2006) Real life applications of nanotechnology in electronics. OnBoard Technol 2006:28
103. Rai A, Singh A, Ahmad A, Sastry M (2006) Role of halide ions and temperature on the morphology of biologically synthesized gold nanotriangles. Langmuir 22(2):736–741
104. Ramesh P, Saravanan K, Manogar P, Johnson J, Vinoth E, Mayakannan M (2021) Green synthesis and characterization of biocompatible zinc oxide nanoparticles and evaluation of its antibacterial potential. Sens Bio-Sens Res 31:100399
105. Ramos AP, Cruz MAE, Tovani CB, Ciancaglini P (2017) Biomedical applications of nanotechnology. Biophys Rev 9:79–89
106. Ravinayagam V, Jermy BR (2020) Nanomaterials and their negative effects on human health. Appl Nanomater Hum Health: 249–273
107. Richter M (2013) Functional diversity of organic molecule enzyme cofactors. Nat Prod Rep 30(10):1324–1345
108. Roberts MF (2013) Alkaloids: Biochemistry, ecology, and medicinal applications. Springer Science & Business Media
109. Rogers MA (2016) Naturally occurring nanoparticles in food. Curr Opin Food Sci 7:14–19
110. Ruangtong J, Jiraroj T, T-Thienprasert, N. P. (2020) Green synthesized ZnO nanosheets from banana peel extract possess anti-bacterial activity and anti-cancer activity. Mater Today Commun 24:101224
111. Ruckenstein E, Kong XZ (1999) Control of pore generation and pore size in nanoparticles of poly (styrene-methyl methacrylate-acrylic acid). J Appl Polym Sci 72(3):419–426
112. Saha S, Sarkar J (2010) In Chattopadhyay D, Patra S, Chakraborty A, Acharya K (eds) 887–895
113. Saka R, Chella N (2021) Nanotechnology for delivery of natural therapeutic substances: A review. Environ Chem Lett 19(2):1097–1106

114. Salunkhe RB, Patil SV, Salunke BK, Patil CD, Sonawane AM (2011) Studies on silver accumulation and nanoparticle synthesis by Cochliobolus lunatus. Appl Biochem Biotechnol 165(1):221–234
115. Sanchez S, Demain AL (2008) Metabolic regulation and overproduction of primary metabolites. Microb Biotechnol 1(4):283–319
116. Sanders WC (2018) Basic principles of nanotechnology. CRC Press
117. Sarathy V, Tratnyek PG, Nurmi JT, Baer DR, Amonette JE, Chun CL, Penn RL, Reardon EJ (2008) Aging of iron nanoparticles in aqueous solution: Effects on structure and reactivity. J Phys Chem C 112(7):2286–2293
118. Saravanan P, SenthilKannan K, Vimalan M, Tamilselvan S, Sankar D (2020) Biofriendly and competent domestic microwave assisted method for the synthesis of ZnO nanoparticles from the extract of Azadirachta indica leaves. Mater Today: Proc 33:3160–3163
119. Schwarzer D, Finking R, Marahiel MA (2003) Nonribosomal peptides: From genes to products. Nat Prod Rep 20(3):275–287
120. Schwille P, Spatz J, Landfester K, Bodenschatz E, Herminghaus S, Sourjik V, Erb TJ, Bastiaens P, Lipowsky R, Hyman A (2018) MaxSynBio: Avenues towards creating cells from the bottom up. Angew Chem Int Ed 57(41):13382–13392
121. Sedaghat S, Agbolag AE (2016) Biosynthesis of silver nanoparticles using pennyroyal water extract as a green route. J Nanostructure Chem 6(1):25–27
122. Shah M, Fawcett D, Sharma S, Tripathy SK, Poinern GEJ (2015) Green synthesis of metallic nanoparticles via biological entities. Materials 8(11):7278–7308
123. Sharma VK, Yngard RA, Lin Y (2009) Silver nanoparticles: Green synthesis and their antimicrobial activities. Adv Coll Interface Sci 145(1–2):83–96
124. Sharma P, Pandey V, Sharma MMM, Patra A, Singh B, Mehta S, Husen A (2021) A review on biosensors and nanosensors application in agroecosystems. Nano Res Lett 16:136. https://doi.org/10.1186/s11671-021-03593-0
125. Siddiqi KS, Husen A (2017) Recent advances in plant-mediated engineered gold nanoparticles and their application in biological system. J Trace Elem Med Biol 40:10–23
126. Shvedova AA, Kagan VE, Fadeel B (2010) Close encounters of the small kind: Adverse effects of man-made materials interfacing with the nano-cosmos of biological systems. Annu Rev Pharmacol Toxicol 50:63–88
127. Singh AK (2022) A review on plant extract-based route for synthesis of cobalt nanoparticles: Photocatalytic, electrochemical sensing and antibacterial applications. Curr Res Green Sustain Chem 100270
128. Singh S, Kumar V, Romero R, Sharma K, Singh J (2019) Applications of nanoparticles in wastewater treatment. In Nanobiotechnology in bioformulations. pp 395–418, Springer
129. Singh SP, Bhargava CS, Dubey V, Mishra A, Singh Y (2017) Silver nanoparticles: Biomedical applications, toxicity, and safety issues. Int J Res Pharm Pharm Sci 4(2):1–10
130. Somorjai GA, Park JY (2008) Colloid science of metal nanoparticle catalysts in 2D and 3D structures. Challenges of nucleation, growth, composition, particle shape, size control and their influence on activity and selectivity. Top Catal 49(3): 126–135
131. Soni N, Prakash S (2011) Factors affecting the geometry of silver nanoparticles synthesis in Chrysosporium tropicum and Fusarium oxysporum. Am J Nanotechnol 2(1):112–121
132. Souza VGL, Fernando AL (2016) Nanoparticles in food packaging: Biodegradability and potential migration to food—A review. Food Packag Shelf Life 8:63–70
133. Stoytcheva M, Zlatev R (2020) Applications of nanobiotechnology. BoD—Books on Demand
134. Strieker M, Tanović A, Marahiel MA (2010) Nonribosomal peptide synthetases: Structures and dynamics. Curr Opin Struct Biol 20(2):234–240
135. Sunny NE, Kaviya A, Kumar SV (2022) Mechanistic approach on the synthesis of metallic nanoparticles from microbes. In Agri-Waste and Microbes for Production of Sustainable Nanomaterials. pp 577–602, Elsevier
136. Süssmuth RD, Mainz A (2017) Nonribosomal peptide synthesis—Principles and prospects. Angew Chem Int Ed 56(14):3770–3821

137. Thakkar KN, Mhatre SS, Parikh RY (2010) Biological synthesis of metallic nanoparticles. Nanomedicine: Nanotechnol, Biol Med 6(2): 257–262
138. Tong WY (2013) Biotransformation of terpenoids and steroids. Springer, Natural Products, pp 2733–2759
139. Tran QH, Le A-T (2013) Silver nanoparticles: Synthesis, properties, toxicology, applications and perspectives. Adv Nat Sci: Nanosci Nanotechnol 4(3):033001
140. Turner AB (1969) Terpenoids and steroids. Annu Rep Sect "B" (Org Chem) 66: 389–422
141. Vickery ML, Vickery B (1981a) Polyketides. In Secondary Plant Metabolism. pp 88–111, Springer
142. Vickery ML, Vickery B (1981b) Secondary plant metabolism. Macmillan Press
143. Wanigasekara J, Witharana C (2016) Applications of nanotechnology in drug delivery and design-an insight. Curr Trends Biotechnol Pharm 10(1):78–91
144. Weber H (2002) Fatty acid-derived signals in plants. Trends Plant Sci 7(5):217–224
145. Weiss J, Takhistov P, McClements DJ (2006) Functional materials in food nanotechnology. J Food Sci 71(9):R107–R116
146. Wuithschick M, Birnbaum A, Witte S, Sztucki M, Vainio U, Pinna N, Rademann K, Emmerling F, Kraehnert R, Polte J (2015) Turkevich in new robes: Key questions answered for the most common gold nanoparticle synthesis. ACS Nano 9(7):7052–7071
147. Yacamán MJ, Ascencio JA, Liu HB, Gardea-Torresdey J (2001) Structure shape and stability of nanometric sized particles. J Vac Sci & Technol B: Microelectron Nanometer Struct Process, Meas, Phenom 19(4):1091–1103
148. Yu A-Q, Pratomo Juwono NK, Leong SSJ, Chang MW (2014) Production of fatty acid-derived valuable chemicals in synthetic microbes. Front Bioeng Biotechnol 2:78
149. Zafar N, Madni A, Khalid A, Khan T, Kousar R, Naz SS, Wahid F (2020) Pharmaceutical and biomedical applications of green synthesized metal and metal oxide nanoparticles. Curr Pharm Des 26(45):5844–5865
150. Zaynab M, Fatima M, Sharif Y, Zafar MH, Ali H, Khan KA (2019) Role of primary metabolites in plant defense against pathogens. Microb Pathog 137:103728
151. Zhang L, Gu FX, Chan JM, Wang AZ, Langer RS, Farokhzad OC (2008) Nanoparticles in medicine: Therapeutic applications and developments. Clin Pharmacol Ther 83(5):761–769
152. Zhao Y, Wu Y, Wang M (2015) Bioactive substances of plant origin 30. Handb Food Chem 967:967–1008
153. Zheng B, Kong T, Jing X, Odoom-Wubah T, Li X, Sun D, Lu F, Zheng Y, Huang J, Li Q (2013) Plant-mediated synthesis of platinum nanoparticles and its bioreductive mechanism. J Colloid Interface Sci 396:138–145

Various Metabolites and Bioactive Compounds from Fruits, and Their Use in Nanoparticles Synthesis and Applications

Arshi Siddiqui, Pragyesh Dixit, Hira Moid, and Uzma Afaq

Abstract Commercially valuable bioactive compounds find a vast range of applications in the medical, pharmaceutical, cosmetic, agriculture and food industry. Nanotechnology is a promising and rapidly emerging field of science. Due to their vast range of applications in the medical, pharmaceutical, cosmetic, agriculture and food industry, bioactive compounds (e.g., flavonoids, phenolic acids, alkaloids, and carotenoids) are commercially valuable goods. Controlled elicitation is one of the promising techniques for enhancing the production of bioactive chemicals in plants. Nanoparticles (NPs) are novel elicitors of bioactive chemicals in plants and could impact the plant's secondary metabolism. The biological production of nanoparticles is becoming more widely recognized as a rapid, environmentally benign, and easy to scale up method. Metallic nanoparticles synthesized from microorganisms and plant extracts are stable and monodispersed when synthesis parameters including pH, temperature, incubation time, and mixing ratio are well controlled. The goal of this chapter is to outline fruit extract NPs synthesis and their various applications.

Keywords Green synthesis · Nanoparticles · Plant extraction · Phytochemicals · Bioactive compounds

1 Introduction

Nanotechnology is a multidisciplinary technical pool that includes physics, chemistry, biology, medicine, and material science, with applications ranging from material and medical science to personal care goods [14-18, 33, 34]. Nanoparticles (NPs)

A. Siddiqui (✉) · H. Moid · U. Afaq
Department of Biosciences, Integral University, Uttar Pradesh, Dasauli Kursi Road, Lucknow 226026, India
e-mail: arshi.apda@gmail.com; sarshi@iul.ac.in

P. Dixit
Dioscuri Centre for Physics and Chemistry of Bacteria, Institute of Physical Chemistry (ICHF), Polish Academy of Sciences, 44/52 Kasprzaka, 01-224 Warsaw, Poland

A. Husen (ed.), *Nanomaterials from Agricultural and Horticultural Products*, Smart Nanomaterials Technology,
https://doi.org/10.1007/978-981-99-3435-5_11

have a great scientific interest because they act as a bridge between bulk materials structures and molecules at atomic level [31]. The utilization of NPs in treatment of cancer [7] and HIV [40] has been enhanced by their reduced size. The biological source and biochemical processing determine the size of the particles along with the impact of external factors like temperature. An increasing interest in creating NPs from various plant parts, specifically from fruits, is on the rise. The main benefit of creating plant-based nanoparticles is that the processes are simple, affordable, sustainable, and environmentally sound.

Many phytochemicals are bioactive substances found in plants. Bioactive chemicals have a number of health benefits, including the prevention of chronic diseases like cancer and diabetes [30]. These health advantages are linked to fruits, and are attributable to the food's bioactive chemicals' synergistic interactions. Along with fruits, vegetables, and grains, over 5000 phytochemicals have been extracted and identified. Vitamin C, folate, provitamin A, potassium, calcium, magnesium, flavonoids, phenolic acids, alkaloids, carotenoids, and fibres are among the most important bioactive substances [30].

Consuming fruit and its products not only enhances one's health but also lowers one's risk of developing a variety of diseases, including age-related macular degeneration, cardiovascular disease, cancer, cataracts of the eye, weakened immune systems, gastrointestinal disorders, high blood pressure, and high cholesterol (LDL) [35]. In order to prevent chronic illnesses (such as cancer, diabetes, heart disease, and obesity) and reduce micronutrient deficiency, FAO (Food and Agriculture Organization of the United Nations)/WHO (World Health Organization) recommends consumption of at least 400 g of vegetables and fruits daily. Fruits get their colour from compounds called carotenoids. Approximately 50 of the 600 carotenoids that have been discovered so far can be converted into vitamin A. In addition, carotenoids are also known for their antioxidant capabilities and to lower the risk of diseases such as cancer, cataracts, cardiovascular disease, and macular degeneration.

Among the above orange coloured fruits are the main source of carotene and have the highest provitamin A. Lycopene, a red carotenoid with antioxidant activity, is mostly found in pink grapefruit, tomato, watermelon, and other red-coloured foods. Studies have shown that lycopene helps in preventing breast, brain, cervix, colon, and prostate cancers [5, 6, 11].

Furthermore, bioactive chemicals are employed in agro-alimentary, fragrances, flavour, colour, and pharmaceutical preparations due to their range of actions. However, secondary metabolites have a critical role in plant defence (against insects, animals, and pathogens) as well as their survival under various biotic and abiotic challenges [30]. Metallic nanoparticles may be a viable alternative for mitigating abiotic stress, as they have been shown to have several favourable effects on crop growth and development. Previous research has demonstrated that these issues can be solved by employing water-soluble nanocurcumin, which has greatly increased biological potential. By loading curcumin onto PLGA (poly-lactic-co-glycolic acid) nanospheres using the nanoprecipitation technique, researchers were able to boost the substance's bioavailability and demonstrate that it has the capacity to pass the blood-brain barrier.

2 Green Synthesis of Nanoparticles by Using Fruit Extract

Previously *Artocarpus heterophyllus* fruit latex is used to make silver and gold NPs [24]. In phytochemical screening, the crude extract of AHL (*Artocarpus heterophyllus*) revealed the random existence of bioactive compounds such as alkaloids, flavonoids, phenolics, anthraquinones, tannins, and proteins. In the synthesis and stabilization of nanoparticles, bioagents operate as capping and self-reducing agents. UV–Visible, FT-IR (Fourier transform infrared spectroscopy), and FESEM (Field emission scanning electron microscopy) are used to examine the produced bimetallic nanoparticles. The morphological study revealed spherical silver and gold NPs with a mean particle size of 15 nm [24]. Phyto-nanotechnology has created new options for nanoparticle synthesis and is an environmentally benign, easy, fast, stable, and cost-effective process. Biocompatibility, scalability and medical usefulness of producing nanoparticles utilizing water as a reducing medium are all advantages of phyto-nanotechnology. Fruit-derived nanoparticles, which are made from easily available plant components and are harmless, are thus appropriate for meeting the rising demand for nanoparticles with biological and environmental applications. Recently, effective gold [39] nanoparticles were manufactured utilizing fruit extracts from the *Aegle marmelos, Eugenia jambolana, and Soursop*, as well as silver nanoparticles from *Benincasa hispida* fruit [3]. Metal NPs have also been synthesized using a variety of plant parts, including leaves, fruits, stems and roots [37]. *Murraya koenigii* leaf extract was used to synthesize and stabilize silver and gold nanoparticles. According to reports, different plant species have diverse processes for generating nanoparticles [37]. Eugenol, *Cinnamomum zeylanisum* primary terpenoid, was identified to have a major role in the formation of gold and silver nanoparticles. Dicot plants, in particular, possess a large number of secondary metabolites that could be used to synthesize nanoparticles [37].

NPs made from bioactive compounds contained in fruit peels have prospective implications for nutraceutical and pharmaceutical delivery that have yet to be explored. Employing fruit peel extract to investigate the biosynthesis of metallic NPs such as silver (AgNPs), gold (AuNPs), zinc oxide, iron, copper, palladium and titanium have been studied. Heat is an important source of energy for the reaction system. The process continues until the plant extracts activate the capping agent, which eventually stops the growth of high-energy atomic growth planes. This will also lead to the creation of NPs of a certain type. The reducing agents usually give electrons to the metal ions and convert them to NPs throughout the synthesis. These NPs are in a high-surface-energy state and tend to aggregate against one other to convert to their low-surface-energy conformations. As a result, the presence of more reducing and stabilizing chemicals reduces nanoparticle aggregation and promotes the formation of smaller NPs.

Fruit extracts are known to contain a significant amount of reducing agents. Fruits such as blueberries, blackberries, grapes, *Terminalia arjuna, Cornus mas* L., *Citrullus lanatus*, and *Punica granatum* L., for instance, contain significant amounts of anthocyanins, ascorbic acid, phenolic compounds, flavonoids, saccharides, and other vitamins [25].

Benincasa Hispida fruit proteins (0.33 mg/mL protein) were incubated with 1.0 mM silver nitrate salt for five days at 37 °C to create the AgNPs. To standardize the synthesis of the desired sized AgNPs, several reactions in various combinations of different temperatures (size of AgNPs increases with decreased instability by increasing the reaction temperature) and other incubation durations were carried out. Fruit protein extract has the ability to reduce $AgNO_3$ with oxidation state " +1" to Ag oxidation state "0" in an aqueous solution. The synergistic action of several proteins and reducing enzymes including serine proteases and angiotensin-converting enzyme (ACE) present in the fruit protein extract is what causes the reduction of $AgNO_3$ to Ag [3].

Dragon Fruit (DF) extract functions as a capping and reducing agent. The distinctive surface plasmon resonance peak developed at 560 nm for the DFAuNPs made with DF extract throughout various time periods. The formation of the SPR band is influenced by a number of variables, including the particle's size and shape. After mixing 1 mL of DF extract with 1.5 mM of $HAuCl_4$, the peak intensity was attained after a reaction of 2 h [8]. The DF-AuNPs' zeta potential, which was measured to be −25.881.41 mV at 25.5 °C, validated the nanoparticles' stability [8]. The NPs derived from fruits have been listed along with various applications in Table 1.

3 Significance of Biological Nanoparticles

The elimination of adverse by-products created during metal nanoparticle synthesis is a major benefit of green synthesis. The nanoparticles made using a green technology have a variety of biological activity. In the world of medicine, tiny particles with a big surface area exhibit good activity. The nanoparticles created are also effective against leishmanial diseases [22]. For nanoparticles with biological uses, biocompatibility, such as reduced metal cytotoxicity, is necessary. When compared to physiochemically generated nanoparticles, biogenic nanoparticles are devoid of hazardous contamination from by-products that attach to nanoparticles during physiochemical synthesis, limiting the biomedical potential of the resulting nanoparticles [37]. When biological nanoparticles come into interaction with complicated biological fluids, their surfaces gradually and selectively absorb biomolecules, generating a corona that interacts with biological systems. These corona layers exceed bare biological nanoparticles in terms of efficacy [37]. The researchers synthesized AuNPs from the peels of two East Asian fruits, oriental melon and peach, and then compared their prospective applications. AuNPs were synthesized using $AuCl_3$ as a precursor and OMP and peach extracts as reducing agents. The involvement of bioactive organic compounds from OMP (oriental melon peel) and peach extracts in the reduction and stabilization of NPs was demonstrated by the weight reduction of the produced AuNPs [28] (Fig. 1).

Table 1 Applications of NPs synthesized from various fruit varieties

Fruit variety	Types of NPs synthesized	Applications	Key references
Syzygium alternifolium	Copper oxide	Antiviral activity against Newcastle Disease Virus (NDV)	[41]
Capparis spinosa	Copper oxide	Antibacterial activity against *S. aureus, Bacillus cereus* Antibacterial activity against *E. coli, Klebsiella pneumoniae*	[9]
Punica granatum	Copper oxide	*Psuedomonas aeruginosa* MTCC 424, *Salmonella enterica* MTCC 1253 and *Enterobacter aerogenes* MTCC 2823	[21]
	Zinc oxide	*Enterococcus faecalis*; Cytotoxicity against HCT116 (colorectal cancer cell line)	[38]
	Silver NP	Antibacterial activity against *E. coli, S. aureus, P. aeruginosa*	[32]
Vtis vinifera	Gold NP	Apoptotic activity against human epidermoid carcinoma A431 cell line	[26]
Citrullus lanatus	Gold NP	Antibacterial activity against *Bacillus cereus* ATCC 13,061, *E. coli* ATCC 43,890, *Listeria monocytogenes* ATCC 19,115, *S, aureus* ATCC 49,444, *Salmonella* *typhimurium* ATCC 43,174; Antioxidant activity;	[27]
Malus pumila	Silver NP	Antibacterial activity against *E. coli, S. aureus, P. aeruginosa* and methicillin-resistant *S. aureus*	[1]
Dimocarpus longan	Silver NP	Antibacterial activity against *E. coli, S. aureus, P. aeruginosa,* *B. subtilis*; Antifungal activity against *Candida albicans*; Cytotoxicity against PC-3 (prostate cancer cell line)	He et al. [12]
Garcinia mangostana	Zinc oxide	Photocatalytic activity against malachite green	[2]

4 Characterization of Green Synthesized Nanoparticles

Characterization is a group of techniques for learning about the creation, characteristics, and uses of nanoparticles. UV–Vis, FTIR, SEM, and AFM analyses were used to evaluate silver nanoparticles generated using the aqueous extract. The change in colour during AgNPs synthesis can be seen with the naked eye at first. The synthesis of

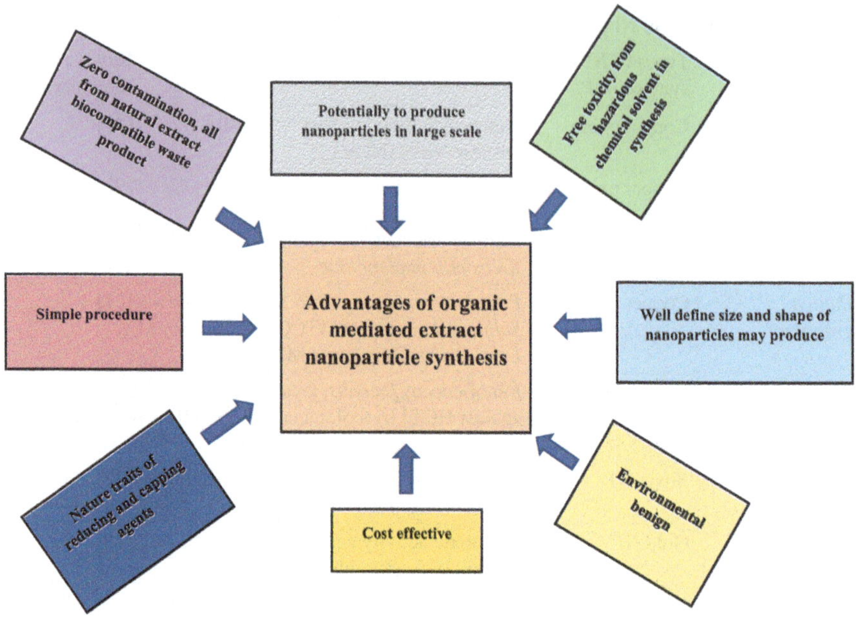

Fig. 1 Significance of green synthesized nanoparticles

AgNPs is usually indicated by the reaction mixture's dark brown colour. UV–visible spectrophotometry is then used to validate the production of AgNPs. In UV–visible spectrophotometry, synthesized AgNPs had a prominent peak of about 400–470 nm. The morphology, size, and form of biosynthesized AgNPs influenced the absorption spectra [13]. The hydrodynamic size and polydispersity index of manufactured nanoparticles are investigated using dynamic light scattering (DLS). Before being sonicated for one minute, the sample was put in a 1.5-mL low-volume disposable cuvette (DTS0112) and calibrated to a 0.5 per cent weight/volume concentration in de-ionized water. The mean particle size was determined using the average of three measurements per sample [3].

The Zeta potential must be measured to ensure the stability of AgNPs in aqueous solutions. AgNPs having a Zeta potential of less than 25 mV or greater than +25 mV are usually stable [13].

SEM and TEM are effective instruments for analysing nanoparticles XRD is an analytical technique for determining the structural characteristics of nanoparticles, such as crystallinity and particle sizes [13]. The shape, size, and particle size distribution of the green-produced AgNPs were shown using TEM. A single drop (10 L) of AgNPs suspension was applied to the carbon-coated copper grid for TEM examination, allowed to dry for two hours at ambient temperature (24 °C), and then put into the sample chamber. Analysis was then carried out at 120 kV accelerating voltage (Bramley, 2020). XRD examination can reveal the precise nature of the silver nanoparticles. Fruit extract-derived green-produced AgNPs have diffraction peaks at 2 on 32.1 °C and 64.4 °C in their XRD patterns, respectively. Thus, the crystalline AgNPs produced by the full reduction of Ag^+ ions by the aqueous fruit extract of *Diospyros malabarica* were clearly visible in the XRD patterns. The crystallization of silver nanoparticles and the organic moieties or contaminants that were linked to their surface resulted in the other unassigned peak [4].

5 Main Contribution of the Green Synthesis of NPs are Flavonoids

Flavonoids include flavone, flavonol, flavanone, flavanonol, and isoflavone derivatives, which are all natural polyphenolic chemicals. Flavonoids have a skeleton made up of two phenyl rings (A and B) joined by an oxygenated heterocycle ring (C), which is hydroxylated in many places. Because they engage in the response to biotic and abiotic stressors, these chemicals play a vital function in plants. Natural flavonoids have received a lot of interest because of their chelating and antioxidative effects, which are necessary for plant physiology and desirable for human health. ROs and free radicals may target a flavonoid-metal complex. This compound, on the other hand, could be a Fenton reaction catalyst, with the ligand moiety acting as a hydroxyl radical acceptor. The antioxidant activity of flavonoid-metal complexes has been found to be higher than that of free ligands [13].

6 Application of Green Synthesized Nanoparticle

In the field of nanobiotechnology, biological nanoparticle synthesis has increased. It develops new materials that are environmentally benign, cost-effective, and have a wide range of applications [23]. The characterized nanoadsorbent's applicability and dependability have been successfully examined for various water samples spiked with CV, including tap water and industrial wastewater [23]. The synthesis of NPs can take either a "Top-Down" or a "Bottom-Up" approach. The "Top-Down" strategy produces NPs by reducing their size using a variety of physical and chemical approaches. In "Bottom-Up" synthesis, NPs are made from small entities such as atoms and molecules, using reduction/oxidation as the primary process. This method produces NPs with fewer flaws and a more uniform chemical composition [19].

Green synthesized nanomaterials play an important role in the use of nanotechnology in a variety of industries. Green nanotechnology is the creation of green nanoproducts and their application to achieve long-term development. Medicines, therapeutic uses, and in vitro diagnostic applications all benefit from green-produced NPs. Greenly produced nanoparticles have good antibacterial, antifungal, and antiparasitic properties [19].

According to reports, Cu_2ONPs made from *Ziziphus spina-christi* (L.) extract have antibacterial action against *Staphylococcus aureus* instead of *Escherichia coli.* In contrast to *E. coli* and *Klebsiella pneumoniae*, Cu_2ONPs made from fruit extract of *Capparis spinosa* were reported in another investigation to exhibit robust antibacterial action against *Bacillus cereus* and *S. aureus* [25]. AgNPs made from *Crataegus pentagyna* fruit extract were found to be effective against *Acinetobacter baumannii*, at minimum inhibitory concentrations and bactericidal concentrations of 0.11, 0.22, 0.11, 0.44, 0.11, 1.7, and 0.11, 7.1 g/mL, respectively. *P. aeruginosa*, *S. aureus*, *E. coli*, and *Enterococcus faecalis*, comparatively [10].

Citrus macroptera (CM) juice was used to generate AuNPs, and the anticancerous potential of these particles was examined in vitro using $HepG_2$ liver cancer cells. The study's findings revealed that the IC_{50} value for CM-AuNPs was 70.2 ng/mL. This research was done to reveal the true potential of CM, which is an ancient treatment for liver problems among the tribal people of Tripura [25] (Fig. 2).

Using fluorescent gold 750 nanocluster for in vivo self-bioimaging of cancer cells. It was made intracellular by turning a chloroauric acid solution into gold nanoparticles in the cytoplasm of malignant cells ($HepG_2$; human hepatocarcinoma cell line, K562; leukaemia cell line) (AuNP). This AuNP was in a clustered form with fluorescence, allowing in vivo fluorescence imaging to preferentially diagnosis malignant cells. In a previous experiment, researchers created carbon nanoparticles using an enhanced green technology (hydrothermal treatment of grape juice), which produced blue-coloured photoluminescence nanoparticles. Low cytotoxicity, biocompatibility, excellent water solubility, and high quantum yield were discovered in this nanoparticle. It fluoresced green at 488 nm and showed cellular uptake by Human HeLa cells, making it a viable option for cellular imaging [29].

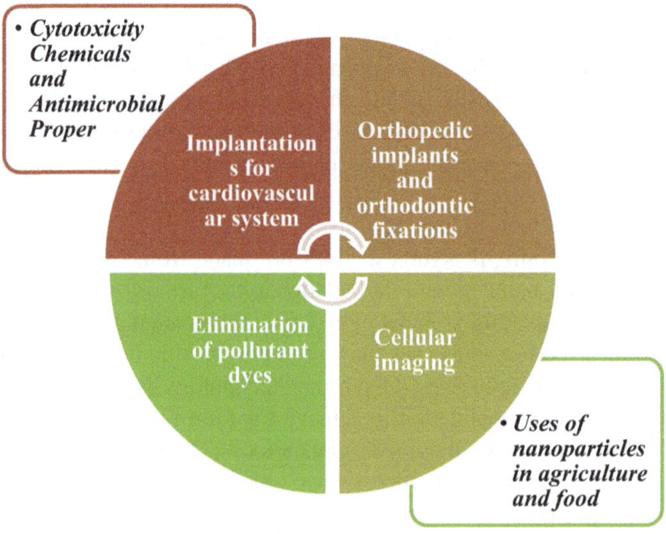

Fig. 2 Application of green synthesized nanoparticles

7 Conclusion

This technique was reasonably inexpensive, quick, and simple, and it did not involve any harmful chemicals. Nanobiotechnology is a rapidly growing field of nanotechnology. The need for biocompatible materials for numerous applications in areas such as health, medicine, water treatment and purification, biosensors, industrial food, and others has drawn increased attention to this subject in recent years. Green approach design, on the other hand, has become a future requirement for the industry. Further study on green nanoparticle synthesis could be very beneficial. In the field of nanotechnology, these environmentally friendly nanoparticles are being assessed for a variety of applications. A further advantage of using fruit extracts for nanoparticle synthesis over other biological methods is that they don't take as long and don't require the upkeep of microbial cultures to maintain the actual potency during the nanoparticle production. Therefore, using plant extracts to create nanoparticles could have a significant impact in the upcoming years. For the creation of nanoparticles using fruit extracts, numerous works can be noted.

References

1. Ali M, Kim B, Belfield KD, Norman D, Brennan M, Ali GS (2016) Green synthesis and characterization of silver nanoparticles using *Artemisia absinthium* aqueous extract—a comprehensive study. Mater Sci Eng C 58:359–365
2. Aminuzzaman M, Ying LP, Goh WS, Watanabe A (2018) Green synthesis of zinc oxide nanoparticles using aqueous extract of Garcinia mangostana fruit pericarp and their photocatalytic activity. Bull Mater Sci 41:50
3. Baker A, Iram S, Syed A, Elgorban AM, Bahkali AH, Ahmad K, Khan MS, Kim J (2021) Fruit derived potentially bioactive bioengineered silver nanoparticles. Int J Nanomed 16:7711
4. Bharadwaj KK, Rabha B, Pati S, Choudhury BK, Sarkar T, Gogoi SK, Kakati N, Baishya D, Kari ZA, Edinur HA (2021) Green synthesis of silver nanoparticles using *Diospyros malabarica* fruit extract and assessments of their antimicrobial, anticancer and catalytic reduction of 4-nitrophenol (4-NP). Nanomaterials 11(8):1999
5. Bramley PM (2000) (2000) Is lycopene beneficial to human health? Phytochemist 54:233–236
6. Clinton SK (1998) Lycopene: chemistry, biology, and implications for human health and disease. Nutr Rev 56:35–51
7. Damoah OY, Wang R, Zhou J, Ding Y, Versatile Y (2016) Nanosystem-based cancer theranostics: design inspiration and predetermined routing. Theranostics 986–1003
8. Divakaran D, Lakkakula JR, Thakur M, Kumawat MK, Srivastava R (2019) Dragon fruit extract capped gold nanoparticles: synthesis and their differential cytotoxicity effect on breast cancer cells. Mat Lett 236:498–502
9. Ebrahimi K, Shiravand S, Mahmoudvand H (2017) Biosynthesis of copper nanoparticles using aqueous extract of *Capparis spinosa* fruit and investigation of its antibacterial activity. Marmara Pharm J 21:866–871
10. Ebrahimzadeh MA, Naghizadeh A, Amiri O, Shirzadi-Ahodashti M, Mortazavi-Derazkola S (2020) Green and facile synthesis of Ag nanoparticles using *Crataegus pentagyna* fruit extract (CP-AgNPs) for organic pollution dyes degradation and antibacterial application. Bioorg Chem 94:103425
11. Giovannucci E, Ascherio A, Rimm EB, Stampfer MJ, Colditz GA, Willett WC (1995) Intake of carotenoids and retinol in relation to risk of prostate cancer. J Natl Cancer Inst 87:1767–1776

12. He Y, Du Z, Ma S, Cheng S, Jiang S, Liu Y, Li D, Huang H, Zhang K, Zheng X (2016) Biosynthesis, antibacterial activity and anticancer effects against prostate cancer (PC-3) cells of silver nanoparticles using Dimocarpus Longan Lour. peel extract. Nanoscale Res Lett 11:1–10
13. Huq MA, Ashrafudoulla M, Rahman MM, Balusamy SR, Akter S (2022) Green synthesis and potential antibacterial applications of bioactive silver nanoparticles: a review. Polymers 14(4):742
14. Husen A (2022) Engineered nanomaterials for sustainable agricultural production, soil improvement and stress management. Elsevier Inc., 50 Hampshire St., 5th Floor, Cambridge, MA 02139, USA. https://doi.org/10.1016/C2021-0-00054-7
15. Husen A, Iqbal M (2019) Nanomaterials and plant potential: an overview. In: Husen A, Iqbal M (eds) Nanomaterials and plant potential, vol 11, 6330th edn. Springer International Publishing AG, Gewerbestrasse, Cham, pp 3–29. https://doi.org/10.1007/978-3-030-05569-1_1
16. Husen A, Jawaid M (2020) Nanomaterials for agriculture and forestry applications. Elsevier Inc., 50 Hampshire St., 5th Floor, Cambridge, MA 02139, USA. https://doi.org/10.1016/C2018-0-02349-X
17. Husen A, Siddiqi KS (2014) Phytosynthesis of nanoparticles: concept, controversy and application. Nanoscale Res Lett 9(229):1–24
18. Husen A, Siddiqi KS (2022) Advances in smart nanomaterials and their applications. Elsevier Inc., 50 Hampshire St., 5th Floor, Cambridge, MA 02139, USA. https://doi.org/10.1016/C2021-0-02202-1
19. Hussain I, Singh NB, Singh A, Singh H, Singh SC (2016) Green synthesis of nanoparticles and its potential application. Biotechnol Lett 38(4):545–560
20. Ingle AP, Shende S, Gupta I, Rai M (2020) Recent trends in the development of nano-bioactive compounds and delivery systems. In: Biotechnological production of bioactive compounds, pp 409–431. Elsevier
21. Kaur P, Thakur A, Chaudhary A (2016) Biogenesis of copper nanoparticles using peel extract of *Punica granatum* and their antimicrobial activity against opportunistic pathogens. Green Chem Lett Rev 9:33–38
22. Khan ZUH, Khan A, Chen Y, Shah NS, Muhammad N, Khan AU, Tahir K, Khan FU, Murtaza B, Hassan SU, Qaisrani SA (2017) Biomedical applications of green synthesized Nobel metal nanoparticles. J Photochem Photobiol B Biol 173:150–164
23. Khani R, Roostaei B, Bagherzade G, Moudi M (2018) Green synthesis of copper nanoparticles by fruit extract of *Ziziphusspina-christi* (L.) Willd: application for adsorption of triphenylmethane dye and antibacterial assay. J Mol Liq 255:541–549
24. Krishnan Sundarrajan S, Pottail L (2021) Green synthesis of bimetallic Ag@ Au nanoparticles with aqueous fruit latex extract of Artocarpus heterophyllus and their synergistic medicinal efficacies. Appl Nanosci 11(3):971–981
25. Kumar H, Bhardwaj K, Dhanjal DS, Nepovimova E, Şen F, Regassa H, Singh R, Verma R, Kumar V, Kumar D, Bhatia SK (2020) Fruit extract mediated green synthesis of metallic nanoparticles: a new avenue in pomology applications. Int J Mol Sci 21(22):8458
26. Nirmala JG, Akila S, Nadar MSAM, Narendhirakannan RT, Chatterjee S (2016) Biosynthesized *Vitis vinifera* seed gold nanoparticles induce apoptotic cell death in A431 skin cancer cells. RSC Adv 6:82205–82218
27. Patra JK, Baek KH (2015) Novel green synthesis of gold nanoparticles using *Citrullus lanatus* rind and investigation of proteasome inhibitory activity, antibacterial, and antioxidant potential. Int J Nanomed 10:7253–7264
28. Patra JK, Das G, Baek KH (2016) Phyto-mediated biosynthesis of silver nanoparticles using the rind extract of watermelon (*Citrullus lanatus*) under photo-catalyzed condition and investigation of its antibacterial, anticandidal and antioxidant efficacy. J Photochemist Photobiol B: Biol 161:200–210
29. Rana A, Yadav K, Jagadevan S (2020) A comprehensive review on green synthesis of nature-inspired metal nanoparticles: mechanism, application and toxicity. J Clean Product 272:122880
30. Rivero-Montejo SDJ, Vargas-Hernandez M, Torres-Pacheco I (2021) Nanoparticles as novel elicitors to improve bioactive compounds in plants. Agriculture 11(2):134

31. Sarkar S, Sarkar SC (2019) Chapter-4 application of nanotechnology in medicine. Med Sci 117:49
32. Shanmugavadivu M, Kuppusamy S, Ranjithkumar R (2014) Synthesis of pomegranate peel extract mediated silver nanoparticles and its antibacterial activity. Am J Adv Drug Deliv 2:174–182
33. Shu J, Tang D (2017) Current advances in quantum-dots-based photoelectrochemical immunoassays. Chem Asian J 12:2780–2789
34. Siddiqi KS, Husen A (2017) Recent advances in plant-mediated engineered gold nanoparticles and their application in biological system. J Trace Elem Med Biol 40:10–23
35. Sidhu JS, Zafar TA (2018) Bioactive compounds in banana fruits and their health benefits. Food Qual Saf 2:183–188
36. Singh J, Dutta T, Kim KH, Rawat M, Samddar P, Kumar P (2018) 'Green'synthesis of metals and their oxide nanoparticles: applications for environmental remediation. J Nanobiotechnol 16(1):1–24
37. Singh P, Kim YJ, Zhang D, Yang DC (2016) Biological synthesis of nanoparticles from plants and microorganisms. Trend Biotechnol 34(7):588–599
38. Sukri S, Shameli K, Wong MMT, Teow SY, Chew J, Ismail NF (2019) Cytotoxicity and antibacterial activities of plant-mediated synthesized zinc oxide (ZnO) nanoparticles using *Punica granatum* (pomegranate) fruit peels extract. J Mol Struct 1189:57–65
39. Vijayakumar S (2019) Eco-friendly synthesis of gold nanoparticles using fruit extracts and in vitro anticancer studies. J Saudi Chem Soc 23(6):753–761
40. Vijayakumar S, Ganesan S (2012) Gold nanoparticles as an HIV entry inhibitor. Curr HIV Res 10:643–646
41. Yugandhar P, Vasavi T, Rao YJ, Devi PUM, Narasimha G, Savithramma N (2018) Cost effective, green synthesis of copper oxide nanoparticles using fruit extract of *Syzygium alternifolium* (Wt.) Walp., characterization and evaluation of antiviral activity. J Cluster Sci 29:743–755

Various Agriculture Crop Plant-Based Bioactive Compounds and Their Use in Nanomaterial Synthesis and Applications

Anil Patani, Ashish Patel, Dharmendra Prajapati, Noopur Khare, and Sachidanand Singh

Abstract Plants, microalgae, fungi, bacteria, and seaweeds are all involved in the biological synthesis of nanoparticles. Agricultural crop plants contain various natural substances, including alkaloids, flavonoids, saponins, steroids, tannins, and others. These natural compounds are derived from various plant parts, such as the leaves, stems, roots, shoots, flowers, bark, and seeds, among others. Plant extracts have been shown to be a potential precursor for the non-hazardous synthesis of nanomaterial in recent studies. As the plant extract is rich in secondary metabolites, it is a reducing and stabilizing agent in the bioreduction reaction that produces unique metallic nanoparticles. Despite the fact that non-biological procedures (chemical and physical) are utilized in the production of nanoparticles, which are extremely harmful and highly toxic to living things, nanoparticles are synthesized. In addition, the biological synthesis of metallic nanoparticles is a cheap, single-step, and environmentally benign process. Various greener nanoparticles, including copper, silver, gold, palladium, cobalt, platinum, zinc oxide, and magnetite, are synthesized using plants. Biosynthesized nanomaterials have efficiently controlled several endemic diseases with minimal side effects. Agricultural crop plant metabolites, their role in nanoparticle synthesis and nanomaterial applications are discussed in this chapter.

Keywords Crops · Nanomaterials · Metabolites · Nanoparticles

A. Patani · D. Prajapati · S. Singh (✉)
Department of Biotechnology, Smt. S.S. Patel Nootan Science and Commerce College, Sankalchand Patel University, Visnagar, Mehsana, Gujarat 384315, India
e-mail: drsachinbioinfo@gmail.com

A. Patel
Department of Life Sciences, Hemchandracharya North Gujarat University, Patan, Gujarat 384265, India

N. Khare
Faculty of Biotechnology, Institute of Bio-Sciences and Technology, Shri Ramswaroop Memorial University, Lucknow, Deva Road, UP, India

A. Husen (ed.), *Nanomaterials from Agricultural and Horticultural Products*, Smart Nanomaterials Technology,
https://doi.org/10.1007/978-981-99-3435-5_12

1 Introduction

Discovering biological substances to synthesis nanomaterials with the desired size distribution and morphologies for various applications in all fields of medicine and industry has become more important in the search for environmentally acceptable and safer ways for nanomaterial creation. The potential medical application of metal nanoparticles (MNPs) motivates scientists to investigate more applications in this field [37–40, 42, 48, 116]. Plants have attracted the most interest among several natural resource groups, such as fungi, bacteria, algae, and marine creatures, for the green synthesis of MNPs [5, 86, 94].

Popular "green factories" for the synthesis of typically non-toxic nanoparticles (NPs) are now plants [111]. Phytonanotechnology is concerned with the plant-based synthesis of MNPs, as well as their characterization, optimization, and wide array of applications [72, 106, 133]. Phytonanotechnology allows for the controlled synthesis of MNPs with precisely defined sizes, shape, and chemical composition [88]. Several plant extracts have been utilized for the green synthesis of several MNPs [55]. This research documented the utilization of several plant components for MNP production, including latex, bark, fruit, root, stem, bud, and leaf. MNP production via phytosynthesis is preferable over microbial synthesis. Because the microbiological method necessitates strict aseptic conditions, a time-consuming process, and lengthy incubation periods [5]. Phytosynthesis appears simpler, more cost-effective, energy-efficient, and ecologically benign than traditional physicochemical approaches for MNP synthesis [88]. Phytonanotechnology has the potential to transform therapies in a wide range of diseases. Particularly, plant-based MNPs have shown significant promise for use in efficient drug delivery to malignant cells, which can enhance the management of cancer patients' treatment [30, 57]. Aqueous solutions of a metal salt and a plant extract are combined to form plant-based MNPs, which are then incubated under various conditions, including pH, salt concentration, plant extract concentration, incubation time, and others [55]. These influencing parameters can change the dimensions and shape of MNPs, which will change their physicochemical and biological characteristics. The production of plant-mediated MNPs is a two-step process that begins with the reduction of metal ions to nanometal nuclei and ends with the stabilization and agglomeration of these nuclei to create a system of colloidal MNPs [88]. It is believed that phytochemicals play a dual role as lowering and stabilizing agents through MNP phytosynthesis [87]. Phytosynthesis of MNPs does not require additive reducers and/or stabilizers, in contrast to conventional chemical processes that do. The alcohol, aldehyde, phenol, and flavonoid present in the plant extract are converted into, respectively, aldehyde, carboxylic acid, ketone, and flavone during the production of MNPs by plants [40].

The investigations reported a vast array of possible biomedical and pharmacological applications for MNPs synthesized by botanical means. For instance, phytosynthesized MNPs were investigated for targeted medication delivery systems as nanocarriers. *Limonia acidissima* L. fruit extract was used for green production of gold nanoparticle (AuNPs) in a study. The fruit peel extract of *Punica granatum* was successfully utilized in the production of AuNPs. *Aloe vera* leaf extracts were utilized to produce 5–50 nm cubic In_2O_3 nanoparticles [67]. Silver nitrate is used as a substrate by *Brassica juncea* and *Medicago sativa*, which turn it into 50 nm AgNPs [35]. Despite the fact that the phytosynthesis of MNPs is still in the laboratory stage, this green strategy is a substantial alternative to conventional methods for large-scale MNP production.

2 Agricultural Crop Plant Metabolites and Their Role in Synthesis of Nanoparticles

It is known that the crude extracts of different plants contain a wide range of primary and secondary metabolite substances, from proteins to different low molecular weight substances, such as flavonoids, phenolic acid, alkaloids, terpenoids, alcoholic substances, amino acids, glutathiones, polysaccharides, antioxidants, organic acids (oxalic, ascorbic, malic, tartaric, protocatechuic acid), and quinones. These metabolites participate in redox reaction pathways, which is general knowledge [11]. They are in responsible of converting metal ions into metallic nanoparticles. It has been hypothesized that the bio-reduction of silver begins with the electrostatic capture of silver ions on the surface of proteins in the plant extract, even though the components involved in the green synthesis of nanoparticles and the mechanism influencing the bio-reduction of ions are not fully understood [75]. The silver ions are then reduced by the proteins, which changes their secondary structure and results in the production of silver nuclei. Silver ions are reduced further, and their concentration at the nuclei results in the formation of silver nuclei [63]. It has also been hypothesized that secondary metabolites (such as proteins, sugars, polyphenols, alkaloids, terpenoids, phenolic acids, and polyphenols) have a role in stabilizing nanoparticles formed as a result of the reduction of metal ions [56, 69]. Table 1 represents plant metabolites responsible for synthesis of nanoparticles in different crop plant species.

2.1 Nanoparticles Based on Phenolic Compounds

A benzene ring that has one or more hydroxy groups on it makes up a phenolic compound. A polyphenolic substance also contains its glycosides and esters, as well as at least two benzene rings with single or multiple hydroxyl substituents. The polyphenol family includes organic substances like phenolic acids with a phenolic ring and a carboxylic acid functional group [87]. According to the findings, phenolic compounds can serve as environmentally friendly reducing agents in the production of MNPs. In a study, spherical silver nanoparticles (AgNPs) in the range of 3–10 nm were created utilizing caffeic acid (3,4-dihydroxy-cinnamic

Table 1 Plant metabolites responsible for synthesis of nanoparticles in different crop plant species

Plants used	Nanoparticles	Parts of plant	Size (nm)	Shapes	Plant metabolites involved in bioreduction	Citation
Nigella sativa	ZnO	Seeds	24	Spherical	Alkaloids, quinones, phenolics, flavonoids	Awan et al. [14]
Morinda citrifolia	Ag	Seeds	3	Spherical	Flavonoids, organic acids	Morales-Lozoya et al. [81]
Curcuma pseudomontana	Au	Rhizomes	20	Spherical	Flavonoid, alkaloids, saponins	Muniyappan et al. [83]
Tinospora cordifolia	Ag	Leaves	34	Spherical	Phenolic compound	Jayaseelan et al. [46]
Trigonellafoenum graecum	Au	Seed	15–25	Spherical	Flavonoids	Aromal and Philip [11]
Withania somnifera	Ag	Leaves	5–40	Irregular, spherical	Methyl 7-oxooctadecanoate	Nagati et al. [85]
Cinnamomum tamala	Cu	Leaves	5–12	Spherical	Caryophyllene, eugenol	Roy and Ghosh [108]
Bacopa monnieri	Cu	Leaves	50–60	Spherical	Triterpenoid, saponins, alkaloids	Thakur et al. [123]
Asparagus adscendens	Cu	Leaves	10–15	Spherical	Saponins, glycosides	Thakur et al. [123]
Cuminum cyminum	Cu	Seeds	25	Spherical	Cuminaldehyde, pinene, limonene, cymene, 1,8-cineole, terpinene, linanool, safranal	Rajesh et al. [103]
Sesamum indicum	Cu	Seeds	n.a	n.a	Phenolic components	Sirisha et al. [119]
Aloe vera	Cu	Flower	40	Spherical	Tannins, flavonoids, alkaloids, carotenoids	Karimi and Mohsenzadeh [52]
Solanum lycopersicum (tomato)	Cu	Fruit extract	40–70	n.a	Ascorbic acid	Batoool and Masood [15]

acid), a phenolic acid [32]. The ecologically friendly production of bimetallic silver-selenium nanoparticles using a combination of quercetin [2-(3,4-dihydroxyphenyl-3,5,7-trihydroxychromen-4-one] and gallic acid (3,4,5-trihydroxybenzoic acid was also reported in the work. The bimetallic nanoparticles produced were between 30 and 35 nm in size and capped with flavonoids and phenolics [79]. In a study, gallotannin (also known as gallotannic acid) ($C_{76}H_{52}O_{46}$) was used to create AgNPs with spherical morphology that ranged in size from 50 to 100 nm [68]. A phenolic molecule called gallotannin is made up of polymers of gallic acid that are joined together by a dipside (hydroxyl-carboxyl) bond [44]. Additionally, curcumin (diferuloylmethane) was used to create spherical AuNPs with a hydrodynamic diameter of 63.4 ± 0.2 nm on average [71]. Curcumin, a polyphenol, is the principal secondary metabolite of *Curcuma longa* and other *Curcuma* species [26, 60]. Importantly, in a study, high-performance liquid chromatography (HPLC) analysis of the phytosynthesized AgNPs showed that there was various phenolic chemicals coupled to AgNPs [75].

2.2 Nanoparticles Based on Flavonoids

The secondary metabolites known as flavonoids are obtained from plants and contain polyphenolic compounds called flavones, flavanonols, flavanones, flavanols or catechins, anthocyanins, neoflavonoids, chalcones, and isoflavones, among others [92]. These secondary metabolites, in accordance with the findings, aid in the green synthesis of MNPs. Epicatechin, a polyphenolic monomeric flavanol, was used effectively as both a reducing and a capping agent during the production of AgNPs, for instance. AgNPs were created with spherical morphologies between 11 and 30 nm [82]. In a study, extremely monodisperse, spherical AuNPs with an average diameter of 18.24 nm were synthesized using the flavonoid kaempferol as a reducing and capping agent [34]. In a study, the flavonoid quercetin was successfully used to create AgNPs with an average size of 11.35 nm and nearly spherical shape [45]. In addition, quercetin was employed as a capping agent in synthesizing superparamagnetic magnetite [58] and gold nanoparticles [90, 91]. MNPs are capped by the hydroxyl and oxo functional groups of quercetin molecules [87]. Hesperidin, naringin, and diosmin were used as reducing and capping agents, respectively, in a test of the synthesis of AgNPs. AgNPs were successfully produced in the 5–40, 20–80, and 5–50 nm size ranges [110]. Similarly, spherical AgNPs in the range of 20–40 nm in diameter were produced using hesperidin [121]. Apigenin, a flavone, was also utilized in synthesizing AuNPs with an average diameter of 20–30 nm as a reducing and capping agent [47]. In addition, the green synthesis of anisotropic AuNPs and quasispherical AgNPs utilizing apiin as the stabilizing and reducing agent has been described. AuNPs and AgNPs were discovered to have an average size of 21 and 39 nm, respectively [53]. Moreover, naringin was successfully employed in the manufacturing of AuNPs, resulting in the synthesis of spherical AuNPs with an average size of 23 nm [118].

2.3 Nanoparticles Based on Terpenoids

Thymol ($C_{10}H_{14}O$) and eugenol ($C_{10}H_{12}O_2$), both phenolic monoterpenoids, were successfully used to create AgNPs with average diameters of 150 and 230 nm, respectively [2]. Thymol was also successfully used to synthesize AgNPs as a capping agent [74]. Additionally, a study reported that thymol was used to synthesize silver and copper nanoparticles as a reducing and capping agent [4, 6, 25]. Using the monoterpene geraniol ($C_{10}H_{18}O$), AgNPs were synthesized with homogeneous size and shape in the range of 1–10 nm, with an average size of 6 nm [109]. A study described a straightforward and unique technique for the green synthesis of AgNPs employing beta-caryophyllene, a natural bicyclic sesquiterpene derived from *Murraya koenigii* leaf extract, as a reducing and stabilizing agent. he AgNPs were produced in spherical forms with a size ranging from 5 to 100 nm and an average diameter of 29.42 nm [51]. Similar to this, sesquiterpenoids from the flower buds of *Tussilago farfara* were effectively used to make AgNPs and AuNPs [59].

2.4 Organic Acid Based Nanoparticles

Different *Morinda citrifolia* plant components, like EMF, EML, and EMDS, were individually used in the green synthesis of AgNPs. The decrease and subsequent stabilization of the AgNPs are caused by certain phytochemicals, flavonoids, and organic acids with –OH and –COOH groups, which are found in various portions of the *M. citrifolia* plant [81]. Anbuvannan et al. [9] described the synthesis of ZnO utilizing carboxylic acid as the reducing and capping agent [9]. Kaviya et al. [54] produced Ag by reducing and capping it with ascorbate [54].

2.5 Alkaloid Based Nanoparticles

Alkaloids are ubiquitous in nature. In addition to plants, terrestrial animals, marine species, and microbes such as bacteria, fungi, and insects can also create them [93]. Most alkaloids are biosynthetically generated from the amino acids lysine, ornithine, tryptophan, tyrosine, and phenylalanine [131]. Using a green synthesis technique, silver nanoparticles (AgNPs) were made from potato alkaloids. The phytopathogenic fungi *Rhizoctonia solani, Alternaria alternata, Botrytis cinerea,* and *Fusarium oxysporum f. sp. lycopersici* were inhibited from growing on their mycelia by potato alkaloids and their nanoparticles at low inhibitory and fungicidal concentrations [7]. Karimi and Mohsenzadeh [52] described the synthesis of Cu nanoparticles employing alkaloids as a capping and reducing agent [52].

2.6 Carotenoids-Based Nanoparticles

Carotenoids are biologically and chemically significant bioactive chemicals. Carotenoids are fat-soluble pigments that give vegetables their orange, yellow, and red colours. The chemicalscarotene, lutein, lycopene, zeaxanthin, astaxanthin, and bixin are found in nature [27]. Karimi and Mohsenzadeh [52] fabricated Cu nanoparticles utilizing carotenoids as a reducing and capping agent [52]. Andal et al. [10] revealed that in the phytosynthesis of AgNPs, beta-carotene functions as both an antioxidant and a capping agent. They extracted beta-carotene from carrots using a solvent extraction technique. Beta-carotene was coupled with AgNPs using a green synthesis approach [10].

2.7 Nanoparticles Based on Protein

For the green synthesis of AgNPs in the size range of 25–45 nm, *Limonia acidissima* (wood apple) purple acid phosphatase was successfully employed as a reducing and capping agent [97]. Purple acid phosphatases, which are members of the family of binuclear metallohydrolases, are known to be crucial for a number of biological activities in both plants and animals [112]. Using the isolated superoxide dismutase (SOD) enzyme from *Papaver somniferum* L., AgNPs were synthesized [50]. The production of spherical AgNPs and AuNPs used two peptides as reducing and stabilizing agents: peptide-1 (NH2-Leu-Aib-Trp-OMe) with a free N-terminus and peptide-2 (Boc-Leu-Aib-Trp-OH) with a tryptophan residue at the C-terminus (Si and Mandal 2007). Glutathione was successfully used to produce AgNPs [129]. 1-valine-based oligopeptides were used to produce AgNPs in a study [73]. In addition, AuNPs were successfully produced by independently combining ornithine, isoleucine, serine, and histidine [13]. In another investigation, the synthesis of AuNPs utilizing histidine was also confirmed [77]. Similarly, AgNPs were produced by utilizing tyrosine and tryptophan as reducing and capping agents, respectively [113]. Aspartic acid was also employed in the production of AgNPs [102]. It has also been reported that very stable and homogeneous AgNPs can be produced using casein [12].

2.8 Nanoparticles Based on Carbohydrates

Galactomannan extracted from the fruit rind of *Punica granatum* was successfully applied to the synthesis of AgNPs with an average diameter of approximately 30 nm. This biopolymer worked as a capping agent as well as a reducing agent [89]. In a separate investigation, heparin and chitosan were successfully used to create AuNPs and AgNPs [36]. Chitosan was used as a reducing and capping agent to create almost spherical AgNPs and AuNPs by Wei and Qian [128]. Researchers produced stable

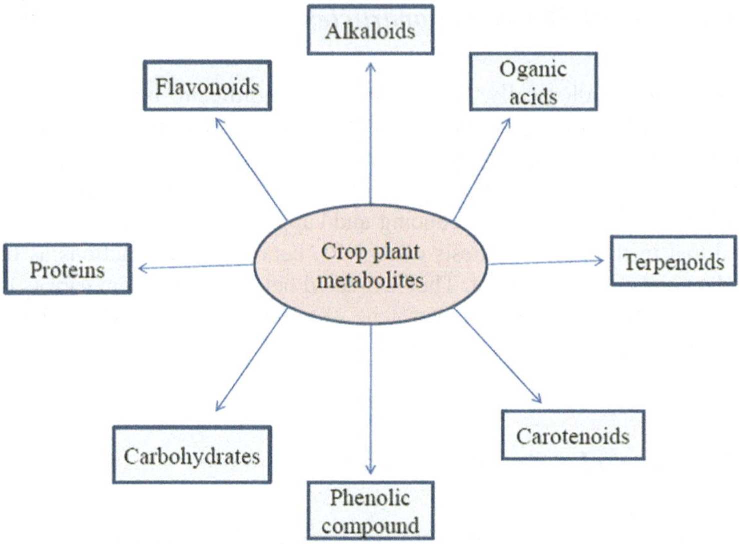

Fig. 1 Plant metabolites from various agricultural crops

and uniform starch-stabilized AgNPs with a mean size of about 14.4 nm by reducing silver nitrate by d-glucose using ultrasound [126]. Remarkably, plants produce enormous amounts of the main metabolites glucose and sucrose during photosynthesis. According to a study, d-glucose was more efficient than sucrose at producing AgNPs [2]. Additionally, the capability for producing AgNPs from glucose and fructose was examined, and glucose was found to be more effective, while fructose had a relatively low reduction potential [31]. AgNPs were also produced using a fucoidan isolate from *Spatoglossum asperum* with spherical to oval morphology and a size range of 20–46 nm [105] (Fig. 1).

3 Application of Nanomaterials

In particular, nanoparticles made of silver, gold, zinc, etc., have special physicochemical properties that are very appealing for biomedical applications, whereas platinum is used for energy storage [3]. It is becoming more common to use silver nanoparticles' optical characteristics as the functional element in many goods and sensors. Unlike most dyes and pigments, silver nanoparticles' colour depends on the size and shape of the particle [124]. It has become more common as an antibacterial agent in clothing and bandages, medical equipment, and home appliances like refrigerators and washing machines [107]. Carbon-based NMs are widely used in various industries, including medication delivery, biosensors, enzyme immobilization, bioimaging, and pollutant removal [23].

3.1 Catalysis

Nanoparticles are used for catalysis. Aluminium, iron, titanium dioxide, clays, and silica have been employed as nanocatalysts for decades. Nanocatalysis is a new science due to its high activity, selectivity, and productivity. 1–10 nm metal nanoparticles have a better catalytic activity to equivalent metal complexes. High surface-to-volume ratio, geometric surface effect, electronic effect, and quantum size effect contribute to nanocatalyst activity. Metal nanoparticles in solution are utilized as effective heterogeneous catalysts due to their environmental friendliness, which is a result of easy product isolation, simple recovery, and excellent recyclability. Metal nanoparticle size affects nanocatalyst conversion and selectivity [62]. Catalytic reagents can reduce a transformation's temperature, eliminate reagent-based waste, and boost a reaction's selectivity, avoiding undesirable side reactions and creating a green technology. Without catalysts, pharmaceuticals, fine chemicals, polymers, fuels, fibres, paints, and lubricants couldn't be made. Catalysts make production more cost-effective, environmentally friendly, and sustainable. Carbon nanotubes are used in photocatalytic processes and as fuel cell, synthetic ammonia, and methane catalysts.

3.2 Water Treatment

There is a lot of interest in the use of nanoparticles in the treatment of wastewater and water. Due to their small size and high specific surface areas, nanomaterials have high adsorption capabilities and reactivity. By using different nanomaterials, it has been reported that heavy metals [122], organic pollutants [130], inorganic anions [64], and microorganisms [49] can all be removed [65]. TiO_2 nanoparticles have been successfully used to apply photocatalytic degradation to the breakdown of pollutants in water and wastewater. Due to its non-toxicity, chemical stability, commercial availability, and strong photoactivity, among other attributes, TiO_2 has been the focus of much research [100, 132]. Contaminants can be converted into CO_2, H_2O, and anions such as NO^{3-}, $PO_4{}^{3-}$, and Cl^- in light and a catalyst [19].

3.3 Sensors

The proliferation of NMs has cemented their use in designing high-performance electrochemical sensing devices for environmental, medical diagnostics, and food safety [96]. According to the review [70], various nanomaterials have been synthesized for the electrochemical analysis of some common additives and contaminants, such as hydrazine, malachite green, bisphenol A, ascorbic acid, caffeine, caffeic acid, sulfite, and nitrite.

Particularly in terms of their high sensitivity and selectivity as well as the shrinking of sensor devices, biosensors have been significantly impacted by nanotechnology. In this situation, new nanostructured glucose biosensors have been created using fluorescent nanomaterials and nanostructures. Diabetes patients frequently employ the electrochemical approach for measuring glucose in the form of a blood glucose metre [22]. Molecular detection devices, such as gas sensors, tiny molecular detectors, electrochemical detectors, and chromatographic applications, may also be made with carbon nanotubes [120]. The addition of nanoparticles and other nanomaterials to the macroelectrode lowers the limit of detection and improves the sensitivity and selectivity of measurements, according to [18].

3.4 Energy Storage

Platinum nanoparticles are promising in energy-related and environmental catalysis. Platinum (Pt) has outstanding catalytic activity in auto exhaust gas treatment, fuel cells, petroleum refining, organic synthesis, and hydrogen production [28]. Pt-based NMs improve the specific surface area and the number of active sites, increasing Pt atom usage [115]. Rice University researchers utilize carbon nanotube sheets to inhibit lithium anode dendrite formation. This method may lead to lithium metal batteries with higher capacity and charge rate than lithium-ion batteries. Lithium metal charges faster and holds ten times more energy per volume than lithium-ion electrodes in most modern devices, including phones and cars. Multiple-walled carbon nanotubes were used to cover lithium metal foil. Rice University researchers grew carbon nanotubes atop graphene to construct high-surface-area, low-resistance electrodes. Researchers created carbon nanotubes using graphene grown on copper. The nanotube-graphene structure is a single molecule with a large surface area since each nanotube's base is atom-by-atom connected to the graphene sheet.

3.5 Nanomedicine

The antibacterial activity of silver nanoparticles, both broad-spectrum and potent, is intensively researched. Nanosilver is utilized in various consumer products, including disinfecting medical devices and household appliances and treating water. Biocompatible gold nanoparticles have sparked a lot of attention in recent years due to their intriguing size-dependent chemical, electrical, and optical properties for possible uses in nanomedicine. The discipline of biomedicine, which is the focus of extensive study, is being revolutionized by some of its applications, including photothermal therapy, medication delivery, photodynamic therapy, gene therapy, biolabelling, and biosensing, among others [84]. AuNPs are non-cytotoxic by nature and have a large surface area, making their surfaces amenable to modification with targeting molecules. This makes them superior to other nanoparticles for various biomedical applications. Since only the affected cells or tissues can be targeted, targeted drug delivery is the most effective treatment. This reduces the adverse effects of medications. This is beneficial in the treatment of cancer, where drugs can be delivered directly to the affected cells without harming healthy cells. Quantum dots [101], Fe_3O_4 [24], and ZnO [104] are effective for targeted drug delivery. Bioimaging, biosensing, and labelling can be performed with gold nanoparticles. Gold nanoparticles have been used as contrast agents in cellular or molecular imaging for decades [43]. Ma and colleagues [66] developed a novel sensor for the colourimetric detection of *Salmonella typhimurium* based on the colour change effect of gold nanoparticles. Currently, gold nanoparticles are widely used in biosensing applications [29]. Biosensing can be used to precisely measure blood glucose [17], detect bacteria [98], and viruses, detect pollutants [8], and monitor pathogens [61].

3.6 Agriculture

Nanotechnology in agriculture can offer a wide range of applications for sustainable development through the creation of nano-fertilizers, nano-pesticides, and nano-herbicides [117]. Silver nanoparticle-based fertilizers have been created to regulate the release of nutrients during plant absorption. By minimizing nutrient loss, soil and groundwater contamination, and chemical interactions between water, soil, and microbes that result in ineffective or toxic chemicals for plants, this system aids in preserving or maintaining soil fertility [76, 78, 125, 127]. Plant pathogens, such as bacteria and fungi that cause disease in plants, can be controlled by directly spraying nanoparticle solutions on seeds, grains, or leaves [16, 33, 95, 99]. The nanostructure shields and increases the active substances from early degradation, resulting in better solubility, stability, specificity, and permeability in the nanopesticides and nanoherbicides, which also deliver active pest control ingredients for extended periods of time [21].

3.7 Food Safety and Packaging

Microbial contamination and food spoiling are significant issues in the food industry, especially in light of the effects of food-borne illnesses on public health. To address these problems, active food packaging with antimicrobial traits and biocidal ingredients is required. This packaging can enhance product quality by preventing food spoiling and microbial contamination. The food sector uses organic acids, enzymes, and polymers as packaging materials (biodegradable and non-degradable). Metal or metallic oxide nanoparticles have recently been developed, and they have benefits over organic and inorganic acids because they can withstand the most demanding processing conditions, such as exposure to high temperatures [20]. Because of this, using nanoparticles in the food packaging sector may offer viable solutions to the issue presented by perishable items, improving their quality and avoiding microbial adhesion. Copper oxide, silver, magnesium oxide, zinc oxide, cadmium selenite/tellurite, titanium, and gold are used in the food sector because they have antibacterial qualities, according to AbdelRahim et al. [1].

4 Conclusion

A potentially eco-friendly, non-toxic, and economical method of producing nanomaterials is offered by the "green synthesis" of such materials. It is possible to create nanoparticles using a variety of crop plant extracts. This chapter focused on the metabolites in crop plants used in agriculture, their function in manufacturing nanoparticles, and applications of nanomaterials. It is well known that a wide variety of organic substances found in plant extracts can act as reducing and stabilizing agents in the production of nanoparticles. Plant-mediated nanoparticles are also stable and anti-aggregative since they contain natural capping agents like proteins. Additionally, nanoparticles made through green synthesis may find use in catalysis, water treatment, sensors, energy storage, nanomedicine, agriculture, food safety, and food packaging. As a result, there are many benefits to the green synthesis of nanomaterials utilizing plant extracts from agricultural crops, including eco-friendliness, biocompatibility, and cost-effectiveness. In light of their distinctive properties, it is concluded that green production of nanomaterials will be vital in many nanotechnology-based processes.

References

1. AbdelRahim K, Mahmoud SY, Ali AM, Almaary KS, Mustafa AEZM, Husseiny SM (2017) Extracellular biosynthesis of silver nanoparticles using *Rhizopus stolonifer*. Saudi J Biol Sci 24(1):208–216
2. Abed MS, Abed AS, Othman FM (2019) Green synthesis of silver nanoparticles from natural compounds: glucose, eugenol and thymol. J Adv Res Fluid Mech Therm Sci 60(1):95–111
3. Aghajanzadeh TA, Prajapati DH, Burow M (2020) Differential partitioning of thiols and glucosinolates between shoot and root in Chinese cabbage upon excess zinc exposure. J Plant Physiol 244:153088
4. Aghajanzadeh TA, Prajapati DH, Burow M (2020) Copper toxicity affects indolic glucosinolates and gene expression of key enzymes for their biosynthesis in Chinese cabbage. Arch Agron Soil Sci 66(9):1288–1301
5. Ahmed RH, Mustafa DE (2020) Green synthesis of silver nanoparticles mediated by traditionally used medicinal plants in Sudan. Int Nano Lett 10(1):1–14
6. Alavi M, Karimi N (2019) Biosynthesis of Ag and Cu NPs by secondary metabolites of usnic acid and thymol with biological macromolecules aggregation and antibacterial activities against multi drug resistant (MDR) bacteria. Int J Biol Macromol 128:893–901
7. Almadiy AA, Nenaah GE (2018) Ecofriendly synthesis of silver nanoparticles using potato steroidal alkaloids and their activity against phytopathogenic fungi. Braz Arch Biol Technol 61

8. Alvarez-Puebla RA, dos Santos Jr DS, Aroca RF (2007) SERS detection of environmental pollutants in humic acid–gold nanoparticle composite materials. Analyst 132(12):1210–1214
9. Anbuvannan M, Ramesh M, Viruthagiri G, Shanmugam N, Kannadasan N (2015) *Anisochilus carnosus* leaf extract mediated synthesis of zinc oxide nanoparticles for antibacterial and photocatalytic activities. Mater Sci Semicond Process 39:621–628
10. Andal P, Tamilselvy S, Priyatharesini PI (2018) Green synthesis of silver nanoparticles from carrot. Res J Pharm Technol 11(7):2757–2760
11. Aromal SA, Philip D (2012) Green synthesis of gold nanoparticles using *Trigonella foenum-graecum* and its size-dependent catalytic activity. Spectrochim Acta Part A Mol Biomol Spectrosc 97:1–5
12. Ashraf S, Abbasi AZ, Pfeiffer C, Hussain SZ, Khalid ZM, Gil PR, ..., Hussain I (2013) Protein-mediated synthesis, pH-induced reversible agglomeration, toxicity and cellular interaction of silver nanoparticles. Coll Surf B: Biointerf 102:511–518
13. Atta AH, El-Shenawy AI, Refat MS, Elsabawy KM (2013) Preparation and characterization of some gold nanometreic compounds with simple organic materials as precursor: spectroscopic, biological and anti-cancer assessments. J Mol Struct 1039:51–60
14. Awan S, Shahzadi K, Javad S, Tariq A, Ahmad A, Ilyas S (2021) A preliminary study of influence of zinc oxide nanoparticles on growth parameters of *Brassica oleracea* var italic. J Saudi Soc Agric Sci 20(1):18–24
15. Batoool M, Masood B (2017) Green synthesis of copper nanoparticles using *Solanum lycopersicum* (tomato aqueous extract) and study characterization. J Nanosci Nanotechnol 1:1–5
16. Bhatt MJ, Patel AD, Bhatti PM, Pandey AN (2008) Effect of soil salinity on growth, water status and nutrient accumulation in seedlings of *Ziziphus mauritiana* (Rhamnaceae). J Fruit Ornamental Plant Res 16(1):383–401
17. Bhumkar DR, Joshi HM, Sastry M, Pokharkar VB (2007) Chitosan reduced gold nanoparticles as novel carriers for transmucosal delivery of insulin. Pharm Res 24(8):1415–1426
18. Brainina K, Stozhko N, Bukharinova M, Vikulova E (2018) Nanomaterials: electrochemical properties and application in sensors. Phys Sci Rev 3(9)
19. Bratovčić A (2019) Photocatalytic degradation of organic compounds in wastewaters. Technologica Acta: Sci/Prof J Chem Tech 11(2):17–23
20. Carbone M, Donia DT, Sabbatella G, Antiochia R (2016) Silver nanoparticles in polymeric matrices for fresh food packaging. J King Saud Univ-Sci 28(4):273–279
21. Castillo-Henríquez L, Alfaro-Aguilar K, Ugalde-Álvarez J, Vega-Fernández L, Montes de Oca-Vásquez G, Vega-Baudrit JR (2020) Green synthesis of gold and silver nanoparticles from plant extracts and their possible applications as antimicrobial agents in the agricultural area. Nanomaterials 10(9):1763
22. Chen L, Hwang E, Zhang J (2018) Fluorescent nanobiosensors for sensing glucose. Sensors 18(5):1440
23. Chen M, Qin X, Zeng G (2017) Biodegradation of carbon nanotubes, graphene, and their derivatives. Trends Biotechnol 35(9):836–846
24. Chertok B, Moffat BA, David AE, Yu F, Bergemann C, Ross BD, Yang VC (2008) Iron oxide nanoparticles as a drug delivery vehicle for MRI monitored magnetic targeting of brain tumors. Biomaterials 29(4):487–549
25. De Kok LJ, Neves MI, Prajapati DH, Shahbaz M, Aghajanzadeh T, Stuiver CEE (2016) The significance of sulfur nutrition and metabolism in the detoxification of excessive levels of copper, zinc and molybdenum in plants. Berichte aus dem Julius Kühn-Institut 184:4
26. Dickinson DA, Iles KE, Wigley AF, Forman HJ (2004) Analysis of transcription factor remodeling in phase II gene expression with curcumin. Methods Enzymol 378:302–318
27. Dos Santos PP, Andrade LDA, Flôres SH, Rios ADO (2018) Nanoencapsulation of carotenoids: a focus on different delivery systems and evaluation parameters. J Food Sci Technol 55(10):3851–3860
28. Duan S, Du Z, Fan H, Wang R (2018) Nanostructure optimization of platinum-based nanomaterials for catalytic applications. Nanomaterials 8(11):949

29. Dykman L, Khlebtsov N (2012) Gold nanoparticles in biomedical applications: recent advances and perspectives. Chem Soc Rev 41(6):2256–2282
30. Ganeshkumar M, Sathishkumar M, Ponrasu T, Dinesh MG, Suguna L (2013) Spontaneous ultra fast synthesis of gold nanoparticles using *Punica granatum* for cancer targeted drug delivery. Colloids Surf, B 106:208–216
31. González Fá AJ, Juan A, Di Nezio MS (2017) Synthesis and characterization of silver nanoparticles prepared with honey: the role of carbohydrates. Anal Lett 50(5):877–888
32. Guo D, Dou D, Ge L, Huang Z, Wang L, Gu N (2015) A caffeic acid mediated facile synthesis of silver nanoparticles with powerful anti-cancer activity. Colloids Surf, B 134:229–234
33. Gupta G, Paul S, Singh S, Pietramellara G, Pathan SI, Danish S, Jha PN (2022) Exploring functional diversity and community structure of diazotrophic endophytic bacteria associated with *Pennisetum glaucum* growing under field in a semi-arid region. Land 11(7):991
34. Halder A, Das S, Bera T, Mukherjee A (2017) Rapid synthesis for monodispersed gold nanoparticles in kaempferol and anti-leishmanial efficacy against wild and drug resistant strains. RSC Adv 7(23):14159–14167
35. Harris AT, Bali R (2008) On the formation and extent of uptake of silver nanoparticles by live plants. J Nanopart Res 10(4):691–695
36. Huang H, Yang X (2004) Synthesis of polysaccharide-stabilized gold and silver nanoparticles: a green method. Carbohyd Res 339(15):2627–2631
37. Husen A (2022) Engineered nanomaterials for sustainable agricultural production, soil improvement and stress management. Elsevier Inc., Cambridge, MA, USA. https://doi.org/10.1016/C2021-0-00054-7
38. Husen A, Iqbal M (2019) Nanomaterials and plant potential: an overview. In: Husen A, Iqbal M (eds) Nanomaterials and plant potential. Springer International Publishing AG, Cham, pp 3–29. https://doi.org/10.1007/978-3-030-05569-1_1
39. Husen A, Jawaid M (2020) Nanomaterialsfor agriculture and forestry applications. Elsevier Inc., Cambridge, MA, USA. https://doi.org/10.1016/C2018-0-02349-X
40. Husen A, Siddiqi KS (2014) Phytosynthesis of nanoparticles: concept, controversy and application. Nanoscale Res Lett 9(1):1–24
41. Husen A (2023) Secondary metabolites based green synthesis of nanomaterials and their applications. Springer Nature Singapore Pte Ltd., 152 Beach Road, #21-01/04 Gateway East, Singapore 189721, Singapore. https://doi.org/10.1007/978-981-99-0927-8
42. Husen A, Siddiqi KS (2022) Advances in smart nanomaterials and their applications. Elsevier Inc., Cambridge, MA, USA. https://doi.org/10.1016/C2021-0-02202-1
43. Hutter E, Maysinger D (2011) Gold nanoparticles and quantum dots for bioimaging. Microsc Res Tech 74(7):592–604
44. Jackson RS (2014) Chemical constituents of grapes and wine. In: Jackson RS (ed) Wine Science, 4th edn. Academic Press, San Diego, pp 347–426
45. Jain S, Mehata MS (2017) Medicinal plant leaf extract and pure flavonoid mediated green synthesis of silver nanoparticles and their enhanced antibacterial property. Sci Rep 7(1):1–13
46. Jayaseelan C, Rahuman AA, Rajakumar G, Vishnu Kirthi A, Santhoshkumar T, Marimuthu S, Elango G (2011) Synthesis of pediculocidal and larvicidal silver nanoparticles by leaf extract from heartleaf moonseed plant, *Tinospora cordifolia* Miers. Parasitol Res 109(1):185–194
47. Jiang J, Mao Q, Li H, Lou J (2017) Apigenin stabilized gold nanoparticles increased radiation therapy efficiency in lung cancer cells. Int J Clin Exp Med 10(9):13298–13305
48. Jin-Chul K, Madhusudhan A, Husen A (2021) Smart nanomaterials in biomedical applications. Springer Nature Switzerland AG, Cham, Switzerland. https://doi.org/10.1007/978-3-030-84262-8
49. Kalhapure RS, Sonawane SJ, Sikwal DR, Jadhav M, Rambharose S, Mocktar C, Govender T (2015) Solid lipid nanoparticles of clotrimazole silver complex: an efficient nano antibacterial against *Staphylococcus aureus* and MRSA. Colloids Surf, B 136:651–658
50. Kalyankar VK, Gadale-Dagade S, Waghmode S (2015) Biosynthesis of Silver nanoparticles using isolated superoxide dismutase enzyme from novel source *Papaverum somniferum* L. Res J Chem Environ 19(3):44–48

51. Kamaraj C, Balasubramani G, Siva C, Raja M, Balasubramanian V, Raja RK, ..., Perumal P (2017) Ag nanoparticles synthesized using β-caryophyllene isolated from *Murraya koenigii*: antimalarial (*Plasmodium falciparum* 3D7) and anticancer activity (A549 and HeLa cell lines). J Cluster Sci 28(3):1667–1684
52. Karimi J, Mohsenzadeh S (2015) Rapid, green, and eco-friendly biosynthesis of copper nanoparticles using flower extract of *Aloe vera*. Synth React Inorg, Met-Org, Nano-Met Chem 45(6):895–898
53. Kasthuri J, Veerapandian S, Rajendiran N (2009) Biological synthesis of silver and gold nanoparticles using apiin as reducing agent. Colloids Surf, B 68(1):55–60
54. Kaviya S, Santhanalakshmi J, Viswanathan B, Muthumary J, Srinivasan K (2011) Biosynthesis of silver nanoparticles using *Citrus sinensis* peel extract and its antibacterial activity. Spectrochim Acta Part A Mol Biomol Spectrosc 79(3):594–598
55. Khatua A, Priyadarshini E, Rajamani P, Patel A, Kumar J, Naik A, ..., Meena R (2020) Phytosynthesis, characterization and fungicidal potential of emerging gold nanoparticles using *Pongamia pinnata* leave extract: a novel approach in nanoparticle synthesis. J Cluster Sci 31(1):125–131
56. Kohli A, Ahmad T, Singh S (2022) *Aegle marmelos* (Bael) and *Annona squamosa* (Sugar Apple). In: Herbs shrubs, and trees of potential medicinal benefits. CRC Press, pp 339–364
57. Kumar CS, Mahesh A, Antoniraj MG, Vaidevi S, Ruckmani K (2016) Ultrafast synthesis of stabilized gold nanoparticles using aqueous fruit extract of *Limonia acidissima* L. and conjugated epirubicin: targeted drug delivery for treatment of breast cancer. RSC Adv 6(32):26874–26882
58. Kumar SR, Priyatharshni S, Babu VN, Mangalaraj D, Viswanathan C, Kannan S, Ponpandian N (2014) Quercetin conjugated superparamagnetic magnetite nanoparticles for in-vitro analysis of breast cancer cell lines for chemotherapy applications. J Colloid InterfSci 436:234–242
59. Lee YJ, Song K, Cha SH, Cho S, Kim YS, Park Y (2019) Sesquiterpenoids from *Tussilago farfara* flower bud extract for the eco-friendly synthesis of silver and gold nanoparticles possessing antibacterial and anticancer activities. Nanomaterials 9(6):819
60. Lestari ML, Indrayanto G (2014) Curcumin. Profiles of drug substances, excipients and related methodology 39:113–204
61. Li F, Zhao Q, Wang C, Lu X, Li XF, Le XC (2010) Detection of Escherichia coli O157: H7 using gold nanoparticle labeling and inductively coupled plasma mass spectrometry. Anal Chem 82(8):3399–3403
62. Li G, Jin R (2013) Catalysis by gold nanoparticles: carbon-carbon coupling reactions. Nanotechnol Rev 2(5):529–545
63. Li S, Shen Y, Xie A, Yu X, Qiu L, Zhang L, Zhang Q (2007) Green synthesis of silver nanoparticles using *Capsicum annuum* L. extract. Green Chem 9(8):852–858
64. Liu F, Yang J, Zuo J, Ma D, Gan L, Xie B, ..., Yang B (2014) Graphene-supported nanoscale zero-valent iron: removal of phosphorus from aqueous solution and mechanistic study. J Environ Sci 26(8):1751–1762
65. Liu H, Simmons CA, Sun Y (2016) A microfabricated platform with on-chip strain sensing and hydrogel arrays for 3D mechanical stimulation of cells. In: 2016 IEEE 29th international conference on micro electro mechanical systems (MEMS). IEEE, pp 267–270
66. Ma X, Song L, Zhou N, Xia Y, Wang Z (2017) A novel aptasensor for the colorimetric detection of *S. typhimurium* based on gold nanoparticles. Int J Food Microbiol 245:1–5
67. Maensiri S, Laokul P, Klinkaewnarong J, Phokha S, Promarak V, Seraphin S (2008) Indium oxide (In_2O_3) nanoparticles using *Aloe vera* plant extract: synthesis and optical properties. J Optoelectron Adv Mater 10(3):161–165
68. Mahajan PG, Dige NC, Vanjare BD, Phull AR, Kim SJ, Lee KH (2019) Gallotannin mediated silver colloidal nanoparticles as multifunctional nano platform: rapid colorimetric and turn-on fluorescent sensor for Hg^{2+}, catalytic and in vitro anticancer activities. J Lumin 206:624–633
69. Makarov VV, Love AJ, Sinitsyna OV, Makarova SS, Yaminsky IV, Taliansky ME, Kalinina NO (2014) "Green" nanotechnologies: synthesis of metal nanoparticles using plants. Acta Nat 6(1 (20)):35–44

70. Manikandan VS, Adhikari B, Chen A (2018) Nanomaterial based electrochemical sensors for the safety and quality control of food and beverages. Analyst 143(19):4537–4554
71. Manju S, Sreenivasan K (2012) Gold nanoparticles generated and stabilized by water soluble curcumin–polymer conjugate: blood compatibility evaluation and targeted drug delivery onto cancer cells. J Colloid Interface Sci 368(1):144–151
72. Mano Bala J, Jannet Vennila J, Ganesh S, Sing S (2011) Analysis of active components in *Acanthus ilicifolius* leaves for their potential chemo protective activity against hepatocellular carcinoma: an in silico approach. J Pharm Res 4:1809–1812
73. Mantion A, Guex AG, Foelske A, Mirolo L, Fromm KM, Painsi M, Taubert A (2008) Silver nanoparticle engineering via oligovaline organogels. Soft Matter 4(3):606–617
74. Manukumar HM, Umesha S, Kumar HN (2017) Promising biocidal activity of thymol loaded chitosan silver nanoparticles (TC@ AgNPs) as anti-infective agents against perilous pathogens. Int J Biol Macromol 102:1257–1265
75. Marslin G, Siram K, Maqbool Q, Selvakesavan RK, Kruszka D, Kachlicki P, Franklin G (2018) Secondary metabolites in the green synthesis of metallic nanoparticles. Materials 11(6):940
76. Martínez-Fernández D, Barroso D, Komárek M (2016) Root water transport of *Helianthus annuus* L. under iron oxide nanoparticle exposure. Environ Sci Poll Res 23(2):1732–1741
77. Maruyama T, Fujimoto Y, Maekawa T (2015) Synthesis of gold nanoparticles using various amino acids. J Colloid Interface Sci 447:254–257
78. Matchado MS, Arulmozhi S, Singh S (2018) Unraveling the inhibitory activities of *Gentiana* derivatives against inflammatory response of rheumatoid arthritis using molecular modeling approach. Int J Pharmaceut Chem Res 10(1):1–15
79. Mittal AK, Kumar S, Banerjee UC (2014) Quercetin and gallic acid mediated synthesis of bimetallic (silver and selenium) nanoparticles and their antitumor and antimicrobial potential. J Colloid Interface Sci 431:194–199
80. Morales-Lozoya V, Espinoza-Gómez H, Flores-López LZ, Sotelo-Barrera EL, Núñez-Rivera A, Cadena-Nava RD, …, Rivero IA (2021) Study of the effect of the different parts of *Morinda citrifolia* L.(noni) on the green synthesis of silver nanoparticles and their antibacterial activity. Appl Surf Sci 537:147855
81. Morales-Lozoya V, Espinoza-Gómez H, Flores-López LZ, Sotelo-Barrera EL, Núñez-Rivera A, Cadena-Nava R D, …, Rivero IA (2021) Study of the effect of the different parts of *Morinda citrifolia* L.(noni) on the green synthesis of silver nanoparticles and their antibacterial activity. Appl Surf Sci 537:147855
82. Moulton MC, Braydich-Stolle LK, Nadagouda MN, Kunzelman S, Hussain SM, Varma RS (2010) Synthesis, characterization and biocompatibility of "green" synthesized silver nanoparticles using tea polyphenols. Nanoscale 2(5):763–770
83. Muniyappan N, Pandeeswaran M, Amalraj A (2021) Green synthesis of gold nanoparticles using *Curcuma pseudomontana* isolated curcumin: its characterization, antimicrobial, antioxidant and anti-inflammatory activities. Environ Chem Ecotoxicol 3:117–124
84. Murali K, Neelakandan MS, Thomas S (2018) Biomedical applications of gold nanoparticles. JSM Nanotechnol Nanomed 6(1):1064
85. Nagati VB, Alwala J, Koyyati R, Donda MR, Banala R, Padigya PRM (2012) Green synthesis of plant-mediated silver nanoparticles using *Withania somnifera* leaf extract and evaluation of their antimicrobial activity. Asian Pac J Trop Biomed 2:1–5
86. Neves MI, Prajapati DH, Parmar S, Aghajanzadeh TA, Hawkesford MJ, Kok LJD (2017) Manganese toxicity hardly affects sulfur metabolism in *Brassica rapa*. In: Sulfur metabolism in higher plants-fundamental, environmental and agricultural aspects. Springer, Cham, pp 155–162
87. Ovais M, Khalil AT, Islam NU, Ahmad I, Ayaz M, Saravanan M, Mukherjee S (2018) Role of plant phytochemicals and microbial enzymes in biosynthesis of metallic nanoparticles. Appl Microbiol Biotechnol 102(16):6799–6814
88. Ovais M, Khalil AT, Raza A, Khan MA, Ahmad I, Islam NU, …, Shinwari ZK (2016) Green synthesis of silver nanoparticles via plant extracts: beginning a new era in cancer theranostics. Nanomedicine 12(23):3157–3177

89. Padinjarathil H, Joseph MM, Unnikrishnan BS, Preethi GU, Shiji R, Archana MG, Sreelekha TT (2018) Galactomannan endowed biogenic silver nanoparticles exposed enhanced cancer cytotoxicity with excellent biocompatibility. Int J Biol Macromol 118:1174–1182
90. Pak PJ, Go EB, Hwang MH, Lee DG, Cho MJ, Joo YH, Chung N (2016) Evaluation of the cytotoxicity of gold nanoparticle-quercetin complex and its potential as a drug delivery vesicle. J Appl Biol Chem 59(2):145–147
91. Pal R, Panigrahi S, Bhattacharyya D, Chakraborti AS (2013) Characterization of citrate capped gold nanoparticle-quercetin complex: experimental and quantum chemical approach. J Mol Struct 1046:153–163
92. Panche AN, Diwan AD, Chandra SR (2016) Flavonoids: an overview. J Nutr Sci 5.
93. Patani AV, Patel RR, Prajapati DH, Patel MS, Patel AD (2020) Evaluation of plant growth promoting rhizobacteria to improve tomato (*Lycopersicon esculentum* Mill.) productivity and resilience under salinity stress. J Cell Tissue Res 20(2):6953–6964
94. Patel AD, Lalcheta K, Gill SS, Tuteja N (2013) Salinity tolerance of *Avicennia officinalis* L. (Acanthaceae) from Gujarat coasts of India. Climate change and plant abiotic stress tolerance, pp 189–208
95. Patel AD, Panchal NS, Pandey IB, Pandey AN (2010) Growth, water status and nutrient accumulation of seedlings of *Jatropha curcas* L.(Euphorbiaceae) in response to soil salinity. In Anales de biología, vol 32, No 1. Facultad de Biología, Universidad de Murcia, pp 59–71
96. Patel DJ, Patani AV, Prajapati DH, Patel CA (2020) Design and synthesis of fluorochemosensors and their cation recognition study. Int J Res Eng, Sci Manag 3(6):279–285
97. Pawar O, Deshpande N, Dagade S, Waghmode S, Nigam Joshi P (2016) Green synthesis of silver nanoparticles from purple acid phosphatase apoenzyme isolated from a new source *Limonia acidissima*. J Exp Nanosci 11(1):28–37
98. Phillips RL, Miranda OR, You CC, Rotello VM, Bunz UH (2008) Rapid and efficient identification of bacteria using gold-nanoparticle–poly (para-phenyleneethynylene) constructs. Angew Chem Int Ed 47(14):2590–2594
99. Prajapati DH, Aghajanzadeh T, De Kok LJ (2017) Impact of sulfur nutrition and H_2S exposure on expression and activity of Group 1 sulfate transporters in developing *Brassica pekinensis* seedlings. Berichte aus dem Julius Kühn-Institut 191:40
100. Prajapati DH, Ausma T, De Boer J, Hawkesford MJ, De Kok LJ (2020) Nickel toxicity in *Brassica rapa* seedlings-Impact on sulfur metabolism and mineral nutrient content. Journal fur Kulturpflanzen-J Cultivated plants 72(9 S):73–478
101. Qi L, Gao X (2008) Emerging application of quantum dots for drug delivery and therapy. Expert Opin Drug Deliv 5(3):263–267
102. Rafey A, Shrivastavaa KBL, Iqbal SA, Khan Z (2011) Growth of Ag-nanoparticles using aspartic acid in aqueous solutions. J Colloid Interf Sci 354(1):190–195
103. Rajesh KM, Ajitha B, Reddy YAK, Suneetha Y, Reddy PS, Ahn CW (2018) A facile biosynthesis of copper nanoparticles using *Cuminum cyminum* seed extract: antimicrobial studies. Adv Nat Sci: Nanosci Nanotech 9(3):035005
104. Rasmussen JW, Martinez E, Louka P, Wingett DG (2010) Zinc oxide nanoparticles for selective destruction of tumor cells and potential for drug delivery applications. Expert Opin Drug Deliv 7(9):1063–1077
105. Ravichandran A, Subramanian P, Manoharan V, Muthu T, Periyannan R, Thangapandi M, ..., Marimuthu PN (2018) Phyto-mediated synthesis of silver nanoparticles using fucoidan isolated from *Spatoglossum asperum* and assessment of antibacterial activities. J Photochem Photobiol B: Biol 185:117–125
106. Reich M, Shahbaz M, Prajapati DH, Parmar S, Hawkesford MJ, De Kok LJ (2016) Interactions of sulfate with other nutrients as revealed by H_2S fumigation of Chinese cabbage. Front Plant Sci 7:541
107. Rineesh NR, Neelakandan MS, Thomas S (2018) Applications of silver nanoparticles for medicinal purpose. JSM Nanotech Nanomed 6(1):1–7
108. Roy K, Ghosh CK (2018). Measurement of electrical conductivity of thin film composed of green synthesized copper nanoparticles. In: 2018 emerging trends in electronic devices and computational techniques (EDCT). IEEE, pp 1–4

109. Safaepour M, Shahverdi AR, Shahverdi HR, Khorramizadeh MR, Gohari AR (2009) Green synthesis of small silver nanoparticles using geraniol and its cytotoxicity against fibrosarcoma-wehi 164. Avicenna J Med Biotech 1(2):111
110. Sahu N, Soni D, Chandrashekhar B, Satpute DB, Saravanadevi S, Sarangi BK, Pandey RA (2016) Synthesis of silver nanoparticles using flavonoids: hesperidin, naringin and diosmin, and their antibacterial effects and cytotoxicity. Int Nano Lett 6(3):173–181
111. Salama HM (2012) Effects of silver nanoparticles in some crop plants, common bean (*Phaseolus vulgaris* L.) and corn (*Zea mays* L.). Int Res J Biotechnol 3(10):190–197
112. Schenk G, Elliott TW, Leung E, Carrington LE, Mitić N, Gahan LR, Guddat LW (2008) Crystal structures of a purple acid phosphatase, representing different steps of this enzyme's catalytic cycle. BMC Struct Biol 8(1):1–13
113. Shankar S, Rhim JW (2015) Amino acid mediated synthesis of silver nanoparticles and preparation of antimicrobial agar/silver nanoparticles composite films. Carbohyd Polym 130:353–363
114. Si S, Mandal TK (2007) Tryptophan-based peptides to synthesize gold and silver nanoparticles: a mechanistic and kinetic study. Chem–Eur J 13(11):3160–3168
115. Sichert JA, Tong Y, Mutz N, Vollmer M, Fischer S, Milowska KZ, Feldmann J (2015) Quantum size effect in organometal halide perovskite nanoplatelets. Nano Lett 15(10):6521–6527
116. Siddiqi KS, Husen A (2017) Recent advances in plant-mediated engineered gold nanoparticles and their application in biological system. J Trace Elem Med Biol 40:10–23
117. Singh A, Sengar RS, Singh AK, Singh S, Kumar A, Ahmad K (2019) Impact of genetic engineering on agriculture food crops. Biotech Today: An Int J Biol Sci 9(1):17–24
118. Singh B, Rani M, Singh J, Moudgil L, Sharma P, Kumar S, …, Kaura A (2016) Identifying the preferred interaction mode of naringin with gold nanoparticles through experimental, DFT and TDDFT techniques: insights into their sensing and biological applications. RSC Adv 6(83):79470–79484
119. Sirisha D, Gandhi N, Hasheena M, Smita A (2014) Eco-friendly method for synthesis of copper nanoparticles and application for removal of aqueous sulphur dioxide (SO_2) and Nitrogen Dioxide (NO_2). Int J Eng Res Techn 3(11):1253–1263
120. Snijesh VP, Singh S (2014) Molecular modeling and network based approach in explaining the medicinal properties of *Nyctanthes arbortristis, Lippia nodiflora* for rheumatoid arthritis. J Bioinform Intell Control 3(1):31–38
121. Stephen A, Seethalakshmi S (2013) Phytochemical synthesis and preliminary characterization of silver nanoparticles using hesperidin. J Nanosc 2013:1–6
122. Tang WW, Zeng GM, Gong JL, Liang J, Xu P, Zhang C, Huang BB (2014) Impact of humic/fulvic acid on the removal of heavy metals from aqueous solutions using nanomaterials: a review. Sci Total Environ 468:1014–1027
123. Thakur S, Rai R, Sharma S (2014) Study the antibacterial activity of copper nanoparticles synthesized using herbal plants leaf extracts. Int J Bio-Technol Res 4:21–34
124. Tungare K, Shahu R, Zambare V, Agarwal P, Jobby R, Nisar N, Jha P (2022) Toxicity mitigation of textile dye reactive blue 4 by hairy roots of *Helianthus annuus* and testing its effect in in vivo model systems. BioMed Res Int 2022
125. Vaghela PM, Patel AD, Pandey IB, Pandey AN (2010) Implications of calcium nutrition on the response of *Salvadora persica* (Salvadoraceae) to soil salinity. Commun Soil Sci Plant Anal 41(22):2644–2660
126. Vasileva P, Donkova B, Karadjova I, Dushkin C (2011) Synthesis of starch-stabilized silver nanoparticles and their application as a surface plasmon resonance-based sensor of hydrogen peroxide. Colloids Surf, A 382(1–3):203–210
127. Verma N, Sao P, Srivastava A, Singh S (2021) Physiological and molecular responses to drought, submergence and excessive watering in plants. In: Harsh environment and plant resilience. Springer, Cham, pp 305–321
128. Wei D, Qian W (2008) Facile synthesis of Ag and Au nanoparticles utilizing chitosan as a mediator agent. Colloids Surf, B 62(1):136–142

129. Wu Q, Cao H, Luan Q, Zhang J, Wang Z, Warner JH, AR Watt A (2008) Biomolecule-assisted synthesis of water-soluble silver nanoparticles and their biomedical applications. Inorganic Chem 47(13):5882–5888
130. Yan J, Han L, Gao W, Xue S, Chen M (2015) Biochar supported nanoscale zerovalent iron composite used as persulfate activator for removing trichloroethylene. Biores Technol 175:269–274
131. Yang L, Stöckigt J (2010) Trends for diverse production strategies of plant medicinal alkaloids. Nat Prod Rep 27(10):1469–1479
132. Zhao Z, Sun J, Xing S, Liu D, Zhang G, Bai L, Jiang B (2016) Enhanced Raman scattering and photocatalytic activity of TiO_2 films with embedded Ag nanoparticles deposited by magnetron sputtering. J Alloy Compd 679:88–93
133. Zuidersma EI, Ausma T, Stuiver CEE, Prajapati DH, Hawkesford MJ, De Kok LJ (2020) Molybdate toxicity in Chinese cabbage is not the direct consequence of changes in sulphur metabolism. Plant Biol 22(2):331–336

Fruit and Vegetable Peels for Nanoparticles Synthesis and Applications

Samandeep Kaur, H. K. Chopra, and P. S. Panesar

Abstract Nanoparticle synthesis using chemical, physical, and biosynthesis methods is a fundamental part of nanotechnology-based applications. The conventional methods used for the synthesis of nanoparticles are expensive, require higher energy, involve the utilization of hazardous chemicals, and produce substantial amounts of by-products. Due to this, the synthesis of NPs by utilizing natural sources is gaining attention as it offers an easy, non-toxic, economic, and sustainable approach to the synthesis of NPs. Fruit and vegetable peels are generally discarded by food manufacturers, but these are profuse with bioactive compounds such as polyphenols, flavonoids, antioxidants, carotenoids, essential oils, etc., which aid in the reduction and stabilization of metal ions during the green synthesis of NPs. Peels of banana, papaya, orange, tomatoes, gourd, etc., have been utilized for the synthesis of metallic NPs such as silver, gold, zinc oxide, iron, copper, palladium, and titanium. The present chapter focuses on different NPs synthesized using fruit and vegetable peels as well as their biological activities such as anti-cancerous, antimicrobial, antioxidant, and catalytic efficiency associated with these NPs.

Keywords Bioactive compounds · Fruit peels · Vegetable peels · NPs · Green synthesis

List of Abbreviations

Calcium Oxide	CaO
Cobalt	Co

S. Kaur · P. S. Panesar
Department of Food Engineering and Technology, Sant Longowal Institute of Engineering and Technology, Longowal, Punjab 148106, India

H. K. Chopra (✉)
Department of Chemistry, Sant Longowal Institute of Engineering and Technology, Longowal, Punjab 148106, India
e-mail: hk67@rediffmail.com

A. Husen (ed.), *Nanomaterials from Agricultural and Horticultural Products*, Smart Nanomaterials Technology,
https://doi.org/10.1007/978-981-99-3435-5_13

Cobalt Chloride	$CoCl_2$
Copper	Cu
Escherichia Coli	*E. Coli*
Gold	Au
Grams	g
Iron	Fe
Iron Oxide	Fe_3O_4
Klebsiella pneumoniae	*K. pneumoniae*
Milliliter	mL
Microgram	μG
Nickel Ferrite	$NiFe_2O_4$
Nanometer	nm
Nanoparticles	NPs
Silver	Ag
Sulfur	S
Seconds	s
Selenium	Se
Silicon dioxide	SiO_2
Thiourea	CH_4N_2S
Titanium dioxide	TiO_2
Zinc Oxide	ZnO
Zirconium	Zr

1 Introduction

In comparison to bulk materials, nanoparticles (NPs) exhibit completely new or improved physical, chemical, and biological properties, due to certain modification in size, morphology, and surface area [45] and has shown applications in various fields such as food, medicine, water treatment, solar energy conversion, and catalysis. The NPs can be synthesized from natural raw materials such as proteins and polysaccharides as well as from inorganic precursors such as metals, and salts [54]. Metal NPs have gained attention due to their antibacterial, antiviral, and anticancer properties. Metal NPs have been used in various biomedical applications such as photo-thermal therapy [20, 27, 69], drug delivery [47, 66], photodynamic therapy [18, 19], gene therapy [26], and biosensing [49].

NPs can be synthesized using various physical, chemical, and biological methods. Since NP synthesis using physical methods is expensive, and use of toxic chemicals in chemical synthesis limit the biomedical applications of NPs, more ecofriendly methods such as green synthesis using microorganisms (such as fungi, yeast) and plant-based extract (such as leaves, root, fruit, etc.) have been employed. In comparison to microorganisms, the plant sources such as fruits and vegetable peels have gained the interest of manufacturers as they do not require any isolation or culture maintenance.

Fruit and vegetable processing industry discard enormous amount of waste into wastelands. This waste is rich in phytochemicals such as polyphenols, flavonoids, antioxidants, carotenoids, essential oils, etc. [35], which have capping abilities, and aid in the reduction as well as stabilization of metal ions during the green synthesis of NPs [54]. Studies have also proven that the bioactive potential of the peels is comparatively higher than the edible fruit. Recently, various fruit and vegetable peels have been utilized for the synthesis of metal NPs, such as synthesis of gold NPs from pomegranate peels [3, 39], onion peels [44], citrus peels [70, 71], etc.; synthesis of silver NPs from banana peels [8, 29], citrus peels [7, 36], pomegranate peels [17], synthesis of iron NPs from jackfruit peels [32], apple peels [65], pear peels [55], etc. The NPs synthesized from these peels exhibit various benefits such as anti-inflammatory [11, 22], anti-cancer [38, 62, 72], antioxidant [14], catalytic activity [5, 73], antimicrobial activity [21, 59] etc. This chapter aims to summarize different studies on nano-particle synthesis using fruits and vegetable peels.

2 Methods for the Synthesis of Nanoparticles

The NPs are synthesized mainly using two approaches (1) top-down approach, in which the macromolecules are reduced to nanoscale using attrition or milling techniques. (2) bottom-up approach, in which chemical and biological processes used to aid assembling of atoms to nanoscale particles [63]. Depending on the size, shape, equipment, and stability of NPs, the NP synthesis can be broadly classified as physical, chemical, and biological methods. Various methods used for synthesis of NPs have been presented in Fig. 1.

2.1 Physical Methods

The physical methods are mainly dependent on the type of instrument used for the synthesis. Various physical methods used for NP synthesis include ball milling, laser ablation which involves breakdown of macromolecules using a laser beam, tube furnace, flame pyrolysis, arc discharge, etc. Ball milling is the simplest method which produces NPs by attrition, where kinetic energy is transferred from moving balls to the material being reduced [30]. The efficiency of ball milling can be improved by coupling it with ultrasound [67]. The drawbacks faced in ball milling include size irregularities and crystal defects. Another common technique used for synthesis of NPs at a commercial scale is flame pyrolysis. In this technique, the precursor (liquid/gaseous) is heated using flame/laser/plasma, transferred to furnace under high pressure, and NPs are recovered [33].

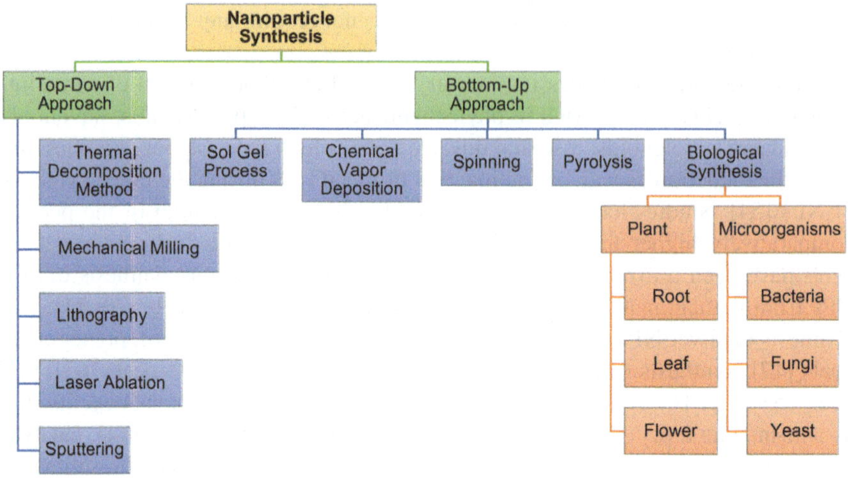

Fig. 1 Various methods used for the synthesis of nanoparticles

2.2 *Chemical Methods*

The major chemical methods used for the synthesis of NPs include sol-gel, solvothermal processes, precipitation etc. In sol-gel method, the precursor which is a chemical solution (mainly metal oxides/chlorides) is dispersed in solvent by sonication, stirring, or shaking. From the resultant solution, the solid and liquid phases are separated using centrifugation, filtration, or sedimentation techniques [51]. In solvothermal synthesis of NPs, chemical reactions occur in sealed containers where the solvents are heated under pressure (temperatures >= boiling points). The reaction temperature, time, type of solvent, reductant as well as surfactant can be changed to modify the shape, size, and crystallinity of NPs [50]. The NP synthesis reaction can further be accelerated by coupling the chemical methods with other techniques such as ultrasound assistance [10], microwave [64], etc.

2.3 *Biological Methods*

The NPs have been synthesized using various physical, chemical, and biological methods. The physical methods used are less economical as they require more space, as well as high temperature and pressures; chemical methods use toxic chemicals hazardous for human and environment [30]. To overcome the drawbacks of conventional techniques, studies have been conducted to develop various cost-effective, efficient, sustainable, and environment-friendly techniques for the synthesis of NPs. The biological methods use plant extracts or microorganisms such as bacteria, fungi, and yeast for the synthesis of NPs. Microorganisms can be considered nano-factories due to the presence of various reductase enzymes which detoxify heavy metals and aid in reduction of metal salts for the synthesis of NPs. Even though microorganisms are eco-friendly, non-toxic, and efficient source of NP synthesis, they require isolation, media preparations, and culture maintenance, which can be time consuming [28]. To overcome this drawback, the plant-based extracts have been used for the synthesis of NPs. Various parts of the plant such as root, leaves, stem, fruits, vegetables, flowers, etc., have been utilized for the synthesis of NPs. Various phytochemicals such as polyphenols, antioxidants, heterocyclic compounds, flavonoids, etc., present in these plant-based extracts help in reduction of metal salts and hence aid in formation of various NPs [30].

3 Synthesis of Nanoparticles Using Fruits and Vegetable Peels

Fruit and vegetable peels are generally discarded by food manufacturers, but these are profuse with bioactive compounds such as polyphenols, flavonoids, antioxidants, carotenoids, essential oils, etc., which aid in the reduction and stabilization of metal ions during the green synthesis of NPs. Due to this, the fruits and vegetable peels having biomedical potential are gaining attention for green synthesis of NPs. Peels of banana, papaya, orange, tomatoes, gourd, etc., have been utilized for the synthesis of metallic NPs such as silver, gold, zinc oxide, iron, copper, palladium, and titanium [63]. In addition to plant extracts, other green techniques such as ultrasound, can be used to assist the NP synthesis as they reduce the particle size of matrix effectively at ambient temperatures and after a short reaction time [9].

During synthesis of NPs from fruits and vegetable peels, the peels are dried, and ground to powder form and solvent extraction of peels is conducted. In the next step, the metal salt (mainly oxide or nitrite) is added to the extract and physical/ thermal treatment (if needed) is provided. The mixture is left undisturbed for some time till a color change is observed in the solution. This mixture is centrifuged and filtered to recover synthesized NPs [23]. The process flowchart for synthesis of NPs from peels of fruits and vegetables has been presented in Fig. 2.

Various studies have been conducted on synthesis of NPs using fruits and vegetable peels (Table 1). Sharma et al. [61] used peels of five different vegetables, i.e., *Lagenaria siceraria, Solanum lycopersicum, Luffa cylindrica, Solanum melongena,* and *Cucumis sativus* to synthesize silver NPs. The synthesized AgNPs around 20 nm in diameter and spherical shape exhibited excellent bactericidal effect against *E. coli,* and *K. pneumoniae.* Studies have also proven that the NPs synthesized from plant extracts are non-toxic and safe for food/ biomedical applications. Gold NPs synthesized using peel extract of pineapple and passion fruit showed no biological cytotoxicity even at concentrations as high as 400 μg/mL [45].

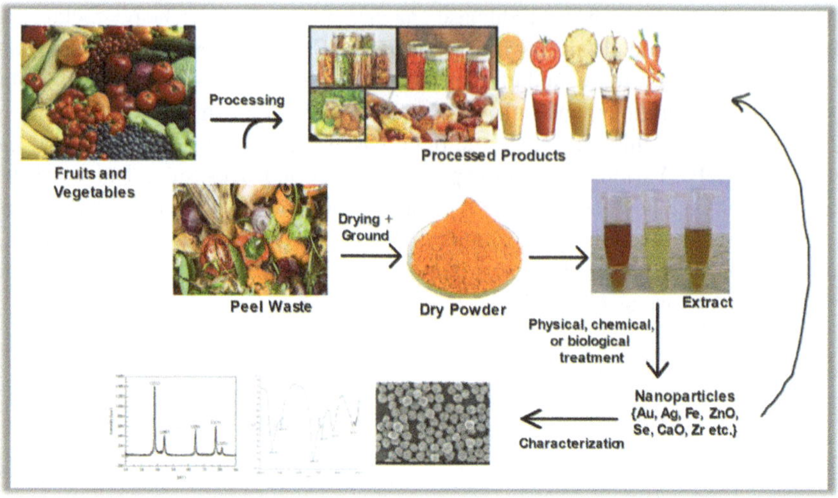

Fig. 2 Process flow for utilization of fruits and vegetable peels for NP synthesis

4 Applications of F&V Derived Nanoparticles

4.1 Desalination of Water

Lack of potable water is one of the major problems faced by human civilization today. The problem of pure water scarcity can be solved to some extent by water desalination. The commercial desalination units operate either on membrane purification techniques, i.e., reverse osmosis, reverse electrodialysis, or on thermal purification techniques, i.e., multi-stage flashing and vapor compression distillation [16]. Solar distillation has been proven to be one of the efficient and cheapest means for water desalination. Studies have been conducted to improve the efficiency of solar distillation systems by utilization of NPs. [6] combined TiO_2 NPs synthesized from jackfruit peel with mat black paint to coat the sides of solar still and added nanofluid (comprise $CoCl_2$, CH_4N_2S, and SiO_2) in the basin. The solar still coated with NPs provided a higher productivity (50.55%) than conventional solar still. In another study, silver NPs synthesized from watermelon peels were utilized for desalination of seawater. It has been observed that silver NPs at 0.05 g loading presented excellent evaporation performance, with an evaporation rate and efficiency of 1.37 kg/m^{2}h and 89.3% respectively [16].

Table 1 Synthesis of nanoparticles from fruits and vegetable peels

Name of the source	Type of NP	Synthesis technique	Average size	Morphology	Reference(s)
Annona squamosa (sugar apple) peels	Silver (Ag)	Microwave-assisted green synthesis	18–35 nm	Spherical shaped	[25]
Citrus paradisi peels	Iron Oxide	Biosynthesis	28–32 nm	Mostly spherical	[37]
Citrus peel extract	Au	Ultrasound-assisted green synthesis	13.65–16.80 nm	Spherical shaped	[22]
Garlic peel	Cobalt (Co)	Green biosynthesis	$\approx$15 nm	Semi-spherical	[73]
Jackfruit	Titanium dioxide (TiO_2)	Green Synthesis using water		Spherical shape (approx.)	[6]
Kiwifruit (*Actinidia deliciosa*) peel	Sulfur (S)	Green synthesis	90–130 nm		[48]
Lemon peel	TiO_2	Green synthesis	80–140 nm	Spherical shape	[43]
Manilkara zapota	Au	Surface functionalized AuNPs synthesis		Spherical shaped	[13]
Onion Peel	Zinc Oxide (ZnO)	Mediated-Green Synthesis	20–80 nm	Spherical shape	[40]
Pomegranate (*Punica granatum* L.) peel	Iron Oxide (Fe_3O_4)	Green biosynthesis	21–23 nm	Cubical shaped	[12]
Pomegranate (*Punica granatum* L.) peel	ZnO	Microwave synthesis	45 nm		[24]
Pomegranate (*Punica granatum* L.) peel	Fe_3O_4	Microwave assisted green synthesis	17.8 ± 6.5 nm	Cubical structure	[42]
Pomelo peel	Ag	Green synthesis + ultrasonication	35–40 nm	Spherical at room and high temperatures, cubical with ultrasonication	[9]
Watermelon peel	ZnO	Biosynthesis	40–88 nm	Spherical shape	[5]

4.2 Antimicrobial Activities

The NPs synthesized from plant sources exhibit excellent antimicrobial properties against several fungi, yeast, gram-negative, and gram-positive bacteria. The studies on antimicrobial activity exhibited by NPs synthesized from fruits and vegetables have been presented in Table 2. The calcium oxide (CaO) NPs fabricated using peel extract of dragon fruit and $CaCl_2$ exhibited antimicrobial activity against *Candida albicans* [52]. The silver NPs synthesized using pomelo peels and $AgNO_3$ inhibited the growth and multiplication of *Bacillus Cereus* (MTCC 430) and *Escherichia Coli* (MTCC 443), and therefore can be utilized for development of antimicrobials [9]. Copper NPs synthesized from pomegranate peels exhibited antimicrobial activity against pathogens such as *Micrococcus luteus* MTCC 1809, *Pseudomonas aeruginosa* MTCC 424, *Salmonella enterica* MTCC 1253, and *Enterobactor aerogenes* MTCC 2823 in vitro. The Copper NPs exhibited higher antimicrobial activity than the peels extract as well as commercial antibiotic as larger zone of inhibition was observed in Copper NPs sample [34].

4.3 Anticancer Activities

NPs synthesized from plant sources have proven to exhibit various anticancer activities making them a suitable material for several biomedical applications. The peels of fruits and vegetables are profuse with various polyphenols and other bioactive compounds which exhibit several antioxidant and anticancer properties. Various studies on anticancer activity exhibited by NPs synthesized from fruits and vegetable peels have been presented in Table 3. The in-vitro cytotoxicity studies of Fe_3O_4 NPs synthesized from *Garcinia mangostana* fruit peel extract exhibited higher cytotoxicity against HCT116 colon cancer cells than CCD112 colon normal cells. And Fe_3O_4 NPs containing extract (10% w/w) showed an IC50 value of 99.80 μg/mL in HCT116 colon cancer cell line which was comparatively lower than CCD112 colon normal cell line, i.e., 140.80 μg/mL [72]. Gold NPs synthesized using peels of *Spondias dulcis* exhibited significant cytotoxicity activity against breast cancer cells (MCF-7) by the production of reactive oxygen species [46].

4.4 Catalytic Activity

The catalytic activity is referred to the ability of any catalyst to increase the rate of a chemical reaction. The extract of fruits and vegetable peels has been utilized for the synthesis of ecofriendly nano-catalysts, which can be recovered/recycled easily and effectively from reacting medium. Zayed et al. [73] utilized garlic peels for the development of cobalt NPs, which effectively catalyzed the degradation of organic

Table 2 Antimicrobial activity exhibited by nanoparticles synthesized from fruits and vegetable peels

Name of the source	Type of NP	Synthesis technique	Average size	Morphology	Application	Key reference
Banana peels and neem leaves	Silver (Ag)	Biosynthesis	168.7 nm from banana peel and 206.4 nm	Spherical shaped	Antimicrobial and dye degrading agent for wastewater treatment	[59]
Musa coccinea	Iron (Fe)	Synthesis of poly-dispersed and non-homogenous MC-ZVI NPs		Spherical shaped	Antimicrobial and antioxidant	[14]
Orange peel	Ag	Photo-biosynthesis	32–47 nm	Spherical shape	Antifungal activity	[41]
Orange peel	Ag	Green biosynthesis		Spherical shaped	Antibacterial Composite Filler Applications	[53]
Orange peel waste	Selenium (Se)	Green Biosynthesis	16–95 nm	Spherical and polydisperse	Antibacterial and Antibiofilm agents against Multidrug-Resistant Bacteria	[57]
Plantain peel	(Zinc Oxide) ZnO	Green synthesis	20 nm	Spherical shape	Antibacterial activity	[31]
Pomegranate peels	CaO	Green synthesis			Antimicrobial activities against *Bacillus* sp., and *Arthrographis cuboidea*	[2]
Pomegranate peels and nano chitosan	Selenium (Se)	Biosynthesis	≈82 nm	Spherical shaped	Antifungal application	[58]
Pomegranate (*Punica granatum*) peel	Zirconium (Zr)		20–60 nm		Antimicrobial and antioxidant potency	[15]

(continued)

Table 2 (continued)

Name of the source	Type of NP	Synthesis technique	Average size	Morphology	Application	Key reference
Pomegranate (*Punica granatum* L.) peel	ZnO	Green synthesis	20–40 nm	Spherical and hexagonal shapes	Antimicrobial activity	[60]
Punica granatum peels	Copper (Cu)	Biogenesis	15–20 nm		Antimicrobial activity against opportunistic pathogens	[34]
Waste vegetable peels	Ag	Green synthesis	20 nm	Spherical	Bactericidal against *E. coli*, and *K. pneumoniae*	[61]

Table 3 Anticancer activity exhibited by nanoparticles synthesized from fruits and vegetable peels

Name of the source	Type of NP	Synthesis technique	Average size	Morphology	Application	Key reference
Garcinia mangostana fruit peel extract	Fe_3O_4	Co-precipitation method	13.42 ± 1.58 nm	Spherical	Anticancer activity against HCT116 colon cells	[72]
Hylocereus undatus (dragon fruit) peel	Ag	Green synthesis	10–50 nm	Spherical shape	Anticancer activity on liver carcinoma (HepG2) cell lines	[62]
Lemon peel	Ag	Green synthesis	20 nm	Different shapes	Antimicrobial test on the lung cancer cell	[21]
Lime peel	$NiFe_2O_4$	Green synthesis	31–35 nm	Spherical	Anticancer effects and inactivation of carcinogen	[38]
Spondias dulcis peel extract	Au	Biogenic synthesis (photosynthesis)	36.75 ± 11.36 nm	Spherical	Cytotoxic activity in human breast cancer cells	[45]

Ag—Silver; Au—Gold; Fe_3O_4—Iron Oxide; $NiFe_2O_4$—Nickel Ferrite

Table 4 Catalytic activity exhibited by nanoparticles synthesized from fruits and vegetable peels

Name of the source	Type of NP	Synthesis technique	Average size	Morphology	Application	Key reference
Citrus macroptera peels	Silver (Ag)	Biogenic synthesis	11 nm	Face centered cubic crystals + spherical	Catalytic activity	[56]
Garlic peel	Cobalt (Co)	Green biosynthesis	≈ 15 nm	Semi-spherical	Catalytic activity	[73]
Lemon peel	TiO_2	Green synthesis	80–140 nm	Spherical shape	Optical and photocatalytic properties	[43]
Onion peels	Ag	Biosynthesis	12.5 nm	Spherical	Catalytic and antioxidant properties	[1]
Pomegranate (*Punica granatum* L.) peel	Iron Oxide (Fe_3O_4)	Green biosynthesis	21–23 nm	Cubical shaped	Optoelectronic applications	[12]
Watermelon peel	ZnO	Biosynthesis	40–88 nm	Spherical shape	Photocatalytic degradation of metronidazole (MTZ)	[5]

pollutants such as 4-nitrophenol and bromophenol blue, at an optimal pH of 9 and maximum rate constant of 4.56×10^{-3} s^{-1}. Saha et al. [56] synthesized Ag NPs from *Citrus macroptera* peels and observed that addition of AgNPs catalyzed the reduction of 4-nitrophenol to 4-aminophenol efficiently. The studies on catalytic activity exhibited by NPs synthesized from fruits and vegetable peels have been presented in Table 4.

4.5 Other Applications

Apart from the above discussed applications, the NPs synthesized form peels have also been used in dye removal processes. Akpomie and Conradie [4] impregnated silver NPs on *Solanum tuberosum* peels and utilized the hybrid system for the adsorption of bromophenol dye from simulated wastewater. The adsorption efficiency was significantly improved with the aid of ultrasonication. In another study, Kumar et al. [37] synthesized iron oxide NPs using citrus peels for treatment of various dyes such as methyl rose, methyl orange, and methylene blue and obtained decolorization% of 96.65%, 89.64%, 80.76% respectively. Ag NPs synthesized from *Citrus macroptera* peels, degraded methyl orange and methylene blue dyes in the presence of sodium borohydride [56]. The iron NPs also presented antioxidant activities against DPPH [37]. Abdullah et al. [1] synthesized silver NPs from onion peels and observed that these NPs exhibited higher antioxidant activities in comparison to ascorbic acid and hence can be used as an alternative antioxidant agent. Yang and Li [68] synthesized silver NPs from mango peels and loaded them to non-woven fabric to obtain a fabric that exhibited antibacterial activity.

5 Conclusions

It has been concluded that the fruits and vegetable industry waste have the potential to be a prominent part of nanotechnology industry. The waste discarded by the processing industries can be valorized by synthesis of NPs for biomedical as well as other applications. Several cytotoxicity studies have proven that the NPs synthesized from plant-based sources are non-toxic and are suitable for food-based applications. The properties of NPs are dependent on their size and shape, which can be varied by changing the process parameters or synthesis technique. The NPs can be used for water desalination, to prevent microbial growth, act as antioxidant compounds. The efficiency of synthesis techniques can be improved by further coupling of two or more techniques.

References

1. Abdullah HSTSH, Asseri SNARM, Mohamad WNKW, Kan SY, Azmi AA, Julius FSY, Chia PW (2021) Green synthesis, characterization and applications of silver NP mediated by the aqueous extract of red onion peel. Environ Pollut 271:116295
2. Ahmad W, Kamboj A, Banerjee I, Jaiswal KK (2022) Pomegranate peels mediated synthesis of calcium oxide (CaO) nanoparticles, characterization, and antimicrobial applications. Inorg Nano-Met Chem 1–8
3. Ahmad N, Radheshyam Rai S (2012) Rapid green synthesis of silver and gold nanoparticles using peels of Punica granatum. Adv Mater Lett 3:376–380
4. Akpomie KG, Conradie J (2020) Biogenic and chemically synthesized Solanum tuberosum peel-silver NP hybrid for the ultrasonic aided adsorption of bromophenol blue dye. Sci Rep 10:1–18
5. Al-Gheethi A, Sundram N, Crane R, Alburihi A, Mohamed RMSR, Al-Shaibani, MM, Kamil NAFM (2022) Metronidazole photocatalytic degradation by zinc oxide nanoparticles synthesized in watermelon peel extract; advanced optimization, simulation and numerical models using machine learning applications. Environ Res 113537
6. Alsaiari AO, Shanmugan S, Abulkhair H, Bamasag A, Moustafa EB, Alsulami RA, Ahmad I, Elsheikh A (2022) Applications of TiO_2/jackfruit peel nanocomposites in solar still: experimental analysis and performance evaluation. Case Stud Therm Eng 102292
7. Ayinde WB, Gitari WM, Samie A (2019) Optimization of microwave-assisted synthesis of silver NP by citrus paradisi peel and its application against pathogenic water strain. Green Chem Lett Rev 12:225–234
8. Bankar A, Joshi B, Kumar AR, Zinjarde S (2010) Banana peel extract mediated novel route for the synthesis of silver nanoparticles. Colloids Surf A Physicochem Eng Asp 368:58–63
9. Barbhuiya RI, Singha P, Asaithambi N, Singh SK (2022) Ultrasound-assisted rapid biological synthesis and characterization of silver nanoparticles using pomelo peel waste. Food Chem 385:132602

10. Bhanvase BA, Sonawane SH (2015) Effect of type and loading of surfactant on ultrasound-assisted synthesis of $CaZn_2(PO_4)_2$ nanoparticles by chemical precipitation. Chem Eng Process 95:347–352
11. Bokhary KA, Maqsood F, Amina M, Aldarwesh A, Mofty HK, Al-Yousef HM (2022) Grapefruit extract-mediated fabrication of photosensitive aluminum oxide NP and their antioxidant and anti-inflammatory potential. Nanomaterials 12:1885
12. Bouafia A, Laouini SE, Tedjani ML, Ali GA, Barhoum A (2022) Green biosynthesis and physicochemical characterization of Fe_3O_4 nanoparticles using Punica granatum L. fruit peel extract for optoelectronic applications. Text Res J 92:2685–2696
13. Chandrakar V, Tapadia K, Gupta SK (2022) Greener approach for gold nanoparticles synthesis from fruit peel extract of Manilkara zapota: a fluorometric assay for determination of thiourea. Inorg Nano-Met Chem 1–9
14. Chau TP, Brindhadevi K, Krishnan R, Alyousef MA, Almoallim HS, Whangchai N, Pikulkaew S (2022) A novel synthesis, analysis, and evaluation of Musa coccinea based zero valent iron nanoparticles for antimicrobial and antioxidant. Environ Res 209:112770
15. Chau TP, Veeraragavan GR, Narayanan M, Chinnathambi A, Alharbi SA, Subramani B, Brindhadevi K, Pimpimon T, Pikulkaew S (2022) Green synthesis of Zirconium nanoparticles using Punica granatum (pomegranate) peel extract and their antimicrobial and antioxidant potency. Environ Res 209:112771
16. Cao H, Cui T, Wang W, Li S, Tang X, Wang H, Zhu G (2020) Green-synthesizing Ag nanoparticles by watermelon peel extract and their application in solar-driven interfacial evaporation for seawater desalination. Mater Res Express 7:045005
17. Devanesan S, AlSalhi MS, Balaji RV, Ranjitsingh AJA, Ahamed A, Alfuraydi AA, AlQahtani FY, Aleanizy FS, Othman AH (2018) Antimicrobial and cytotoxicity effects of synthesized silver nanoparticles from Punica granatum peel extract. Nanoscale Res Lett 13:1–10
18. El-Hussein A, Mfouo-Tynga I, Abdel-Harith M, Abrahamse H (2015) Comparative study between the photodynamic ability of gold and silver nanoparticles in mediating cell death in breast and lung cancer cell lines. J Photochem Photobiol B Biol 153:67–75
19. Erdogan O, Abbak M, Demirbolat GM, Birtekocak F, Aksel M, Pasa S, Cevik O (2019) Green synthesis of silver nanoparticles via Cynara scolymus leaf extracts: the characterization, anticancer potential with photodynamic therapy in MCF7 cells. PLoS One 14:e0216496
20. Estelrich J, Busquets MA (2018) Iron oxide nanoparticles in photothermal therapy. Molecules 23:1567
21. Gandhad SS, Jeergal PR, Veena E, Hublikar L, Horakeri LD, Mathad SN, Patil SR (2022) Synthesis and characterization of silver nanoparticles using green route. Int J Adv Sci Technol 8:2252
22. Gao L, Mei S, Ma H, Chen X (2022) Ultrasound-assisted green synthesis of gold nanoparticles using citrus peel extract and their enhanced anti-inflammatory activity. Ultrason Sonochem 83:105940
23. Gautam A, Rawat S, Verma L, Singh J, Sikarwar S, Yadav BC, Kalamdhad AS (2018) Green synthesis of iron nanoparticle from extract of waste tea: an application for phenol red removal from aqueous solution. Environ Nanotechnol Monit Manag 10:377–387
24. George PP (2022) Value-addition of pomegranate fruit peel biowaste: microwave mediated green synthesis, characterization and antimicrobial activity of zinc oxide nanoparticles. Asian Basic Appl Res J 17–24
25. Hassan AMS, Mahmoud AS, Ramadan MF, Eissa MA (2022) Microwave-assisted green synthesis of silver nanoparticles using Annona squamosa peels extract: characterization, antioxidant, and amylase inhibition activities. Rend Lincei Sci Fis Nat 33:83–91
26. Hu K, Chen X, Chen W, Zhang L, Li J, Ye J, Zhang Y, Zhang L, Li C, Yin L, Guan YQ (2018) Neuroprotective effect of gold nanoparticles composites in Parkinson's disease model. Nanomed Nanotechnol Biol Med 14:1123–1136
27. Huang X, Jain PK, El-Sayed IH, El-Sayed MA (2008) Plasmonic photothermal therapy (PPTT) using gold nanoparticles. Lasers Med Sci 23:217–228

28. Hulkoti NI, Taranath TC (2014) Biosynthesis of nanoparticles using microbes—a review. Colloids Surf B Biointerfaces 121:474–483
29. Ibrahim HM (2015) Green synthesis and characterization of silver nanoparticles using banana peel extract and their antimicrobial activity against representative microorganisms. J Radiat Res Appl Sci 8:265–275
30. Ijaz I, Gilani E, Nazir A, Bukhari A (2020) Detail review on chemical, physical and green synthesis, classification, characterization and applications of nanoparticles. Green Chem Lett Rev 13:223–245
31. Imade EE, Ajiboye TO, Fadiji AE, Onwudiwe DC, Babalola OO (2022) Green synthesis of zinc oxide nanoparticles using plantain peel extracts and the evaluation of their antibacterial activity. Sci Afr 16:e01152
32. Jain R, Mendiratta S, Kumar L, Srivastava A (2021) Green synthesis of iron nanoparticles using Artocarpus heterophyllus peel extract and their application as a heterogeneous Fenton-like catalyst for the degradation of Fuchsin basic dye. Curr Res Green Sustain Chem 4:100086
33. Kammler HK, Mädler L, Pratsinis SE (2001) Flame synthesis of nanoparticles. Chem Eng Technol 24:583–596
34. Kaur P, Thakur R, Chaudhury A (2016) Biogenesis of copper nanoparticles using peel extract of Punica granatum and their antimicrobial activity against opportunistic pathogens. Green Chem Lett Rev 9:33–38
35. Kaur S, Panesar PS, Chopra HK (2021) Citrus processing by-products: an overlooked repository of bioactive compounds. Crit Rev Food Sci Nutr 1–20
36. Kaviya S, Santhanalakshmi J, Viswanathan B, Muthumary J, Srinivasan K (2011) Biosynthesis of silver nanoparticles using citrus sinensis peel extract and its antibacterial activity. Spectrochim Acta A Mol Biomol Spectrosc 79:594–598
37. Kumar B, Smita K, Galeas S, Sharma V, Guerrero VH, Debut A, Cumbal L (2020) Characterization and application of biosynthesized iron oxide nanoparticles using citrus paradisi peel: a sustainable approach. Inorg Chem Commun 119:108116
38. Malik AR, Aziz MH, Atif M, Irshad MS, Ullah H, Gia TN, Ahmed H, Ahmad S, Botmart T (2022) Lime peel extract induced $NiFe_2O_4$ NPs: synthesis to applications and oxidative stress mechanism for anticancer, antibiotic activity. J Saudi Chem Soc 26:101422
39. Manna K, Mishra S, Saha M, Mahapatra S, Saha C, Yenge G, Gaikwad N, Pal R, Oulkar D, Banerjee K, Saha KD (2019) Amelioration of diabetic nephropathy using pomegranate peel extract-stabilized gold nanoparticles: assessment of NF-κB and Nrf2 signaling system. Int J Nanomed 14:1753
40. Modi S, Yadav VK, Choudhary N, Alswieleh AM, Sharma AK, Bhardwaj AK, Khan SH, Yadav KK, Cheon J, Jeon BH (2022) Onion peel waste mediated-green synthesis of zinc oxide nanoparticles and their phytotoxicity on mung bean and wheat plant growth. Materials 15:2393
41. Mohamed YMA, Elshahawy IE (2022) Antifungal activity of photo-biosynthesized silver nanoparticles (AgNPs) from organic constituents in orange peel extract against phytopathogenic Macrophomina phaseolina. Eur J Plant Pathol 162:725–738
42. Mundekkad D, Alex AV (2022) Analysis of structural and biomimetic characteristics of the green-synthesized Fe_3O_4 nanozyme from the fruit peel extract of Punica granatum. Chem Pap 76:3863–3878
43. Nabi G, Ain QU, Tahir MB, Nadeem Riaz K, Iqbal T, Rafique M, Hussain S, Raza W, Aslam I, Rizwan M (2022) Green synthesis of TiO_2 nanoparticles using lemon peel extract: their optical and photocatalytic properties. Int J Environ Anal Chem 102:434–442
44. Patra JK, Kwon Y, Baek KH (2016) Green biosynthesis of gold nanoparticles by onion peel extract: synthesis, characterization and biological activities. Adv Powder Technol 27:2204–2213
45. Pechyen C, Ponsanti K, Tangnorawich B, Ngernyuang N (2021) Waste fruit peel-mediated green synthesis of biocompatible gold nanoparticles. J Mater Res Technol 14:2982–2991
46. Pechyen C, Ponsanti K, Tangnorawich B, Ngernyuang N (2022) Biogenic synthesis of gold nanoparticles mediated by Spondias dulcis (Anacardiaceae) peel extract and its cytotoxic activity in human breast cancer cell. Toxicol Rep 9:1092–1098

47. Pham XN, Nguyen TP, Pham TN, Tran TTN, Tran TVT (2016) Synthesis and characterization of chitosan-coated magnetite nanoparticles and their application in curcumin drug delivery. Adv Nat Sci Nanosci Nanotechnol 7:045010
48. Priyadarshi R, Roy S, Rhim JW (2022) Enhanced functionality of green synthesized sulfur nanoparticles using kiwifruit (Actinidia deliciosa) peel polyphenols as capping agents. J Nanostructure Chem 12:389–399
49. Prosposito P, Burratti L, Venditti I (2020) Silver nanoparticles as colorimetric sensors for water pollutants. Chemosensors 8:26
50. Rafienia M, Bigham A, Hassanzadeh-Tabrizi SA (2018) Solvothermal synthesis of magnetic spinel ferrites. J Med Signals Sens 8:108
51. Ramesh S (2013) Sol-gel synthesis and characterization nanoparticles $Ag_{3(2+x)}$ $Al_xTi_{4-x}O_{11+\delta}$ ($0.0 \leq x \leq 1.0$). J Nanosci 13:1–8
52. Ramli M, Rossani RB, Nadia Y, Darmawan TB, Ismail YS (2019) Nanoparticle fabrication of calcium oxide (CaO) mediated by the extract of red dragon fruit peels (Hylocereus Polyrhizus) and its application as inorganic-anti-microorganism materials. IOP Conf Ser Mat Sci Eng 509:012090
53. Rathinavel S, Saravanakumar SS (2022) Synthesis of silver nanoparticles through orange peel powder for antibacterial composite filler applications. J Polym Environ 30:1407–1414
54. Rodríguez-Félix F, Graciano-Verdugo AZ, Moreno-Vásquez MJ, Lagarda-Díaz I, Barreras-Urbina CG, Armenta-Villegas L, Olguín-Moreno A, Tapia-Hernández JA (2022) Trends in sustainable green synthesis of silver nanoparticles using agri-food waste extracts and their applications in health. J Nanomater 2022
55. Rong K, Wang J, Zhang Z, Zhang J (2020) Green synthesis of iron nanoparticles using Korla fragrant pear peel extracts for the removal of aqueous Cr (VI). Ecol Eng 149:105793
56. Saha P, Mahiuddin M, Islam AN, Ochiai B (2021) Biogenic synthesis and catalytic efficacy of silver nanoparticles based on peel extracts of citrus macroptera fruit. ACS Omega 6:18260–18268
57. Salem SS, Badawy MSE, Al-Askar AA, Arishi AA, Elkady FM, Hashem AH (2022) Green biosynthesis of selenium nanoparticles using orange peel waste: characterization, antibacterial and antibiofilm activities against multidrug-resistant bacteria. Life 12:893
58. Salem MF, Abd-Elraoof WA, Tayel AA, Alzuaibr FM, Abonama OM (2022) Antifungal application of biosynthesized selenium nanoparticles with pomegranate peels and nanochitosan as edible coatings for citrus green mold protection. J Nanobiotechnology 20:1–12
59. Sengupta A, Sarkar A (2022) Synthesis and characterization of nanoparticles from neem leaves and banana peels: a green prospect for dye degradation in wastewater. Ecotoxicology 31:537–548
60. Shaban AS, Owda ME, Basuoni MM, Mousa MA, Radwan AA, Saleh AK (2022) Characterization and genotoxic effects of green synthesis zinc oxide nanoparticles mediated by Punica granatum peel extract and its antimicrobial activity
61. Sharma K, Kaushik S, Jyoti A (2016) Green synthesis of silver nanoparticles by using waste vegetable peel and its antibacterial activities. J Pharm Sci Res 8:313
62. Shyamalagowri S, Charles P, Manjunathan J, Kamaraj M, Anitha R, Pugazhendhi A (2022) In vitro anticancer activity of silver nanoparticles phyto-fabricated by Hylocereus undatus peel extracts on human liver carcinoma (HepG2) cell lines. Process Biochem 116:17–25
63. Suhag R, Kumar R, Dhiman A, Sharma A, Prabhakar PK, Gopalakrishnan K, Singh A (2022) Fruit peel bioactives, valorisation into nanoparticles and potential applications: a review. Crit Rev Food Sci Nutr 1–20
64. Swaminathan V, Deheri PK, Bhame SD, Ramanujan RV (2013) Novel microwave assisted chemical synthesis of $Nd_2Fe_{14}B$ hard magnetic nanoparticles. Nanoscale 5:2718–2725
65. Ting ASY, Chin JE (2020) Biogenic synthesis of iron nanoparticles from apple peel extracts for decolorization of malachite green dye. Wat Air Soil Poll 231:1–10
66. Vazquez NI, Gonzalez Z, Ferrari B, Castro Y (2017) Synthesis of mesoporous silica nanoparticles by sol-gel as nanocontainer for future drug delivery applications. Boletín de la Sociedad Española de Cerámica y Vidrio 56:139–145

67. Yang B, Chen D (2017) Synthesis of CuO nanoparticles for catalytic application via ultrasound-assisted ball milling. Process Appl Ceram 11:39–44
68. Yang N, Li WH (2013) Mango peel extract mediated novel route for synthesis of silver nanoparticles and antibacterial application of silver nanoparticles loaded onto non-woven fabrics. Ind Crops Prod 48:81–88
69. Yang W, Liang H, Ma S, Wang D, Huang J (2019) Gold nanoparticle based photothermal therapy: development and application for effective cancer treatment. Sustain Mater Technol 22:e00109
70. Yang B, Qi F, Tan J, Yu T, Qu C (2019) Study of green synthesis of ultrasmall gold nanoparticles using citrus sinensis peel. Appl Sci 9:2423
71. Yuan CG, Huo C, Gui B, Cao WP (2017) Green synthesis of gold nanoparticles using citrus maxima peel extract and their catalytic/antibacterial activities. IET Nanobiotechnol 11:523–530
72. Yusefi M, Shameli K, Yee OS, Teow SY, Hedayatnasab Z, Jahangirian H, Webster TJ, Kuča K (2021) Green synthesis of Fe_3O_4 nanoparticles stabilized by a Garcinia mangostana fruit peel extract for hyperthermia and anticancer activities. Int J Nanomed 16:2515
73. Zayed MF, Eisa WH, Anis B (2022) Garlic peel as promising low-cost support for the cobalt nano-catalyst; synthesis and catalytic studies. J Environ Manage 312:114919

Grass and Their Waste Products for Nanoparticles Synthesis and Applications

Anurag Tiwari, Kajal Pandey, Sachidanand Singh, and Sonam Chawla

Abstract Nanotechnology is a rapidly growing miniaturization technology resulting in various inventions for easing human life and one such example of miniaturization is the synthesis of nanoparticles. Green synthesis of nanoparticles is a technique relying on agriculture and plant waste for synthesizing nanoparticles for various applications of benefit. The use of grass and grass waste for nanoparticle synthesis is one the most sustainable, non-toxic, cheap and eco-friendly methods even during large scale production of nanoparticles. The various grass species used for this purpose are *Cymbopogon citratus*, *Cymbopogon flexuosus*, *Enhalus acoroides*, *Chrysopogon zizanioides*, *Cynodon dactylon*, *Cyperus rotundus*, *Poa annua* and many more. They are used to derive different silver, copper, gold, silica, copper oxide, zinc oxide nanoparticles which have shown numerous applications. In this chapter, we have described the use of various grass species and grass waste for production of different nanoparticles and their applications and properties.

Keywords Nanotechnology · Nanoparticles · Green synthesis · Grass waste · Therapeutic · Industrial applications

1 Introduction

Grass is a green plant with no wood or wood fibers possessing various economical and health benefits. Grasses can be classified as either grass, sedge or of rush families, scientifically classified as Poaceae or Cyperaceae or Juncaceae familes, respectively. Grass and grass waste are emerging as major players in the field of green nanotechnology because of their easy availability, low cost, low toxicity and

A. Tiwari · K. Pandey · S. Chawla (✉)
Department of Biotechnology, Jaypee Institute of Information and Technology, Sector 62, Noida, Uttar Pradesh 201309, India
e-mail: sonam.chawla@jiit.ac.in; sonam9a@gmail.com

S. Singh
Department of Biotechnology, Sankalchand Patel University, Visnagar, Gujarat 384315, India

A. Husen (ed.), *Nanomaterials from Agricultural and Horticultural Products*, Smart Nanomaterials Technology,
https://doi.org/10.1007/978-981-99-3435-5_14

no potential hazard to the environment, i.e. sustainable synthesis [9]. Nanotechnology is a combined branch of science and engineering which includes miniaturization and development of nanoparticles between the range of 1–100 nm and has numerous applications in the field of agriculture, pharmaceuticals, biosensors and medical science [11]. Nano studies lead to the development of recyclable nanoparticles for assessing the human body to treat life threatening diseases like AIDS, cancer, exhibiting antimicrobial activity and for waste water treatment because of its ability to exclude heavy metal content from water [9, 19].

One of the major concerns in green technology is the synthesis of nanoparticles by using ways which do not have any harmful effects to humans as well as the environment. Nanoparticles can be synthesized using plants, algae, bacteria but green synthesis of nanoparticles has comparative advantages to all and dried grass is one of the best options for their synthesis [9]. Microbes are not preferred for the synthesis of nanoparticles because they need special controlled conditions for growth and any changes in temperature, pH or other parameters can hamper their biological activity which can increase the risk of spread of bacterial diseases in human and the large-scale commercial production of nanoparticles through microbes is difficult because of biosafety issues [19, 25]. Living plant or biomass can be used for nanoparticle production but use of plant extract is recommended to ensure the isolation of the compounds which can reduce metal toxic effects and then using them for synthesis of nanoparticles [6, 18, 19]. Grass can be used for the biosynthesis of many metallic nanoparticles such as silica, zinc oxide, silver, gold, palladium and copper nanoparticles. Most of the grasses have the capability to synthesize gold and silver nanoparticles easily but silver nanoparticles are favored because of their antimicrobial activity and also because of their lightness and strength [19]. Zinc oxide nanoparticles are reported to have photocatalytic activity and because of their excellent antimicrobial property, they can be used as replacement for conventional antibiotics as microorganisms can easily develop resistance against them [24].

2 Green Synthesis of Nanoparticles Using Plant Waste

Green nanotechnology is an interdisciplinary field that is a rapidly evolving and favored domain of research, recognized as a technology that is sustainable and non-hazardous in comparison to the chemical and physical methods presently widespread for nanoparticle synthesis. This biogenic path is classified as "green" as it is independent of the requirement for extremely toxic chemicals or high energy inputs [29]. A new area of nanotechnology called "green nanoparticle synthesis" uses materials from plants and microbes as well as complete cells, metabolites, and other ecologically friendly resources to create metallic nanoparticles. As they enhance the properties of nanoparticles in a more sustainable way, biological approaches are now preferred in manufacture of metal and metal oxide nanoparticles of required three dimensional characteristics. The major advantage of the technology is its clean and sustainable nature. The immense biodiversity of plant kingdom in terms of enzymatic pathways and the ability to catalyze the synthesis of a vast range of secondary metabolites like alkaloids, flavonoids, saponins, steroids, tannins from different parts is central to green synthesis of nanoparticles. These enzymes and secondary metabolites confer a reducing/oxidizing capacity as well as act as stabilizing agents for during the synthetic reaction facilitating formation of respective metal nanoparticles [1].

With the above premise on green synthesis, the ability to use grasses and agricultural wastes in place of plants or microorganisms which may have other utility, further makes the process simple and cost effective. An important recommendation in green synthesis is that to simplify the biosynthetic process the plant extract be used instead of live plants wherein the reducing/oxidizing compounds, required to transform the metallic ion, are extracted and utilized in isolation for addition to the reaction mixture to facilitate extracellular metal nanoparticles. The three-dimensional characteristics of the nanoparticles i.e. size/shape can be controlled by controlling the reaction parameters such as pH, concentration of reactants and temperature [19]. It is important to emphasize here that developing technologies based on utilization of agricultural wastes for green synthesis is another facet which makes the process sustainable (elimination of waste by reutilizing) and can be glorified as a "waste to wealth" technology.

3 Nanoparticle Synthesis Using Grass/Grass Waste and Diversity in Size

Metallic nanoparticles display well defined physical properties such as size between 1 nm and 100 nm, delimited electrical, magnetic, catalytic and optical properties, in comparison to the bulk state of the same metal [19]. Grass and grass waste are emerging as key "green" raw material in green nanotechnology due to its ease of availability, low cost and sustainability. The key three dimensional characteristics required for nanoparticles synthesized using grass extracts/agricultural wastes should fulfill the mentioned criterion—particle size should be in the range of 1–100 nm, have large surface-volume relation, shape and compositional features, light, catalytic and have additional functional features such as antimicrobial properties [19, 23, 32]. Kahatami and co-workers demonstrated the synthesis of silver nanoparticles nearly 15 nm particle size at ambient laboratory conditions by exposing silver nitrate solution to grass extract. The dried grass was prepared by washing post-collection, disinfection using alterations between sodium hypochlorite solution and distilled water, twice. The grass was then exposed to 70% alcohol for 2 min followed by thrice washing with autoclaved deionized water, to wash away any residual disinfection agents. Hereafter the grass extract is prepared via addition of 100 ml of deionized water to clean and dry grass waste nearly 20 g and for boiling for 15 min at ambient conditions. The resultant extract is filtered using Whatman filter paper and storing at 4 °C. Henceforth

0.1 ml stock solution of silver nitrate was used to prepare different concentrations of silver nitrates which was then mixed with nearly 15 ml of the grass extract, making up the volume to 45 ml. At the reaction mixture is advised to be stored at 28 °C, the dark [19]. The first indicator for successful synthesis of silver nanoparticles is the transformed color of dried grass extract from light yellow to reddish black. This was an ecofriendly synthesis of silver nanoparticles by utilizing grass wastes as the extract acts as both reducing and capping agents obliterating the use of other chemical entities in the synthetic technology. The synthesized "green" silver nanoparticles can be characterized for various physicochemical properties using UV–Vis spectroscopy, X-ray powder diffraction technology, and transmission electron microscopy.

3.1 Gold Nanoparticles

Gold nanoparticles have immense and diverse applications in biomedical field such as in biosensor design, imaging, photothermal therapy and drug delivery [12, 13, 16, 31]. Various physical, chemical and biological method have evolved for synthesizing gold nanoparticles (Au-NPs) with diversity in three-dimensional characteristics [22]. A distinct characteristic of Au-NPs, distinguishing it from bulk metal (yellow), is the wine red color of the solution. Size and shape of Au-NPs has critical impact on the chemical properties of these nanoparticle [2]. Mahdi and co-workers demonstrated the green synthesis of Au-NPs using grass extract using $HAuCl_4$, deionized water, and the extract of the grass wastes. They reported color changes post incubation with aqueous grass extract from yellow to red indicating the synthesis of Au-NP. The synthesized Au-NPs using lemon grass extract were reported to be triangular and spherical shape as indicated by transmission electron microscope visualization and particle size analysis indicated nearly 13 nm as the average particle size [22].

3.2 Copper Oxide Nanoparticles

Amongst the nano sized metal oxide semiconductors, cupric oxide (CuO) nanoparticles are notable for narrow band gap energy along with higher surface area, hydrophobicity and magnetic properties, large electrical and thermal conductivity, and proficient catalytic activity. The grass extract mediated CuO nanoparticle synthesis requires the grass extract prepared by drying, boiling and filtering *Cynodon* species. The extract was added to $Cu(NO_3)_2 \cdot 3H_2O$ under stirring conditions at 80 °C. Notably the solution color changes from blue to dark green while stirring, confirming the CuO nanoparticle formation. The stirring is continued till a green paste is left and washed and precipitated. The precipitates are added to silica crucibles and calcified at a temperature of 200 °C for upto 60 min. The green color again changes to dark color indication synthesis of CuO nanostructures. Importantly, the size and surface area to volume ratio of CuO nanostructures regulates the bioactivity i.e. antibacterial action of CuO nanoparticles [35].

3.3 Zinc Oxide Nanoparticles

Green methods for the synthesis of nanoparticles are ecologically compatible, obliterating the utilization of toxic chemicals or reagents, besides simplifying the process [14, 33]. In this process of synthesis, grass extract has multiple roles as reducing agent as well as a capping intervention. The preparation of grass extract was completed by drying, boiling and filtering. The heating was done for grass extract and thereafter zinc nitrate hexahydrate was added. The next step was constant stirring at 80 °C until a paste of yellowish-white color was acquired and then heated in the muffle furnace for one hour and powder collected was finely crushed. The zinc nitrate salt was the only chemical utilized in the formation of zinc oxide nanoparticles. The average particle size of ZnONPs is determined as in the range of 25–50 nm [5]. The green synthesis of ZnONPs have been done using lemongrass and *Chrysopogon Zizanioides* grass extract [5, 24]. Thus the synthesis of ZnONPs was achieved using green synthesis principles and excluding the dependence on chemical reagents. Zinc oxide nanoparticles show photocatalytic activity as well as antimicrobial properties [24].

4 Grass Species Used and Their Applications

Green synthesis of nanoparticles is a rapidly popularizing technology as compared to physical/chemical synthetic paths for nanoparticle synthesis. There are numerous varieties of grass species that can be used to develop nanoparticles with different applications as shown in the table given below (Table 1).

The main reason for using grass and grass waste for synthesizing nanoparticles is its biocompatibility and eco-friendly nature. Also, it is a cost-effective method during large scale production of nanoparticles. Some commonly used grass species and their waste are described below along with their applications.

4.1 **Cymbopogon Citratus *and* Cymbopogon Flexuosus**

Lemon grass also known as *Cymbopogon flexuosus* is used to derive silver, gold and silica nanoparticles which have huge industrial potential and the grass is mainly cultivated in Uttar Pradesh, Assam and Kerala [17]. Silica nanoparticles are synthesized from leaf and root extract of lemon grass and then further be used in the treatment of

Table 1 Green synthesis of nanoparticles from grass and their properties/applications

Grass or grass waste used	Nanoparticle synthesized	Properties/applications	Key reference
Cymbopogon citratus	Silver nanoparticles	Antibacterial	[3]
Cymbopogon flexuosus	Silica, silver and gold nanoparticles	Treatment of Osteoporosis	[34]
Poa annua	Silver nanoparticles	Potential drug carrier and therapeutic	[27]
Chrysopogon zizanioides	Zinc oxide nanoparticles	Photo degradation and antimicrobial	[24]
Seagrass: *Enhalus acoroides*	Silver nanoparticles	Benefits diabetes treatment	[30]
Cynodon dactylon	Copper oxide nanoparticles	Antibacterial	[35]
Cyperus rotundus	Copper oxide nanoparticles	Antibacterial	[35]
Pennisetum purpureum	Mesoporous silica nanoparticles	Adsorption of anionic and cationic dyes	[4]
Dried grass: Waste product	Silver nanoparticles	Anticancer, antifungal, antibacterial	[19]
Dry grass	Carbon containing Nanoparticles	Neurotropic, antioxidant	[28]

osteoporosis because of the excellent blending capacity of silica nanoparticles with osteoporotic drug and bone mineralisation properties of silica [26].

Another variety of lemon grass used for green synthesis of nanoparticles is *Cymbopogon citratus.* The leaves of *Cymbopogon citratus* are used to make the alkalinized extract to derive the silver nanoparticles which possess antibacterial properties [3]. The bacterial strains of *E. coli* and *B. cereus* which develop drug resistance can be easily tackled using AgNP derived from *Cymbopogon citratus* [3].

4.2 *Chrysopogon Zizanioides*

Zinc Oxide nanoparticles are synthesized using the grass extracts of *Chrysopogon zizanioides* and are known for their antimicrobial and photocatalytic degradation activity [24]. Textile industries emit a lot of dye contaminants into the water bodies which have harmful effects to the environment. This can be regulated by the use of zinc oxide nanoparticles due to their photocatalytic oxidative degradation. *Chrysopogon zizanioides* derived ZnONPs display potent antimicrobial action against Gram positive, Gram negative as well as fungal species. In a experiment it was observed that zinc oxide nanoparticles exhibit IC 100 for *E. coli* (1 mg/ml), IC 100 for *C. albicans* (0.5 mg/ml) and IC 70 for *S. aureus* (0.5 mg/ml) as inhibitory concentrations [24].

4.3 Enhalus Acoroides

Enhalus acoroides is abundantly found in the Gulf of Mannar and is known widely for its therapeutic use. The grass extracts of *Enhalus acoroides* are used to derive silver nanoparticles which are further used as anti-diabetic agents because of their ability to reduce the activity of α-glucosidase which is a digestive enzyme [30]. The seagrass extract has shown potent pharmacological effect against diabetes but requires more research, in vivo studies and clinical trials for better results.

4.4 Cynodon Dactylon *and* Cyperus Rotundus

The green synthesis of copper oxide nanoparticles using *Cynodon* species is an eco-friendly and sustainable approach considering its wide spread availability across different regions of India. The copper oxide nanoparticles obtained from *Cynodon dactylon* appear like rice spikelets in shape and the ones derived from *Cyperus rotundus* exhibit composite structure [35]. These nanoparticles exhibit significant antibacterial action against several bacterial species, via disruption of cellular membranes and malfunctioning of cellular enzymes [7].

4.5 Dried Grass (Waste Product)

The waste material of grass, that is, the dried grass can be used for silver nanoparticle synthesis. Dried grass extract and silver nitrate are reacted in optimal conditions for producing nanoparticles nearly 15 nm particle size and can be used for its antifungal, anticancer and antimicrobial properties [19]. Significant anti-bacterial activity of silver nanoparticles synthesized using dried grass extract is reported against *Acetobacter baumannii* and *Pseudomonas aeruginosa* [19]. In comparison, the antifungal activity of silver nanoparticles are reported to have the same effect as one of the most common fungicide, amphotericin B and comparatively much more than fluconazole [20]. Carbon containing nanoparticles derived from grass waste exhibit neurotropic activity and antioxidant activity and these grass extracts can be used to inhibit the generation of reactive oxygen species in the termini of nerves [28]. A proposed applications of grass-based nanoparticles in different industries is shown below in the figure (Fig. 1).

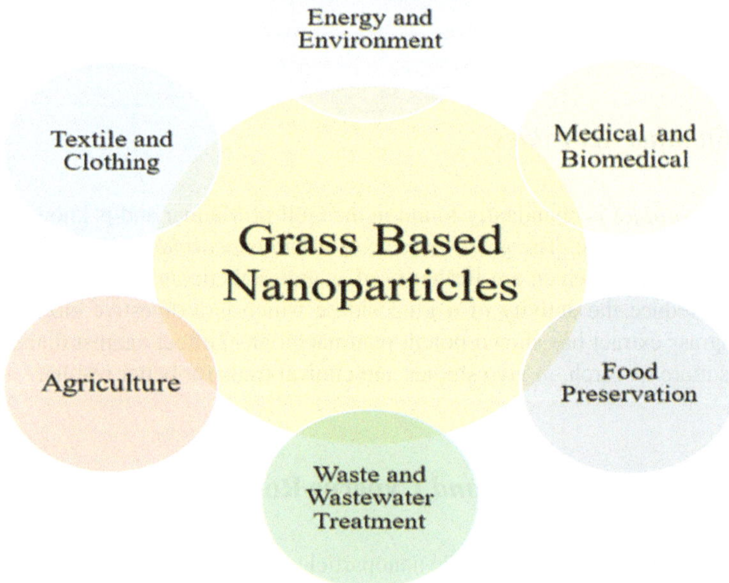

Fig. 1 Applications of grass and grass waste-based nanoparticles

5 Advantages

"Green synthesis" of nanoparticles involves the use of sustainable and environment friendly approaches and raw materials. Green synthesis generated nanoparticles have diversified natures, with larger stability and suitable dimensions as they are synthesized utilizing a one-step method. One of the objectives of research in the area of green technology (eco-friendly) in the area of nanoparticles formation is to reduce possible risks in the production process and utilization for humans, also the environment [15, 19]. Grass and grass waste has great usage potential in the field of green nanotechnology because of their easy availability, low toxicity, low cost and no potential hazard to the environment even when carrying the large-scale production [9]. In contrast to microorganisms or even some other plants because of the controlled parameters for their growth and biosafety issues, the dried grass utilization to synthesize nanoparticles is comparatively safe, easy and cost-effective. Grass and Grass wastes usage can be regarded as sustainable and plentiful natural resources in the direction of green synthesis. It can be utilized for the biosynthesis of many metallic nanoparticles such as silica, zinc oxide, silver, gold, palladium and copper nanoparticles and because of their different properties like antimicrobial, catalytic degradation, antioxidant properties, they can have various applications in the field of biomedical, pharmaceuticals, agriculture etc. Grass wastes can be utilized to convert low-cost waste into health [11].

6 Conclusion

The above overview depicts the use of grass and grass waste as an ideal method for the production of various nanoparticles like gold, silver, silica, zinc/copper oxide, and carbon containing nanoparticles. It is an eco-friendly way to develop nanoparticles which can be further used for wastewater treatment, targeting diseases and many other applications because of their small size, metal degrading potential and other numerous properties like antimicrobial, antioxidant and neurotropic effects. In future, more research is warranted to expand the applications of grass based green synthesis of nanoparticles which is one of the most promising, sustainable and cheap technologies for production of nanoparticles.

References

1. Adelere IA, Lateef A (2016) A novel approach to the green synthesis of metallic nanoparticles: the use of agro-wastes, enzymes, and pigments. Nanotechnol Rev 5(6):567–587
2. Ahmed S, Ikram S (2015) Synthesis of gold nanoparticles using plant extract: an overview. Nano Res Appl 1:1–6
3. Ajayi E, Afolayan A (2017) Green synthesis, characterization and biological activities of silver nanoparticles from alkalinized *Cymbopogon citratus* Stapf. Adv Nat Sci: Nanosci Nanotechnol 8(1):8
4. Akpotu SO, Moodley B (2018) Effect of synthesis conditions on the morphology of mesoporous silica from elephant grass and its application in the adsorption of cationic and anionic dyes. J Environ Chem Eng 6(4):5341–5350
5. Anvekar T, Rajendra V, Kadam H (2017) Green synthesis of zno nanoparticles, its characterization and application. Mater Sci Res India 14(2):153–157
6. Darroudi M, Sarani M, Kazemi Oskuee R, Khorsand Zak A, Amiri MS (2014) Nanoceria: gum mediated synthesis and in vitro viability assay. Ceram Int 40(2):2863–2868
7. Das D, Nath BC, Phukon P, Dolui SK (2013) Synthesis and evaluation of antioxidant and antibacterial behavior of CuO nanoparticles. Colloids Surf B Biointerfaces 101:430–433
8. Devadiga A, Shetty KV, & Saidutta MB (2015). Timber industry waste-teak (Tectona grandis Linn.) leaf extract mediated synthesis of antibacterial silver nanoparticles. Int Nano Lett 5(4):205–214
9. Goutam SP, Saxena G, Roy D, Yadav AK (2019) Green synthesis of nanoparticles and their applications in water and wastewater treatment. Bioremediation Ind Waste Environ Saf: Ind Waste Its Manag 16:349–379
10. Heydari R, Rashidipour M (2015) Green synthesis of silver nanoparticles using extract of oak fruit hull (Jaft): synthesis and in vitro cytotoxic effect on MCF-7 cells. Int J Breast Cancer 2015:1–6
11. Hobson GW (2011) Nanotechnology. Compr Biotechnol (Second Ed) 3:683–697
12. Husen A (2017) Gold nanoparticles from plant system: synthesis, characterization and their application. In: Ghorbanpourn M, Manika K, Varma A (eds) Nanoscience and plant–soil systems, vol.–48, Springer International Publishing AG, Gewerbestrasse 11, 6330 Cham pp 455–479

13. Husen A (2019a) Medicinal plant-product based fabrication nanoparticles (Au and Ag) and their anticancer effect. In: Kintzios SE, Barberaki M, Flampouri (eds) Plants that fight cancer, 2nd edn. Taylor & Francis/CRC Press, pp 133–147
14. Husen A (2019b) Natural product-based fabrication of zinc oxide nanoparticles and their application. In: Husen A, Iqbal M (eds) N*anomaterials* and plant potential, Springer International Publishing AG, Gewerbestrasse 11, 6330 Cham, pp 193–291. https://doi.org/10.1007/978-3-030-05569-1_7
15. Husen A, Iqbal M (2019) Nanomaterials and plant potential: An overview. In: Husen A, Iqbal M (eds) Nanomaterials and Plant Potential, Springer International Publishing AG, Gewerbestrasse 11, 6330 Cham, pp 3–29. https://doi.org/10.1007/978-3-030-05569-1_1
16. Husen A, Rahman QI, Iqbal M, Yassin MO (2019) Plant-mediated fabrication of gold nanoparticles and their applications. In: Husen A, Iqbal M (eds) N*anomaterials* and plant potential, Springer International Publishing AG, Gewerbestrasse 11, 6330 Cham, pp. 71–110. https://doi.org/10.1007/978-3-030-05569-1_3
17. Joy PP, Sumitha M, George M, Ancy J (2006) Lemongrass: the fame of Cochin. Ind J ArecanutSps Med Plts 8:55–64
18. Khatami M, Alijani H, Sharifi I, Sharifi F, Pourseyedi S, Kharazi S et al (2017) Leishmanicidal activity of biogenic Fe_3O_4 nanoparticles. Sci Pharm 85(4):36
19. Khatami M, Sharifi I, Nobre MAL, Zafarnia N, Aflatoonian MR (2018) Waste-grass-mediated green synthesis of silver nanoparticles and evaluation of their anticancer, antifungal and antibacterial activity. Green Chem Lett Rev 11(2):125–134
20. Kim KJ, Sung WS, Moon SK, Choi JS, Kim JG, Lee DG (2008) Antifungal effect of silver nanoparticles on dermatophytes. J Microbiol Biotechnol 18(8):1482–1484
21. Krishnaswamy K, Vali H, Orsat V (2014) Value-adding to grape waste: green synthesis of gold nanoparticles. J Food Eng 142:210–220
22. Mahdi HS, Parveen A (2019) Biosynthesis, characterization and antibacterial activity of gold nanoparticles (AU-NPS) using Black Lemon extract. Mater Today: Proc 18:5164–5169
23. Mishra VK, Husen A, Rahman QI, Iqbal M, Sohrab SS, Yassin MO, Bachheti RK (2019) Plant-based fabrication of silver nanoparticles and their application. In: Husen A, Iqbal M (eds) N*anomaterials* and plant potential, Springer International Publishing AG, Gewerbestrasse 11, 6330 Cham, pp 135–175. https://doi.org/10.1007/978-3-030-05569-1_5
24. More P, Inamdar V, Suresh S et al (2021) Synthesis of zinc oxide nanoparticles using *Chrysopogon Zizanioides* grass extract, its applications in photodegradation and antimicrobial activity. J Mater Sci: Mater Electron 32:20725–20741
25. Mortazavi SM, Khatami M, Sharifi I, Heli H, Kaykavousi K, Sobhani Poor MH, Kharazi S, Lima Nobre MA (2017) Bacterial biosynthesis of gold nanoparticles using *Salmonella enterica* subsp. *Enterica Serovar Typhi* isolated from blood and stool specimens of patients. J Cluster Sci 28:2997–3007
26. Muniraju S, Pradeep S, Madihalli C. et al. (2020) Nanostructured silica particles from *Cymbopogon flexuosus* (Nees ex Steud.): extraction, characterization and study of their blend with osteoporotic drugs. SN Appl Sci 2: 2179
27. Nesrin K, Yusuf C, Ahmet K, Ali SB, Muhammad NA, Suna S, Fatih S (2020) Biogenic silver nanoparticles synthesized from Rhododendron ponticum and their antibacterial, antibiofilm and cytotoxic activities. J Pharm Biomed Anal: 179
28. Paliienko K et al. (2021) Carbon-containing nanoparticles from grass: green synthesis, optical, spectroscopic, oxidative properties and neurotropic action in brain nerve Terminals," In 2021 IEEE 11th International Conference Nanomaterials: Applications & Properties (NAP), pp 1–5
29. Saratale RG, Saratale GD, Shin HS, Jacob JM, Pugazhendhi A, Bhaisare M, Kumar G (2017) New insights on the green synthesis of metallic nanoparticles using plant and waste biomaterials: current knowledge, their agricultural and environmental applications. Environ Sci Pollut Res 25(11):10164–10183
30. Senthilkumar P, Santhosh Kumar DSR, Sudhagar B et al (2016) Seagrass-mediated silver nanoparticles synthesis by *Enhalus acoroides* and its α-glucosidase inhibitory activity from the Gulf of Mannar. J Nanostruct Chem 6:275–280

31. Siddiqi KS, Husen A (2017) Recent advances in plant-mediated engineered gold nanoparticles and their application in biological system. J Trace Elem Med Biol 40:10–23
32. Siddiqi KS, Husen A, Rao RAK (2018) A review on biosynthesis of silver nanoparticles and their biocidal properties. J Nanobiotechnol 16(14):1–28
33. Siddiqi KS, Rahman A, Tajuddin HA (2018) Properties of zinc oxide nanoparticles and their activity against microbes. Nanoscale Res Lett 13(141):1–13
34. Singh J, Dutta T, Kim KH et al (2018) 'Green' synthesis of metals and their oxide nanoparticles: applications for environmental remediation. J Nanobiotechnol 16:84
35. Suresh S et al. (2020) Green synthesis of copper oxide nanostructures using *Cynodon dactylon* and *Cyperus rotundus* grass extracts for antibacterial applications. Ceramics Internationalm Part B 46(8):12525–12537

Future Prospective and Risk Factors Associated with the Use of Nanoparticles

Senari N. Wijesooriya, Nadun H. Madanayake, and Nadeesh M. Adassooriya

Abstract Nanotechnology has been a promising technological application where the usage of nanoparticles (NPs) in commercial and household operations has grown exponentially. The unique and inherent properties of different nanomaterials (NMs) have greatly impacted many fields. However, the increased use of NPs is being impeded by the possible toxic concerns on the environment because of their deliberate and accidental release. The size and form of NMs and their concentration, period of exposure effect, dosage, surface area, and surface functionalization significantly impact their behavior. Therefore, NMs can cause beneficial and harmful effects on niches including plants, animals, and microorganisms as well as on humans by interacting via diverse mechanisms. Hence, nanotechnology has become a double-edged sword that requires assessing its potential risk factors on all forms of life for its sustainable use. Therefore, this chapter elaborates on future prospects and risk factors of frequently used NPs on different trophics in the biosphere.

Keywords Titanium oxide NPs · Copper NPs · Silver nanoparticles · Zinc oxide NPs · Iron oxide NPs

S. N. Wijesooriya
Gonawila (NWP), Department of Food Science and Technology, Wayamba University of Sri Lanka, Kuliyapitiya, Sri Lanka

N. H. Madanayake
Department of Botany, University of Peradeniya, Peradeniya 20400, Sri Lanka

N. M. Adassooriya (✉)
Department of Chemical and Process Engineering, University of Peradeniya, Peradeniya, Sri Lanka
e-mail: nadeeshm@sci.pdn.ac.lk

Department of Chemical Engineering, Massachusetts Institute of Technology, 77 Massachusetts Avenue, Cambridge, MA 02139, USA

A. Husen (ed.), *Nanomaterials from Agricultural and Horticultural Products*,
Smart Nanomaterials Technology,
https://doi.org/10.1007/978-981-99-3435-5_15

1 Introduction

Nanotechnology is at the cutting edge of technology that has enhanced the efficiency of numerous industrial applications to transform our lives in better ways. Fabrication of nanomaterials (NMs) have drastically increased with time. Life cycle assessment of nanoparticles (NPs) and nano-enabled products shows that these materials could release deliberately or accidentally to different environmental matrices. The inherent properties of NMs make them highly reactive candidates which can interact with different environmental components. Although this is a promising approach, many debates and arguments are aroused in the community due to the negative impacts of nano-related implementations. Moreover, long-lasting prevalence and exposure can result in deleterious effects on biological counterparts. Hence, it is utterly important exactly to understand the release behavior, toxicity, and consequences on flora and fauna to manage them more wisely and safely [17]. The NMs can simply be defined as materials with at least one of its dimensions within 1–100 nm. The unique physicochemical properties of NPs including their size, shape, composition, and surface characteristics have a significant impact on their behavior [62].

Effects of NMs can vary depending on the properties of the immediate environment in which they exist. These materials can accumulate, transform, agglomerate, or react with other biological components to develop new types of NPs. NMs can reside in different environments such as soil, water, and air. It has been reported that soil is a major sink which they can settle. Plants, which exist at the intersection of these matrices, can easily uptake them. Therefore, this can lead to tropic transfer which may can have potential negative impacts on the ecosystem [39–41, 73, 82, 83]. Although studies on nanoscience and nanotechnology exponentially increase with time, most of these explorations are centered on fabrication and their potential applications. However, studies on nanotoxicology are lower than 1% and require urgent attention [18]. It is important to focus on more systematic studies on the pros and cons of NMs while targeting more on their risk assessments. Therefore, this chapter will discuss on the future prospects and risk factors associated NPs.

2 Risk-Causing Factors of Nanoparticles

The risks associated with NPs are ultimately due to the high surface area to volume ratio, quantum confinement, and other aspects such as non-conventional mobility [32]. Hence, the toxicological studies are based on their mobility and interaction with the different components in the microenvironment [63]. Fig. 1 summarizes and pictorially represents the factors influencing the NPs toxicity on biotic components.

Grieger et al. [33] reviewed the intrinsic properties of NMs as a secondary factor that influences on their unpredictability in risk assessment. Particle size has been reported as a frequently described character in risks. Primarily, this has been defined using techniques like electron microscopy to measure the primary dimensions. In

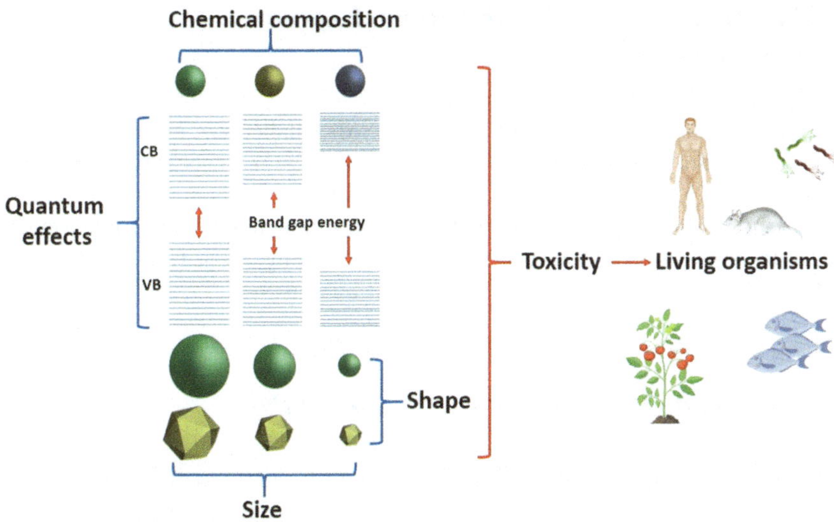

Fig. 1 Physicochemical properties of NMs that lead to toxic effects on living organisms

addition, mean particle size and its distribution are also applied. Furthermore, Organization for Economic Cooperation and Development (OECD) practiced utilizing NPs size distribution in the gaseous and liquid media as an assessment criterion for risk assessment [34]. Moreover, recommendations and the standards proposed by Scientific Committee on Emerging and Newly Identified Health Risks (SCENIHR) defined NP size as a vital parameter in risk analysis. For instance, terms such as "exposure to airborne particles" refers to the inspiration of particles of various sizes [22]. Although NP size or the size distribution is reported as a risk assessment factor certain authors pointed out that particle size cannot be considered as a sole character, determining the mechanisms leading to toxicity. Therefore, it is essential to identify and grade the factors unique to different types of NMs. Also, other physicochemical properties of NPs can be inter-related and become a determinant factor in assessing interactions between particles with biotic and abiotic substrates. It has been reported that NPs surface area has a dose-based response on living organisms. For instance, lung autopsy samples from asbestos miners showed that the degree of fibrosis and identified retained fibers were dependent on mass, quantity, and area. Fibers with variable sizes retained fiber surface area best predicted the extent of fibrosis [94]. Therefore, NPs surface area serves as a dose-metric character for the toxicological activities of NPs.

3 The Risk Associated with Human Exposure to Nanoparticles

Bioavailability and easiness of access due to NMs smaller size and larger surface area can risks the individual's health. Accordingly, there is a necessity to study and assess the entry of NMs to humans via nano-enabled products in cosmetics, foods, pharmaceuticals, etc. [43]. Therefore, this section will highlight on possible dangers caused by most widely utilized NMs synthesized using TiO_2, Cu, Ag, ZnO and Fe_xO_y.

3.1 Titanium Oxide Nanoparticles (TiO_2 NPs)

TiO_2 NPs are widely incorporated in pharmaceutical, and cosmetic products. TiO_2 NPs exhibit distinct physicochemical properties compared to their bulk counterparts [36]. In addition, these materials are implied in electronics, energy and chemical industries. TiO_2 NPs can directly pass in the brain via olfactory pathway. Márquez-Ramírez et al. [53] observed that TiO_2 NPs have cytotoxic effects on human glial cells, showing their impacts on nervous system. It can inhibit the cell proliferation and cause alterations related to reduction in F-actin fiber immune location. Studies using animal models have reported that TiO_2 NPs can induce pulmonic inflammations. Sun et al. [88] elaborated on chronic lung toxicity in mice via the oxidative stress and mechanisms involved in pulmonic inflammations, leading to intra-tracheal instillation. Deposition of TiO_2 NPs in the lung tissues resulted in a noticeable increment in lung indices, inflammation, and bleeding. Also, exposure of NPs drastically enhanced the generation of reactive oxygen species (ROS) and lipid peroxidation. In addition, the antioxidant capacity of lung tissues was reduced. Because of the nanocytotoxicity, authors have reported that TiO_2 NPs upregulate the levels of tumor necrosis factor-α , cyclooxygenase-2, heme oxygenase-1, interleukin-2, interleukin-4, interleukin-6, interleukin-8, interleukin-10, interleukin-18, interleukin-1, and cyp1A1, respectively, while down-regulating the expression of NF-B-inhibiting factor and heat shock protein 70. Thus, TiO_2 NPs-induced pulmonary inflammation in mice is related to induced oxidative stress and the expression of inflammatory cytokines. Chen et al. [13] investigated the potential risks of TiO_2 NPs for longer durations of exposure using *Danio rerio* as a model organism. It was observed that there is a dose and time-dependent retardation in growth and liver weight ratio. Also, it was suggested that TiO_2 NPs can translocate and accumulate in the gills, liver, and brain by crossing the blood–brain and blood-heart barriers. Moreover, TiO_2 NPs are also known to cause kidney damage in mouse models. Furthermore, TiO_2 NPs inducing nephric injuries have been studied at the molecular level [35].

3.2 *Copper Nanoparticles (Cu NPs)*

The toxicological outcomes of Cu NP exposure vary depending on a variety of factors, and the mode of entry identified as the vital parameter for its toxicity [101]. According to Chen et al. [15] Cu NPs mainly get deposited in the respiratory tract via mucociliary clearance, and directly via ingestion to digestive tract. Acidity due to HCl in the human stomach solubilize Cu NPs facilitating the diffusing into other body parts. These interactions of Cu NPs are mainly determined by charge and their potential to distribute systemically. Reacted Cu NPs in the stomach can subsequently enter the small intestine and diffuse into the villi, and this will systemically distribute throughout the body [11]. Also Chen et al. [15] demonstrated that the accumulation of Cu NPs in kidneys, liver, and spleen of rodents had resulted in organ damage. Higher surface area enhances the reactivity of Cu NPs with H^+ ions in gastric juice to produce Cu ions leading to uncontrollable cellular uptake. Higher doses of Cu ions in the hepatic cells disrupt the normal metabolism of the liver causing nephritis and renal tubule necrosis. Furthermore, biomolecule oxidation, complexing with protein and other macromolecules is also possible with Cu NPs. Also, Cu NPs can undergo dissolution, Fenton-like reactions [25]; photo-oxidation, and generation of ROS [30] in living tissues. Most of these reactions make molecular alterations by interacting with Cu-based NPs leading to oxidative stress.

3.3 *Silver Nanoparticles (Ag NPs)*

Exposure to Ag and Ag-based composites can happen dermally, orally, and via breathing. Once Ag NPs enter the body, it can translocate and enter the other organs such as the liver, spleen, and brain [47]. Suresh et al. [89] revealed that different forms of Ag NPs can induce cytotoxic effects on lung tissues in a dose-dependent manner. Deng et al. [20] stated that Ag NPs can induce cytotoxic effects on human lung adenocarcinoma cells by enhancing the expression of gap junctional intercellular communication and connexin43.

The rate of intracellular dissolution and release of Ag ions has been identified as a vital factor in the cytotoxicity of Ag NPs in lung tissues [49]. Also, the persistent release of Ag NPs leads to cause mild pulmonary fibrosis and inflammation which results in sub-chronic injury responses [23, 97]. Exclusively, 10 nm-sized Ag NPs observed to be more cytotoxic to lung tissue than a particle with larger mean sizes. This proves that NP size has a significant impact on potential toxicity [29, 97]. Moreover, bioavailability, and surface coatings, as well as the duration of exposure can influence on lung toxicity. Accumulation of Ag NPs in the liver via inhalation was reported by Takenaka et al. [90]. Several mechanisms have been proposed by researchers on the human liver tissues, leading to oxidative stress, cell damage, and apoptosis. [42, 67]. Ji et al. [42] mentioned cytoplasmic vacuolization and hepatic focal necrosis in liver tissues following exposure to Ag NPs. On the other hand, kidneys are also affected by Ag NPs. Ag^+ ions released from medical devices can enter the bloodstream and accumulate in the kidneys [92]. Degradation of proximal convoluted tubules, capsular and membranous thickening, and mesangial defects have all been reported in response to Ag NPs exposure [45].

3.4 Zinc Oxide Nanoparticles (ZnO NPs)

ZnO NPs are frequently utilized to manufacture consumer products such as cosmetics, food packaging materials, and antimicrobial coatings. There is a huge risk of ZnO NP exposure in humans. ZnO NPs can enter an individual via different portals including dermal, inhalation, and oral passages. Hence, the interaction of ZnO NPs with the body organs has a range of impacts on different organ systems (nervous system, digestive tract, circulatory, respiratory, and reproductive system). Moreover, the effects of ZnO NPs significantly affected by the dose, route of entry, and exposure time. Also, the cytotoxic effects of these NMs are found to be developed in a concentration- and time-dependent manner against various cell types.

Wang et al. [96] and Chen et al. [14] recently reported that the cytotoxic effects of ZnO NPs result in cell inflammation. Safar et al. [74] explained that ZnO NPs can trigger cell inflammation and mitochondrial disfunction in THP-1 human monocytic cell lines. Sizova et al. [85] recently described hepatotoxicity from ZnO NPs in mouse models. Srivastav et al. [86] found that oral application ZnO NPs at 10, 50, and 300 mg kg^{-1} induce hepatotoxicity and nephrotoxicity in mice. Yan et al. [99] explained that intratracheal exposure to ZnO NPs at 1.25, 2.5, and 5 mg kg^{-1} resulted in atherosclerosis in Wistar rats. Choi et al. [16] demonstrated that 0.01, 0.1, 1, and 10 mg kg^{-1} ZnO NPs caused pericardial edema in zebrafish embryos. Moreover, only fewer studies are found with respect to ZnO NPs-induced cardiac toxicity.

Direct ingestion of ZnO NPs reported to cause deleterious effects primarily on the liver, kidneys, and lungs [24]. Sharma et al. [77] confirmed that the mice models exposed to ZnO NPs had resulted in injuries to their livers. Also, cytological changes in liver, pancreas, heart, and stomach specimens of rat models were observed by Pasupuleti et al. [65]. Nounou et al. [61] tested using rat models on the effects of ZnO NPs once they were ingested orally. It was observed that ZnO NPs can develop eosinophilia, and lymphocyte infiltration, followed by damage to lung tissues via oxidative stress, inflammatory responses, and DNA damage. Despite ZnO NPs showing cytotoxic effects, there is no clear evidence on the effect of particle size on toxicity. However, concentration-dependent toxicities were observed with ZnO NPs and results in systemic diseases, alveolar injuries, and inflammatory responses once inspired [5, 12]. Inhalation studies conducted by Adamcakova-Dodd et al. [1] demonstrated that cytotoxic and pulmonary inflammation due to the dissolution of ZnO NPs in the respiratory tract when mice were exposed to 3.5 mgm^{-3} at a rate of 4 h per day for 13 weeks. Therefore, it has been shown that the dose, exposure pathway, and duration of post-exposure has a significant impact on the low sub-chronic toxicity of ZnO NPs.

3.5 *Iron Oxide Nanoparticles (FeO NPs)*

FeO NPs is a prominent NM that has been employed in medical implications especially for theranostics applications. However, direct inhalation FeO NPs were reported to generate oxidative stresses in the lungs, liver, spleen, and brain, by promoting Fenton reactions. Hence this can lead to cause deleterious effects on living tissues via inflammation, reduced viability, and lysis [91]. Mostly, FeO NPs show size and surface functionalization-dependent effects on living tissues. For example, Naqvi et al. [59] investigated the effect of Tween 80 functionalized FeO NPs (30 nm) on murine macrophage (J774) cells. To assess the concentration- and time-dependent toxicity of FeO NPs. It was reported that higher doses of NPs (300–500 μgmL^{-1}) for extended durations (6 h) had significantly reduced the viability of cells to 55–65%. Necrosis-apoptosis assay showed that the majority had undergone apoptosis. Also, elevated doses of FeO NPs resulted in an increment of ROS production causing cell damage and death. Hence, it can be speculated that FeO NPs show time and concentration-dependent oxidative stress on cells. Magdolenova et al. [51] assessed the effect of pristine and oleate functionalized Fe_3O_4 NPs on human lymphoblastoid TK6 cells and human blood cells. It was observed surface functionalized Fe_3O_4 had a concentration-dependent cytotoxic and genotoxic effects, whereas pristine Fe_3O_4 NPs did not cause any cytotoxic or genotoxic effects on tested cell lines under experimental conditions thus provided. Therefore, it clearly manifests that the surface-modified NPs had significantly altered the performance and cellular uptake of the NPs, mimicking pathological morphological changes in the cells.

The effects of particle size and surface coating of FeO NPs in vitro and in vivo systems were examined by Feng et al. [26]. They showed that the cellular uptake, cytotoxicity, distribution, and clearance of FeO NPs are significantly influenced by both size and coating. Feng et al. [26] compared the effect of polyethyleneimine and polyethylene glycol-coated FeO NPs on SKOV-3 human ovarian cancer cells and RAW 264.7 murine macrophages. It was observed that polyethyleneimine-coated FeO NPs showed an enhanced uptake in comparison to polyethylene glycol-coated FeO NPs on macrophages and cancer cells. Also, cytotoxic effects on tested cell lines were mainly caused due to ROS generation and cell apoptosis. In addition, smaller-size (10 nm) polyethylene glycol-coated NPs had shown a higher cellular uptake in comparison to larger particles (30 nm). However, polyethyleneimine-coated FeO NPs did not show any autophagy. Hence, this can be concluded that polyethyleneimine has a protective function against cytotoxic effects incurred by FeO NPs. Moreover, biodistribution studies portrayed that FeO NPs have a trend to accumulate in the liver and spleen. This study shows that FeO particle size and surface coating plays a vital role in their effect on biological applications.

Despite recent advances in nanotoxicology research, scientists have yet to accurately predict the behavior and biokinetics of NPs. Artificial conditions and in vitro or in vivo studies using animal models and cell lines show complicated interactions in dose, size, and surface coating-dependent manner. Also, the inherent properties of NPs may result in the diversity of their distribution, uptake, overcoming barriers, immune response, and metabolism. Therefore, there is a greater limitation of data available to exactly examine the risks of NMs.

4 The Environmental Risk Associated with Nanoparticle Exposure

Following the discharge of NPs into the environment, it can build up as wastes in ecosystems, endangering living organisms. NMs can accumulate in different environments, and it is essential to know how NPs will behave [44, 70]. Accumulation of NMs in soil in relation to deliberate and accidental release can impose significant impacts on abiotic and biotic processes [6]. Hanna et al. [38] proposed that estuarine and marine sediments can act as potential endpoints for several NPs. The higher reactivity of metallic NMs can significantly influence soil microbial activity by disrupting the activity of soil microbial biocatalysts [21, 28, 84, 98]. Additionally, carbon-based NMs are said to be less harmful to soil microbes than metallic NPs [27]. Metal and metal oxide NPs show a detrimental effect on the abundance, activity, and diversity of soil bacteria, fungi, and other microbes even at extremely low doses.

4.1 Effect of ZnO and CuO NPs Exposure on Environment

In terrestrial, and aquatic environments, NPs undergo numerous transformations. Therefore, the toxic characteristics of modified particles differ from their pristine form. Transformation may influence the toxicity of NPs through changes in physicochemical properties, and bioavailability. In addition, synergistic effects of metallic NPs, environmental contaminants, increased release of toxic metal ions, and changed interactions with biomacromolecules also affects.

Shen et al. [80] studied the effect of ZnO NPs on soil microbe's metabolism using a laboratory scale setup. It was shown that soil samples treated with different doses of ZnO NPs significantly reduced the ammonification rate, soil respiration, dehydrogenase activity, and fluorescent diacetate hydrolase (FDAH) activity with time. Also, authors have reported that the ecotoxicity of acidic soil is greater than that of neutral soils, where no significant toxicity on soil microbes was observed in alkaline soils. According to Rajput et al. [70] ammonification was observed to be decreased by up to 37.8 percent in soil treated with 1 mgg^{-1} ZnO NPs on soil for more than a three-month test period, in which respiration was suppressed by 14.2%, and dehydrogenase and FDAH activities followed a similar pattern in soil treated with 1–10 mg ZnO-NPs. Furthermore, ZnO NPs decrease the populations of P-solubilizing, Azotobacter, and K-solubilizing microbes, because of the suppression of enzymatic activities of catalase, and urease [10].

In a study conducted by Zeng et al. [100], the ZnO NPs formed by microemulsion has a broad range of antibacterial uses. Moreover, a similar study revealed that ZnO NPs have antimicrobial properties against bacteria like *Escherichia coli, Staphylococcus aureus, Bacillus subtilis, Pseudomonas fluorescens*, and *Salmonella typhimurium* and anti-fungal activities against *Aspergillus fumigates* and *Aspergillus. flavus* [60]. Parallel studies of Reddy et al. [72] also revealed that ZnO NPs were found to be toxic to both gram-negative (*E. coli*) and gram-positive (*S. aureus*) bacteria. Manzoor et al. [52] discovered antibacterial activity against food-borne and water-borne bacterial pathogens such as *E. coli, Campylobacter jejuni*, and *Vibrio cholera*. It was revealed that ZnO NPs cause changes in cellular shape, finally leading to bacterial mortality.

It is a well known fact that NPs have a momentous impact on the crop development and yield. In plant tissues, which include edible plant components, NPs have been reported to accumulate. The effects of ZnO NPs on plant species are also a concern regarding environmental toxicity [50, 87]. According to Lee et al. [48], high concentrations of ZnO NPs (10–2000 mg/L) on buckwheat (*Fagopyrum esculentum*) led to reduced biomass, negatively affecting root surface cells, and an aberrant ROS defense mechanism. Also, Lin and Xing [50] investigated the toxicological effect of ZnO NPs on ryegrass. It was shown that existence lowered biomass, shrank the root tip and epidermis, and led to cortical cells to become collapse and be severely vacuolated. Moreover, Ghosh et al. [31] examined ZnO NPs (85 nm) for their, genotoxicity, cytotoxicity, and metabolic impacts in *Allium cepa, Nicotiana tabacum*, and *Vicia faba*. They found that exposure to the ZnO NPs which have a diameter of 85 nm, can disrupt the membrane integrity, DNA strand fractures, enhanced chromosome aberrations, micronucleus production, and cell-cycle blockage at the G2/M phase. Parallel studies by Murali et al. [57], found that NPs exposure enhanced intracellular ROS production and lipid peroxidation in *Nicotiana tabacum* and *Vicia faba*. Also, Tripathi et al. [95] reviewed that nitric oxide alleviated ZnO NP-induced phytotoxicity in wheat seed sprouting.

Cu plays a structural role in hormone signaling, mitochondrial respiration, cell wall metabolism, numerous regulatory proteins, photosynthetic electron transport, and oxidative stress responses [54]. However, CuO NPs also show some toxic effects on plants. For instance, CuO NPs strongly negatively affect the shoot and root growth of perennial ryegrass (*Lolium perenne L.*) [4]. According to Shi et al. [81], treatment with CuO NPs reduced duckweed growth by 50%. (*Landoltia punctata*). Adhikari et al. [2] demonstrated that seed sprouting of *Glycine max L.* and *Cicer aretinium L.* is affected by CuO NPs. These toxicity symptoms can be dose-dependent which can lead to a decrease in root extension in primary roots.

CuO NPs have also been shown to be toxic in Syrian barley (*Hordeum vulgare L.*) [78], owing to stress-induced regulation of antioxidative response and a decrease in photosynthetic capabilities. Under CuO NP treatments, antioxidant enzymes of the ascorbate glutathione cycle were variably altered as rising CuO NP concentrations markedly raised levels of H_2O_2. These results showed that CuO NPs are toxic to plant physiological, biochemical, and photosynthetic processes, suggesting that excessive ROS production can disturb the cellular redox system, growing the oxidized forms and resulting in oxidative damage to biological molecules like lipids, proteins, and nucleic acids [37, 79]. Furthermore, CuO NPs treatments have a substantial impact on the growth of barley seedlings; a progressive reduction in shoot weight and length was seen when CuO NPs concentration is increased [78].

Perreault et al. [66] demonstrated the toxicity of CuO NPs in aquatic plants like *Lemna gibba.* According to that study, it was found that the toxicity of CuO NPs mainly depends on their solubility in the media. Core–shell CuO NPs caused to generate four times as many reactive oxygen species than CuO NPs and copper sulfate, as indicated by the accumulated copper concentration in the plants. This, shows the availability of the polymer shell altered the harmful effect caused in *L. gibba.* Similarly, Costa et al. [19] have shown that exposure of 1,000 mg of CuO NP per L^{-1} can cause the full loss of PSII photochemical quenching and also decline the rate of transpiration, photosynthesis in *Oryza sativa*, var. Jyoti. Despite the fact that NMs are already successfully employed as nanopesticides and nanofertilizers, additional research is needed to evaluate their impacts on plants and the environment [69, 75].

4.2 Effect of Ag NPs Exposure on Environment

Ag NPs are frequently used in everyday life, which unavoidably raises the danger of exposure for people and ecosystems. Additionally, Ag NPs may be released into the environment during the production, delivery, erosion, laundering, or disposal of goods containing Ag NPs. Although the long history of silver use has revealed no obvious negative effects, there is concern regarding the possible risks of Ag NPs in the environment. Studies have shown, Ag inhibits soil microbial development at lower dosages than those of other heavy metals due to its possible bactericidal effect [58, 93]. According to the findings of McGee et al. [55], dehydrogenase and urease activities were found to be significantly reduced in soil contacted to 50 mg kg^{-1} of Ag NPs (20 nm) and a significant change in the bacterial community population was also shown with respect by decreasing the Acidobacteria and Verrucomicrobiota count, while increasing the population of Proteobacteria. Also, Ag NPs had a considerable impact on the composition of the fungal community and the prevalence of archeal and bacterial amoA genes. Similarly, Ag-NPs (20 nm) at 660 mg kg^{-1} caused a reduction in signature bacterial fatty acids as well as in the diversity and uniformity of bacterial and fungal DNA sequences in low arctic soil. Also, microbial respiration was found to be 50% lower compared to normal. Furthermore, Ag NPs were toxic to nitrogen-fixing Rhizobia [46]. However, Ag NPs have a significant effect on the microbial population in contrast to silver microparticles. According to the literature, the concentration of Ag NPs (1–10 nm) in forest soils at 10 and 100 mg kg^{-1} caused a significant drop in microbial biomass. Furthermore, an elevation in the metabolic quotient, demonstrating stress caused by environmental changes due to NPs [9]. The biological processes of Ag-NPs (40 nm) at dosages ranging from 0 to 50 mg kg^{-1} were assessed in calcareous soils with varying salinity levels and textures [68]. It is noteworthy to mention that the biological effects of the NPs were based on soil type and Ag concentration.

4.3 *Effect of TiO_2 NPs Exposure on Environment*

Titanium NMs are less toxic compared to other nanometal oxides. At submilligram or liter concentrations, ZnO, CuO, and TiO_2 NPs were found to be the most hazardous respectively. Accordingly, toxicity on algal species *Pseudokirchneriella subcapitata*, by TiO_2 NPs are owing to cell entrapment rather than metal ion dissolution [3]. Many organisms, including fish, cyanobacteria, and algae, have been reported to be toxic to TiO_2 NPs. Although benthic microbial populations are crucial parts of freshwater environments, a piece limited knowledge about how the addition of TiO_2 NPs would affect them. In ecosystem activities like primary production, organic matter degradation, nutrient cycling, and pollution bioremediation, benthic bacterial communities play an important role. In a study evaluating its impact on simulated stream populations, addition of TiO_2 NPs at a concentration of 1 mg L^{-1} results in a sharp decline in bacterial activity, which is quickly reversed within three weeks to control levels [56]. Similarly, Ozaki et al. [64] showed that the incorporation of TiO_2 NPs into artificial streams for a brief time can cause a notable decline in the number of bacterial cell numbers in artificial streams.

4.4 *Effect of Iron-Based Nanoparticles on the Environment*

The crystal structure, as well as charge, size, solubility, and the presence of organic molecules, are claimed to influence NP reactivity and consequently toxicity [7]. Lethal effects of iron-based NPs on bacteria are linked to oxidative stress generated by ROS production. These outcomes could be brought on by the interaction of oxygen with reduced iron species or by the ionic transport chain being disrupted as a result of the NPs' attraction for cell membranes [7, 102]. For example, Shah et al. [76] reported that Zero-valent Iron NPs (ZVFe-NPs) in soil (2–58 nm at 550 mg kg^{-1}) did not cause notable alterations in bacterial population based on pyrosequencing results. Moreover, environmental factors rather than the existence of NPs can induce toxic effects. Cao et al. [8] studied that ZVFe-NPs at 1,000 mg kg^{-1} (to be utilized in rhizoremediation processes) did not show negative impacts on microbiological parameters, but did have indirect harmful effects on plant root development. Similarly, Rashid et al. [71] showed that, when 2,000 mg kg^{-1} was administered after

180 days of incubation, fungal colonies, heterotrophic cultivable bacteria, microbial biomass carbon, and nitrogen mineralization were drastically reduced.

Cao et al. [8] studied the effect of co-precipitated Fe_3O_4-NPs on the arbuscular mycorrhizal (AM) community and plant/fungi-soil ecosystems. They observed the *Zea mays L.* proliferation and the AM fungal community and compared to bulk Fe_3O_4 in a greenhouse pot experiments. In comparison to bulk Fe_3O_4, the authors showed that the maximum studied concentration of Fe_3O_4 NPs was harmful to AM fungi, affecting variety and modifying community structure. Particulate Fe_3O_4 may damage soil fertility by limiting AM fungal nutrition supplies for maize.

5 Conclusion and Future Perspectives

Due to their distinct characteristics, NPs are frequently utilized in consumer products. Nevertheless, there are enough proofs to show that NMs can cause toxic impacts on higher organisms. Hence, attention is being drawn to the toxicological effects of NPs on the environment and human health. The basis for nanotoxicological investigations is the interaction between the biotic and abiotic elements of the environment. However, it is not quite clear how NPs and living things interact with one another. The rising use of NPs clearly illustrates the deleterious influence of these components. It is consequently vital to research the toxicity and behavior of NPs in water, living organisms (biota), soil, and sediments, as well as their toxicity in combination. The unique features of NPs contribute to their toxicity by increasing their use in both residential and commercial activities. The intrinsic properties of NPs, such as size, nature, shape, charge, surface chemistry, medium of synthesis, storage time, stability, aggregation, mobility, and reactivity, influence their toxicity. The study on the toxicity of NPs is widespread, with several toxicity assays being tested for different types of NPs. Based on the available literature, definitive conclusions cannot be drawn. To investigate the toxicity of all types of NPs, standard methods must be developed. The toxicological influence of NPs on the ecosystem necessitates a long-term assessment of the dangers associated with NPs prior to their large-scale manufacture and commercial and medicinal use.

References

1. Adamcakova-Dodd A, Stebounova LV, Kim JS, Vorrink SU, Ault AP, O'Shaughnessy PT, Grassian VH, Thorne PS (2014) Toxicity assessment of zinc oxide nanoparticles using sub-acute and sub-chronic murine inhalation models. Part Fibre Toxicol 11(1):1–15
2. Adhikari T, Kundu S, Biswas AK, Tarafdar JC, Rao AS (2012) Effect of copper oxide nanoparticle on seed germination of selected crops. J Agric Sci Technology A 2(6A): p 815
3. Aruoja V, Pokhrel S, Sihtmäe M, Mortimer M, Mädler L, Kahru A (2015) Toxicity of 12 metal-based nanoparticles to algae, bacteria and protozoa. Environ Sci Nano 2(6):630–644

4. Atha DH, Wang H, Petersen EJ, Cleveland D, Holbrook RD, Jaruga P, Dizdaroglu M, Xing B, Nelson BC (2012) Copper oxide nanoparticle mediated DNA damage in terrestrial plant models. Environ Sci Technol 46(3):1819–1827
5. Bengalli R, Ferri E, Labra M, Mantecca P (2017) Lung toxicity of condensed aerosol from E-CIG liquids: influence of the flavor and the in vitro model used. Int J Environ Res Public Health 14(10):1254
6. Boxall AB, Chaudhry Q, Sinclair C, Jones A, Aitken R, Jefferson B, Watts C (2007) Current and future predicted environmental exposure to engineered nanoparticles. Central Science Laboratory, Department of the Environment and Rural Affairs, London, UK, 89
7. Braunschweig J, Bosch J, Meckenstock RU (2013) Iron oxide nanoparticles in geomicrobiology: from biogeochemistry to bioremediation. New Biotechnol 30(6):793–802
8. Cao J, Feng Y, Lin X, Wang J, Xie X (2017) Iron oxide magnetic nanoparticles deteriorate the mutual interaction between arbuscular mycorrhizal fungi and plant. J Soils Sediments 17(3):841–851
9. Carbone S, Antisari LV, Gaggia F, Baffoni L, Di Gioia D, Vianello G, Nannipieri P (2014) Bioavailability and biological effect of engineered silver nanoparticles in a forest soil. J Hazard Mater 280:89–96
10. Chai H, Yao J, Sun J, Zhang C, Liu W, Zhu M, Ceccanti B (2015) The effect of metal oxide nanoparticles on functional bacteria and metabolic profiles in agricultural soil. Bull Environ Contam Toxicol 94(4):490–495
11. Chang YN, Zhang M, Xia L, Zhang J, Xing G (2012) The toxic effects and mechanisms of CuO and ZnO nanoparticles. Materials 5(12):2850–2871
12. Chen F, Ehlerding EB, Cai W (2014) Theranostic nanoparticles. J Nucl Med 55(12):1919–1922
13. Chen J, Dong X, Xin Y, Zhao M (2011) Effects of titanium dioxide nano-particles on growth and some histological parameters of zebrafish (Danio rerio) after a long-term exposure. Aquat Toxicol 101(3–4):493–499
14. Chen L, Batjikh I, Hurh J, Han Y, Huo Y, Ali H, Li JF, Rupa EJ, Ahn JC, Mathiyalagan R, Yang DC (2019) Green synthesis of zinc oxide nanoparticles from root extract of Scutellaria baicalensis and its photocatalytic degradation activity using methylene blue. Optik 184:324–329
15. Chen Z, Meng H, Xing G, Chen C, Zhao Y, Jia G, Wang T, Yuan H, Ye C, Zhao F, Chai Z, Zhu C, Fang X, Ma B, Wan L (2006) Acute toxicological effects of copper nanoparticles in vivo. Toxicol Lett 163(2):109–120
16. Choi JS, Kim RO, Yoon S, Kim WK (2016) Developmental toxicity of zinc oxide nanoparticles to zebrafish (Danio rerio): a transcriptomic analysis. PLoS ONE 11(8):e0160763
17. Colvin VL (2003) The potential environmental impact of engineered nanomaterials. Nat Biotechnol 21(10):1166–1170
18. Cortez-Jugo C, Czuba-Wojnilowicz E, Tan A, Caruso F (2021) A Focus on "Bio" in bio–nanoscience: the impact of biological factors on nanomaterial interactions. Adv Healthcare Mater 10(16):2100574
19. Da Costa MVJ, Sharma PK (2016) Effect of copper oxide nanoparticles on growth, morphology, photosynthesis, and antioxidant response in Oryza sativa. Photosynthetica 54(1):110–119
20. Deng F, Olesen P, Foldbjerg R, Dang DA, Guo X, Autrup H (2010) Silver nanoparticles up-regulate Connexin43 expression and increase gap junctional intercellular communication in human lung adenocarcinoma cell line A549. Nanotoxicology 4(2):186–195
21. Du W, Sun Y, Ji R, Zhu J, Wu J, Guo H (2011) TiO 2 and ZnO nanoparticles negatively affect wheat growth and soil enzyme activities in agricultural soil. J Environ Monit 13(4):822–828
22. Eduarda W, Heederik D (1998) Methods for quantitative assessment of airborne levels of noninfectious microorganisms in highly contaminated work environments. Am Ind Hyg Assoc J 59(2):113–127
23. El-Sheikh AA, Abd El-Fatah SS, Moawad RS, Yousef DM (2018) Human umbilical cord mesenchymal stem cells ameliorate silver nanoparticles induced sub-chronic lung toxicity in adult male albino rats. Life Sci J 15(10)

24. Esmaeillou M, Moharamnejad M, Hsankhani R, Tehrani AA, Maadi H (2013) Toxicity of ZnO nanoparticles in healthy adult mice. Environ Toxicol Pharmacol 35(1):67–71
25. Fahmy B, Cormier SA (2009) Copper oxide nanoparticles induce oxidative stress and cytotoxicity in airway epithelial cells. Toxicol In Vitro 23(7):1365–1371
26. Feng Q, Liu Y, Huang J, Chen K, Huang J, Xiao K (2018) Uptake, distribution, clearance, and toxicity of iron oxide nanoparticles with different sizes and coatings. Sci Rep 8(1):1–13
27. Frenk S, Ben-Moshe T, Dror I, Berkowitz B, Minz D (2013) Effect of metal oxide nanoparticles on microbial community structure and function in two different soil types. PLoS ONE 8(12):e84441
28. Ge Y, Schimel JP, Holden PA (2011) Evidence for negative effects of TiO_2 and ZnO nanoparticles on soil bacterial communities. Environ Sci Technol 45(4):1659–1664
29. Gliga AR, Skoglund S, Odnevall Wallinder I, Fadeel B, Karlsson HL (2014) Size-dependent cytotoxicity of silver nanoparticles in human lung cells: the role of cellular uptake, agglomeration and Ag release. Part Fibre Toxicol 11(1):1–17
30. Gobin AS, Froude VE, Mathur AB (2005) Structural and mechanical characteristics of silk fibroin and chitosan blend scaffolds for tissue regeneration. J Biomed Mater Res Part A: Off J Soc Biomater, Jpn Soc Biomater, Aust Soc Biomater Korean Soc Biomater 74(3):465–473
31. Ghosh M, Jana A, Sinha S, Jothiramajayam M, Nag A, Chakraborty A, Mukherjee A, Mukherjee A (2016) Effects of ZnO nanoparticles in plants: cytotoxicity, genotoxicity, deregulation of antioxidant defenses, and cell-cycle arrest. Mutat Res/Genet Toxicol Environ Mutagen 807:25–32
32. Grass RN, Limbach LK, Athanassiou EK, Stark WJ (2010) Exposure of aerosols and nanoparticle dispersions to in vitro cell cultures: a review on the dose relevance of size, mass, surface and concentration. J Aerosol Sci 41(12):1123–1142
33. Grieger KD, Hansen SF, Baun A (2009) The known unknowns of nanomaterials: describing and characterizing uncertainty within environmental, health and safety risks. Nanotoxicology 3(3):222–233
34. Grieger KD, Linkov I, Hansen SF, Baun A (2012) Environmental risk analysis for nanomaterials: review and evaluation of frameworks. Nanotoxicology 6(2):196–212
35. Gui S, Zhang Z, Zheng L, Cui Y, Liu X, Li N, Sang X, Sun Q, Gao G, Cheng Z, Cheng J (2011) Molecular mechanism of kidney injury of mice caused by exposure to titanium dioxide nanoparticles. J Hazard Mater 195:365–370
36. Hackenberg S, Friehs G, Froelich K, Ginzkey C, Koehler C, Scherzed A, Burghartz M, Hagen R, Kleinsasser N (2010) Intracellular distribution, geno-and cytotoxic effects of nanosized titanium dioxide particles in the anatase crystal phase on human nasal mucosa cells. Toxicol Lett 195(1):9–14
37. Halliwell B, Gutteridge JM (1985) Free radicals in biology and medicine
38. Hanna SK, Miller RJ, Zhou D, Keller AA, Lenihan HS (2013) Accumulation and toxicity of metal oxide nanoparticles in a soft-sediment estuarine amphipod. Aquat Toxicol 142:441–446
39. Husen A (2020a) Interactions of metal and metal-oxide nanomaterials with agricultural crops: an overview. In: Husen, A, Jawaid M (eds) Nanomaterials for agriculture and forestry applications. Elsevier Inc. 50 Hampshire St., 5th Floor, Cambridge, MA 02139, USA, pp 167–197. https://doi.org/10.1016/B978-0-12-817852-2.00007-X
40. Husen A (2020b) Carbon-based nanomaterials and their interactions with agricultural crops. In: Husen A, Jawaid M (eds), Elsevier Inc. 50 Hampshire St., 5th Floor, Cambridge, MA 02139, USA, pp 199–218. https://doi.org/10.1016/B978-0-12-817852-2.00008-1
41. Husen A (2022) Engineered nanomaterials for sustainable agricultural production, soil improvement and stress management. Elsevier Inc., 50 Hampshire St., 5th Floor, Cambridge, MA 02139, USA. https://doi.org/10.1016/C2021-0-00054-7
42. Ji JH, Jung JH, Kim SS, Yoon JU, Park JD, Choi BS, Chung YH, Kwon IH, Jeong J, Han BS, Shin JH (2007) Twenty-eight-day inhalation toxicity study of silver nanoparticles in Sprague-Dawley rats. Inhalation Toxicol 19(10):857–871
43. Karimi M, Sadeghi R, Kokini J (2018) Human exposure to nanoparticles through trophic transfer and the biosafety concerns that nanoparticle-contaminated foods pose to consumers. Trends Food Sci Technol 75:129–145

44. Keller AA, McFerran S, Lazareva A, Suh S (2013) Global life cycle releases of engineered nanomaterials. J Nanopart Res 15(6):1–17
45. Kim YS, Kim JS, Cho HS, Rha DS, Kim JM, Park JD, Choi BS, Lim R, Chang HK, Chung YH, Kwon IH (2008) Twenty-eight-day oral toxicity, genotoxicity, and gender-related tissue distribution of silver nanoparticles in Sprague-Dawley rats. Inhalation Toxicol 20(6):575–583
46. Kumar N, Palmer GR, Shah V, Walker VK (2014) The effect of silver nanoparticles on seasonal change in arctic tundra bacterial and fungal assemblages. PLoS ONE 9(6):e99953
47. Lankveld DP, Oomen AG, Krystek P, Neigh A, Troost–de Jong, A., Noorlander, C.W., Van Eijkeren, J.C.H., Geertsma, R.E. and De Jong, W.H. (2010) The kinetics of the tissue distribution of silver nanoparticles of different sizes. Biomaterials 31(32):8350–8361
48. Lee S, Chung H, Kim S, Lee I (2013) The genotoxic effect of ZnO and CuO nanoparticles on early growth of buckwheat, Fagopyrum esculentum. Water Air Soil Pollut 224(9):1–11
49. Leo BF, Chen S, Kyo Y, Herpoldt KL, Terrill NJ, Dunlop IE, McPhail DS, Shaffer MS, Schwander S, Gow A, Zhang J (2013) The stability of silver nanoparticles in a model of pulmonary surfactant. Environ Sci Technol 47(19):11232–11240
50. Lin D, Xing B (2007) Phytotoxicity of nanoparticles: inhibition of seed germination and root growth. Environ Pollut 150(2):243–250
51. Magdolenova Z, Drlickova M, Henjum K, Rundén-Pran E, Tulinska J, Bilanicova D, Pojana G, Kazimirova A, Barancokova M, Kuricova M, Liskova A (2015) Coating-dependent induction of cytotoxicity and genotoxicity of iron oxide nanoparticles. Nanotoxicology 9(sup1):44–56
52. Manzoor U, Siddique S, Ahmed R, Noreen Z, Bokhari H, Ahmad I (2016) Antibacterial, structural and optical characterization of mechano-chemically prepared ZnO nanoparticles. PLoS ONE 11(5):e0154704
53. Márquez-Ramírez SG, Delgado-Buenrostro NL, Chirino YI, Iglesias GG, López-Marure R (2012) Titanium dioxide nanoparticles inhibit proliferation and induce morphological changes and apoptosis in glial cells. Toxicology 302(2–3):146–156
54. Marschner C (1996) In Marschner H (ed) Mineral nutrition of higher plants. Academic Press, London (1995), p 889, (ISBN 0-12-473543-6). Price: 29.95 Pound Sterling
55. McGee CF, Storey S, Clipson N, Doyle E (2017) Soil microbial community responses to contamination with silver, aluminium oxide and silicon dioxide nanoparticles. Ecotoxicology 26(3):449–458
56. Miao L, Wang P, Wang C, Hou J, Yao Y, Liu J, Lv B, Yang Y, You G, Xu Y, Liu Z (2018) Effect of TiO_2 and CeO_2 nanoparticles on the metabolic activity of surficial sediment microbial communities based on oxygen microelectrodes and high-throughput sequencing. Water Res 129:287–296
57. Murali M, Gowtham HG, Singh SB, Shilpa N, Aiyaz M, Alomary MN, Alshamrani M, Salawi A, Almoshari Y, Ansari MA, Amruthesh KN (2022) Fate, bioaccumulation and toxicity of engineered nanomaterials in plants: current challenges and future prospects. Sci Total Environ 811:152249
58. Murata T, Kanao-Koshikawa M, Takamatsu T (2005) Effects of Pb, Cu, Sb, In and Ag contamination on the proliferation of soil bacterial colonies, soil dehydrogenase activity, and phospholipid fatty acid profiles of soil microbial communities. Water Air Soil Pollut 164:103–118
59. Naqvi S, Samim M, Abdin MZ, Ahmed FJ, Maitra AN, Prashant CK, Dinda AK (2010) Concentration-dependent toxicity of iron oxide nanoparticles mediated by increased oxidative stress. Int J Nanomed 5:983
60. Navale GR, Thripuranthaka M, Late DJ, Shinde SS (2015) Antimicrobial activity of ZnO nanoparticles against pathogenic bacteria and fungi. JSM Nanotechnol Nanomed 3(1):1033
61. Nounou H, Attia H, Shalaby M, Arafah M (2013) Oral exposure to zinc oxide nanoparticles induced oxidative damage, inflammation and genotoxicity in rat's lung. Life Sci J 10(1):1969–1979
62. Oberdörster G, Maynard A, Donaldson K, Castranova V, Fitzpatrick J, Ausman K, Carter J, Karn B, Kreyling W, Lai D, Olin S (2005) Principles for characterizing the potential human health effects from exposure to nanomaterials: elements of a screening strategy. Part Fibre Toxicol 2(1):1–35

63. Oberdörster G, Stone V, Donaldson K (2007) Toxicology of nanoparticles: a historical perspective. Nanotoxicology 1(1):2–25
64. Ozaki A, Adams E, Binh CTT, Tong T, Gaillard JF, Gray KA, Kelly JJ (2016) One-time addition of nano-TiO_2 triggers short-term responses in benthic bacterial communities in artificial streams. Microb Ecol 71(2):266–275
65. Pasupuleti M, Schmidtchen A, Malmsten M (2012) Antimicrobial peptides: key components of the innate immune system. Crit Rev Biotechnol 32(2):143–171
66. Perreault F, Samadani M, Dewez D (2014) Effect of soluble copper released from copper oxide nanoparticles solubilisation on growth and photosynthetic processes of *Lemna gibba* L. Nanotoxicology 8(4):374–382
67. Piao MJ, Kang KA, Lee IK, Kim HS, Kim S, Choi JY, Choi J, Hyun JW (2011) Silver nanoparticles induce oxidative cell damage in human liver cells through inhibition of reduced glutathione and induction of mitochondria-involved apoptosis. Toxicol Lett 201(1):92–100
68. Rahmatpour S, Shirvani M, Mosaddeghi MR, Nourbakhsh F, Bazarganipour M (2017) Dose–response effects of silver nanoparticles and silver nitrate on microbial and enzyme activities in calcareous soils. Geoderma 285:313–322
69. Raikova RT, Gabriel DA, Aladjov HT (2006) Modeling investigation of learning a fast elbow flexion in the horizontal plane—prediction of muscle forces and motor units action. Comput Methods Biomech Biomed Engin 9(4):211–219
70. Rajput S, Singh LP, Pittman CU Jr, Mohan D (2017) Lead (Pb^{2+}) and copper (Cu^{2+}) remediation from water using superparamagnetic maghemite (γ-Fe2O3) nanoparticles synthesized by Flame Spray Pyrolysis (FSP). J Colloid Interface Sci 492:176–190
71. Rashid MI, Shahzad T, Shahid M, Imran M, Dhavamani J, Ismail IM, Basahi JM, Almeelbi T (2017) Toxicity of iron oxide nanoparticles to grass litter decomposition in a sandy soil. Sci Rep 7(1):1–11
72. Reddy KM, Feris K, Bell J, Wingett DG, Hanley C, Punnoose A (2007) Selective toxicity of zinc oxide nanoparticles to prokaryotic and eukaryotic systems. Appl Phys Lett 90(21):213902
73. Rizwan M, Ali S, Qayyum MF, Ok YS, Adrees M, Ibrahim M, Zia-ur-Rehman M, Sadeghi L, Tanwir F, Babadi VY (2015) In vitro toxicity of iron oxide nanoparticle: oxidative damages on Hep G2 cells. Exp Toxicol Pathol 67(2):197–203
74. Safar R, Doumandji Z, Saidou T, Ferrari L, Nahle S, Rihn BH, Joubert O (2019) Cytotoxicity and global transcriptional responses induced by zinc oxide nanoparticles NM 110 in PMA-differentiated THP-1 cells. Toxicol Lett 308:65–73
75. Selivanov VN, Zorin EV (2001) Sustained Action of ultrafine metal powders on seeds of grain crops. Perspekt Materialy 4:66–69
76. Shah V, Collins D, Walker VK, Shah S (2014) The impact of engineered cobalt, iron, nickel and silver nanoparticles on soil bacterial diversity under field conditions. Environ Res Lett 9(2):024001
77. Sharma V, Singh P, Pandey AK, Dhawan A (2012) Induction of oxidative stress, DNA damage and apoptosis in mouse liver after sub-acute oral exposure to zinc oxide nanoparticles. Mutat Res/Genet Toxicol Environ Mutagen 745(1–2):84–91
78. Shaw AK, Ghosh S, Kalaji HM, Bosa K, Brestic M, Zivcak M, Hossain Z (2014) Nano-CuO stress induced modulation of antioxidative defense and photosynthetic performance of Syrian barley (Hordeum vulgare L.). Environ Exp Bot 102:37–47
79. Shaw AK, Hossain Z (2013) Impact of nano-CuO stress on rice (Oryza sativa L.) seedlings. Chemosphere 93(6): 906–915
80. Shen Z, Chen Z, Hou Z, Li T, Lu X (2015) Ecotoxicological effect of zinc oxide nanoparticles on soil microorganisms. Front Environ Sci Eng 9(5):912–918
81. Shi J, Abid AD, Kennedy IM, Hristova KR, Silk WK (2011) To duckweeds (Landoltia punctata), nanoparticulate copper oxide is more inhibitory than the soluble copper in the bulk solution. Environ Pollut 159(5):1277–1282
82. Siddiqi KS, Husen A (2016) Engineered gold nanoparticles and plant adaptation potential. Nanoscale Res Lett 11(400):1–10

83. Siddiqi KS, Husen A (2021) Plant response to silver nanoparticles: a critical review. Crit Rev Biotechnol 42(7):773–990. https://doi.org/10.1080/07388551.2021.1975091
84. Simonin M, Guyonnet JP, Martins JM, Ginot M, Richaume A (2015) Influence of soil properties on the toxicity of TiO_2 nanoparticles on carbon mineralization and bacterial abundance. J Hazard Mater 283:529–535
85. Sizova E, Miroshnikov S, Nechitailo X (2019) Assessment of the structural reorganization of liver and biochemical parameters of blood serum after the introduction of zinc nanoparticles and its oxides. Environ Sci Pollut Res 26(17):17110–17120
86. Srivastav AK, Dhiman N, Tiwari R, Arjaria N, Prakash J, Jagdale P, Ayanur A, Singh D, Patnaik S, Kumar M (2019) Sub-acute oral exposure of zinc oxide nanoparticles causes alteration in iron homeostasis through acute phase response: a protective effect by surface modification. J Trace Elem Med Biol 52:270–287
87. Stampoulis D, Sinha SK, White JC (2009) Assay-dependent phytotoxicity of nanoparticles to plants. Environ Sci Technol 43(24):9473–9479
88. Sun Q, Tan D, Ze Y, Sang X, Liu X, Gui S, Cheng Z, Cheng J, Hu R, Gao G, Liu G (2012) Pulmotoxicological effects caused by long-term titanium dioxide nanoparticles exposure in mice. J Hazard Mater 235:47–53
89. Suresh AK, Pelletier DA, Wang W, Morrell-Falvey JL, Gu B, Doktycz MJ (2012) Cytotoxicity induced by engineered silver nanocrystallites is dependent on surface coatings and cell types. Langmuir 28(5):2727–2735
90. Takenaka S, Karg E, Roth C, Schulz H, Ziesenis A, Heinzmann U, Schramel P, Heyder J (2001) Pulmonary and systemic distribution of inhaled ultrafine silver particles in rats. Environ Health Perspect 109(suppl 4):547–551
91. Tan KX, Barhoum A, Pan S, Danquah MK (2018) Risks and toxicity of nanoparticles and nanostructured materials. In: Emerging applications of nanoparticles and architecture nanostructures, pp 121–139, Elsevier
92. Tang J, Xiong L, Wang S, Wang J, Liu L, Li J, Yuan F, Xi T (2009) Distribution, translocation, and accumulation of silver nanoparticles in rats. J Nanosci Nanotechnol 9(8):4924–4932
93. Throbäck IN, Johansson M, Rosenquist M, Pell M, Hansson M, Hallin S (2007) Silver (Ag^+) reduces denitrification and induces the enrichment of novel nirK genotypes in soil. FEMS Microbiol Lett 270(2):189–194
94. Timbrell V (1982) Deposition and retention of fibres in the human lung. In Inhaled Particles V, pp 347–369, Pergamon
95. Tripathi DK, Singh S, Singh S, Pandey R, Singh VP, Sharma NC, Prasad SM, Dubey NK, Chauhan DK (2017) An overview on manufactured nanoparticles in plants: uptake, translocation, accumulation, and phytotoxicity. Plant Physiol Biochem 110:2–12
96. Wang M, Wang J, Liu Y, Wang J, Nie Y, Si B, Liu Y, Wang X, Chen S, Hei TK, Wu L (2019) Subcellular targets of zinc oxide nanoparticles during the aging process: role of cross-talk between mitochondrial dysfunction and endoplasmic reticulum stress in the genotoxic response. Toxicol Sci 171(1):159–171
97. Wang X, Ji Z, Chang CH, Zhang H, Wang M, Liao YP, Lin S, Meng H, Li R, Sun B, Winkle LV (2014) Use of coated silver nanoparticles to understand the relationship of particle dissolution and bioavailability to cell and lung toxicological potential. Small 10(2):385–398
98. Xu C, Peng C, Sun L, Zhang S, Huang H, Chen Y, Shi J (2015) Distinctive effects of TiO_2 and CuO nanoparticles on soil microbes and their community structures in flooded paddy soil. Soil Biol Biochem 86:24–33
99. Yan Z, Wang W, Wu Y, Wang W, Li B, Liang N, Wu W (2017) Zinc oxide nanoparticle-induced atherosclerotic alterations in vitro and in vivo. Int J Nanomed 12:4433
100. Zeng X, Zhang F, He N, Zhang B, Liu X, Li X (2016) ZnO nanoparticles of different shapes and their antimycotic property against penicillium and mucor. Nanosci Nanotechnol Lett 8(8):688–694

101. Zhang X, Wang S, Xu L, Feng L, Ji Y, Tao L, Li S, Wei Y (2012) Biocompatible polydopamine fluorescent organic nanoparticles: facile preparation and cell imaging. Nanoscale 4(18):5581–5584
102. Zucolotto V (2013) Nanotoxicology: materials, methodologies, and assessments. Springer Science & Business Media

The manufacturer's authorised representative in the EU is Springer Nature Customer Service Centre GmbH, Europaplatz 3, 69115 Heidelberg, Germany. If you have any concerns regarding our products, please contact ProductSafety@springernature.com

Printed and bound by CPI Group (UK) Ltd, Croydon, CR0 4YY
30/09/2026
02225318-0011